THE SPACE-AGE SOLAR SYSTEM

Joseph F. Baugher

AT&T Corporation
Skokie, Illinois

Illinois Institute of Technology
Chicago, Illinois

JOHN WILEY & SONS
New York   Chichester   Brisbane
Toronto   Singapore

Library of Congress Cataloging in Publication Data:

Baugher, Joseph F.
   The space-age solar system.

   Bibliography: p. 427
   Includes index.
   1. Solar system.   2. Astronomy.   3. Planets.
I. Title.
QB501.B38   1988   523.2   87-6255
ISBN 0-471-85034-9 (pbk.)

Printed in the United States of America

10 9 8 7 6 5 4 3 2 1

To Judith

Preface

This book is an introductory text describing the modern exploration of our Solar System. The emphasis is on the internal structures, surfaces, and histories of the planets and moons of the solar system, in light of the recent discoveries made by interplanetary spacecraft. The book is aimed at a level approximately equivalent to that of *Scientific American*. It is intended to be read by anyone having an armchair interest in astronomy and/or space travel and does not presume any previous knowledge or technical background in the subject.

The book is appropriate for introductory astronomy survey courses in colleges and universities, especially for those courses placing heavy stress on planetary astronomy. The text is also sufficiently lucid and engaging so that students in the liberal arts should have little difficulty in reading it with full understanding and interest. This book should also appeal to those members of the general public who have an interest in astronomy and in space travel.

The book is meant to be a "travelogue" describing the exotic places and fantastic worlds that will in time be explored in detail as humanity expands outward into the Solar System. The discussion is as current and up-to-date as possible; the recent probes to Halley's comet and the Voyager Uranus results are described in detail. The exposition is intended to be largely qualitative and descriptive rather than quantitative or mathematical. There are no complicated equations, and no knowledge of mathematics is required beyond that learned in high school.

The text begins with a historical sketch of the development of modern theories of the Solar System, starting from the ancient Greek models of the universe and extending to the modern notions of Copernicus, Kepler, and Newton. A description of the objects in the Solar System follows. The presentation begins with the Sun at the center of the Solar System and generally proceeds outward. The planets are subdivided into three distinct categories: small rocky worlds such as the Earth, giant hydrogen-rich planets such as Jupiter, and distant icy worlds such as Uranus. Since the moons of Jupiter and Saturn are so different in structure from the planets themselves, these worlds are treated in separate chapters. There is a chapter describing the structure of comets. The book concludes with a discussion of the question of the origin of the Solar System, in which the results of the preceding chapters are used to outline the current knowledge of the mechanism by which the Solar System came into being and subsequently evolved to its present form.

First, the Sun is described, with special emphasis on the ways its internal structure differs from that of a planet. Also, the sources of solar energy are discussed in some detail.

Next, the Earth is described from a planetary perspective. Much of the terminology to be used in the rest of the text is introduced here. Special attention is paid to the origin and evolution of Earth's atmosphere. Recent knowledge confirms that life has played the central role in the history of Earth's atmosphere. This description of the Earth establishes the framework for comparison with the structures, atmospheres, surfaces, and histories of the other planets and moons in the Solar System.

A great deal of attention is paid to the Moon, since so far it is the only other world whose surface human beings have walked upon and studied in person. Next, the torrid planet Mercury is considered, with extensive use of the results obtained by the *Mariner 10* spaceprobe, which flew past that planet in 1974. The cloud-shrouded planet Venus is described from the viewpoint of results obtained by recent automated landing craft and radar-equipped satellites. The results obtained by the Viking landers are used as the basis of a detailed description of the polar caps, ancient river valleys, and volcanoes of the planet Mars.

The giants Jupiter and Saturn are treated in a separate section, since they are so different from the inner terrestrial planets. The results found during the Voyager flybys of these planets are described. The moons of these giant worlds are treated in separate chapters, drawing heavily upon the data gathered during the Voyager mission.

Next, the ice worlds, Uranus, Neptune, and Pluto are described. Preliminary results from the Voyager flyby of Uranus are discussed, and the icy moons of this world are described. Other chapters describe the meager knowledge that we have about the distant worlds Neptune and Pluto. The comets are described in this section, since they are believed to originate from points far beyond the orbit of Pluto.

The final section describes our current knowledge of the origin of the Solar System from the standpoint of meteorites, which are the only samples of extra-terrestrial matter that we have (aside from Moon rocks).

The metric system of units is used throughout the text. That is, distances are given in meters or kilometers, masses in kilograms, and times in seconds, hours, or days, and so on. Temperature is expressed in degrees Celsius ($^\circ$C). Occasionally, the English equivalent of a commonly quoted number is also given. Numbers that are too large (or too small) to be written conveniently as a string of digits are specified in terms of power-of-ten notation. For example, the number 1,000,000 (one million) is written as 10^6. One billion is written as 10^9. The fraction 1/1,000,000 is written as 10^{-6}.

I am deeply indebted to my parents and to many teachers, who encouraged my youthful curiosity about the natural world. I thank Mr. Robert McConnin of Wiley, who has spent so much of his time in assisting me in the development of the final manuscript of this work. Toby Muller and Gilda Stahl were invaluable in checking the manuscript for internal consistency. Charlene Cassimire was essential in supervising the final production of the book. Any errors remaining are, of course, mine.

JOSEPH BAUGHER
Chicago, Illinois

Contents

THE FAMILY OF THE SUN

PART ONE PHOTO:
The liftoff of the *Titan IIIE*-Centaur booster rocket, which carried the spacecraft *Voyager 1* on its momentous journey to the planets of the outer Solar System. (Photograph courtesy of NASA/JPL)

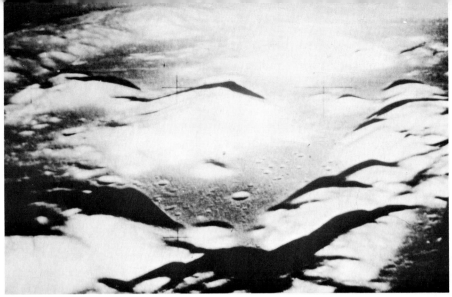

1

Introduction

Throughout most of recorded Western history, the known world was largely confined to the lands of northern Europe and the shores of the Mediterranean Sea. Although it was generally conceded that the Earth was round, much of its surface remained unexplored. Few people ever traveled far from home, and most ocean voyages never ventured very far from land. There were only vague rumors of exotic and fantastic lands far to the south in Africa and to the east in China. The vast continents of the Americas were totally unknown.

In the latter part of the fifteenth century, a remarkable series of exploratory ocean voyages began. Beginning with Christopher Columbus in 1492, ships set sail from Europe to explore the entire Earth. Within a few short decades, sailing vessels had circumnavigated the globe and had visited all corners of the planet. The vast lands of Africa, the Americas, and the Far East were opened up for colonization and exploitation by the nations of Europe. For good or ill, the world was changed forever.

The sudden increase in the size of the known world also led to an expansion of the human mind and spirit. Throughout much of the Middle Ages, the primary goal of human life was the attainment of salvation in the life to come, with worldly concerns being of secondary interest. The natural world was seen as the reflection of the glory of God, and science was largely subordinated to religion. In almost all matters concerning science and the natural world, total

Photograph of the Taurus-Littrow valley on the Moon, the landing site for *Apollo 17*. This is a mountainous region near the edge of Mare Serenitatis. (Photograph courtesy of NASA)

reliance was placed upon ancient authorities such as Aristotle and Plato. The discovery of the New World unfettered the human spirit, leading to new forms of expression in art, music, and literature. Encouraged by the astounding discoveries being made by these ocean voyages, the sciences freed themselves from the constraints imposed by the rigidities of medieval thought. The natural world now became an arena for robust and vigorous study. The technical and scientific revolution that created our modern world was underway.

We may now be poised at the edge of a new age of discovery, one with effects as profound and far-reaching as the age of exploration that opened up the world at the beginning of the sixteenth century. This is the exploration of the moons and planets of the Solar System, made possible by the advent of rocketry and space travel. Within a few short decades in the late twentieth century, we have made the most astounding progress. Our machines have flown past all of the planets within the orbit of Neptune and have sent back detailed photographs of their surfaces. Automated spacecraft have landed on Mars and Venus and have sent back stunning views of their alien landscapes. Spacecraft have flown through the heads of comets, providing incredibly detailed data about their structure and composition. Human beings have actually landed on the surface of the Moon, returning samples of lunar soil and rock to Earth for study. Almost overnight we have acquired a set of detailed maps of over two dozen new worlds. Planets and moons that were heretofore nothing more than distant points of light in the sky have suddenly become places as real to us as the Rocky Mountains, the Amazon River, the jungles of Africa, or the Gobi Desert of China.

Our machines have made the most astounding discoveries about the moons and planets of the Solar System. These worlds have turned out to be far stranger than any writer of science fiction could have ever imagined. A schematic drawing of the relative sizes of the Sun, the nine planets, and the various moons in the Solar System is shown in Figure 1.1.

The Moon sits coldly in the sky — airless, waterless, and sterile. So far, it is the only other world where human beings have walked upon the surface. Once thought to be dead and relatively uninteresting, it now seems that the Moon has as interesting and varied a history as does the Earth itself. The back side of the Moon, totally unknown to us before 1959, is now as well mapped as many areas on Earth. The footprints left behind by the Apollo astronauts still remain in the dusty plains of the Moon, patiently awaiting a return visit.

Tiny Mercury, closest to the Sun, is so hot on one side that lead would melt. It is so cold on the other side that air would turn to liquid. It is a world that looks a lot like the Moon on the outside but is like the Earth on the inside.

Venus, forever veiled under thick clouds, has a surface much like the medieval depiction of Hell, with intense heat and a dense, poisonous, sulfurous atmosphere. Much of its surface still remains a mystery. Is it a vigorously active world like the Earth, with volcanoes and drifting continents? Or is it inactive like the Moon?

The red planet Mars is covered with giant volcanoes, ancient river valleys, vast polar ice caps, and a gigantic rift complex that would stretch across the entire width of the United States. Mars is currently waterless and nearly airless

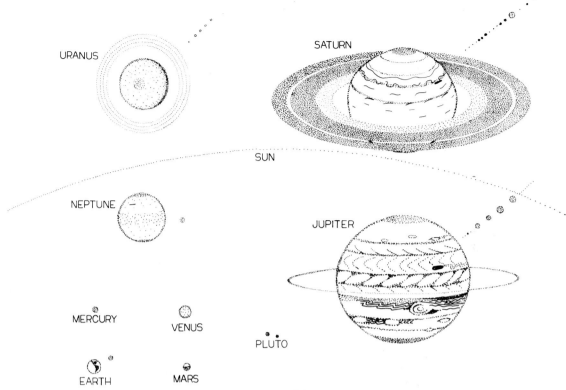

FIGURE 1.1 Relative sizes of the nine planets and their moons.

but at one time it may have been much more like Earth, with a dense atmosphere, flowing rivers, deep oceans, and perhaps even living beings.

Gigantic Jupiter is an immense ball of swirling, turbulent gases, so large that Earth would be swallowed up in one of the minor eddies on its surface. Jupiter's strong magnetic field entraps a zone of radiation so intense that it would kill astronauts within a matter of hours, were they foolish enough to venture too close to the planet.

The moons of Jupiter are just as intriguing as the planet itself. Io, the innermost moon of Jupiter, has numerous active volcanoes that spew forth floods of liquid sulfur. The volcanic activity is so intense that the moon virtually turns itself inside out once every few million years. Europa, the second Jovian moon, may have a subterranean ocean beneath its icy outer crust. Ganymede, the largest moon in the Solar System, is mostly made of ice, but its surface has large continents that drift much like those on Earth. Dark, cold Callisto has a surface scarred by craters and an enormous ringed basin as large as the state of Alaska.

Saturn is a virtual twin of Jupiter but has an exquisite set of rings made up of billions of tiny ice particles. It has an intriguing set of moons. They are worlds composed largely of water ice, battered and scarred by the impacts of gigantic meteors. Many of them are two-faced, with one hemisphere being completely different from the other. Distant Iapetus has one hemisphere as bright as newly fallen snow, but the other is as dark as asphalt pavement. Perhaps the most mysterious of them all is the giant moon Titan, with a dense atmosphere of thick, orange-colored clouds. It may have a surface covered with organic molecules, with rivers of liquid methane slicing their way through hydrocarbon canyons. Life, or at least the precursors of life, may be present.

Uranus and Neptune are twin worlds whose very existence was unknown to ancient astronomers. They are largely made up of frozen water, methane, and ammonia, with a dense atmosphere of hydrogen. Triton, the large moon of Neptune, probably has an appreciable atmosphere. Its frigid surface may be covered with an ocean of liquid nitrogen.

Tiny Pluto and its moon Charon sit at the edge of the Solar System. These two worlds do an amazing celestial dance; they move in such a way that they perpetually turn their same faces toward each other. Even though Pluto is exceedingly cold, it may have a thick atmosphere that alternately appears and disappears once every Plutonian year.

Comets, once thought to be supernatural objects that brought warnings of impending disaster, now are seen as samples of primordial, undisturbed matter that may date back to the very beginning of the Solar System. Surrounding the comet is a vast storm of gas and dust, which shields a central nucleus of frozen ice no larger than a city. For the first time in history spacecraft have penetrated through this cloud of gas and dust to view the hidden nucleus and probe some of its mysteries.

These worlds beckon to us, crying out for exploration. Provided that we humans do not destroy ourselves or lose our nerve, we will shortly venture forth to study their exotic and fascinating landforms and exploit their resources. Perhaps human beings shall eventually colonize their surfaces. In the next few hundred years there may be a Solar-System-wide civilization encompassing billions of people inhabiting a dozen worlds. The opening up of the Solar System to exploration may be the beginning of a new growth in science, religion, art, and philosophy, one as profound as the renaissance in art and literature that brought the medieval stagnation to an end.

This book is intended as a description of what we will find when we begin our expansion into the Solar System, a sort of tour guide for the future.

2

From Flat Earth to Spaceflight: Development of Modern Theories of the Solar System

When I was a small child, I lived on a farm on the lower Eastern Shore of Maryland. At that time there were two sawmills located at the edge of the farm, one to the east and the other to the west. Like many other children I was fascinated by the engines, saws, gears, and noise of these machines. I also happened to notice that the Sun rose over the eastern sawmill every morning. Every evening it set over the western sawmill. Quite naturally, I suppose, I pictured the Sun as being attached to a gigantic chain that passed across the sky between these two sawmills. Once every day, the engines and gears of these sawmills pulled the Sun across the sky!

Gradually, my horizons broadened and I found that my youthful cosmology could not stand the test of experiment. The center of the universe was not really located at my house. I suppose that ancient, primitive peoples approached the movement of the Sun, the Moon, and the stars in much the same way as I had done as a child. Our early ancestors must have looked up at the night sky in wonder and amazement. The heavenly bodies must be very far away; when people climbed trees, the stars seemed no closer. The stars always

Photograph of a crescent Earth taken by the crew of *Apollo 12* while in orbit around the Moon. (Photograph courtesy of NASA)

seemed to keep the same fixed locations in the sky in relation to each other. The stellar patterns never changed, and the same groupings of stars rose in the sky again and again, year after year. The stellar pattern seemed to be so eternally unchanging that prominent groupings of stars were given names. These groups (or *constellations*) were imagined to have been placed there by the intervention of divine beings and were named after gods, animals, or mythical, heroic figures.

Our early ancestors also noted that the entire stellar pattern rotated as a unit once a day about a fixed axis. This axis pointed toward the North Star, located in the constellation Ursa Minor (the Small Bear). It was perhaps only natural to imagine that the stars were simply tiny points of light attached to the inside of an enormous hollow sphere, with the Earth sitting at the very center.

However, there were a few celestial objects that did change their positions in the sky from one night to the next. The Sun and the Moon appear to Earth-bound observers to travel across the celestial sphere, moving in systematic and regular paths from one constellation of stars to another. The ancients also noted that there were five bright "stars" that moved in relation to the other (fixed) stars. These five objects were given the names of deities — Mercury, Venus, Mars, Jupiter, and Saturn. All seven of these mobile celestial objects were collectively referred to as *planets,* the Greek word for "wanderer."

Very much later, it was discovered that there was an intrinsic difference between the Sun and the other planets. The Sun (as well as the fixed stars) emits its own light. The Moon and the other planets, however, shine only by the light that is reflected from them by the Sun. Consequently, the term *planet* is now understood to refer only to those bodies in the Solar System that shine via reflected light.

CRYSTAL SPHERES AND PLANETARY EPICYCLES

Perhaps the first recorded attempt to describe the structure of the universe in any systematic manner was made by the ancient Greek astronomer Anaximander of Miletus (610 – 546 B.C.). He pictured the entire universe as a series of concentric transparent spherical shells, all centered on the Earth. Such a model of the universe is termed *geocentric,* as it places the Earth at the center of all things. A schematic view of Anaximander's model of the universe is shown in Figure 2.1. The outermost celestial sphere carried the fixed stars. It turned on its invisible axis once every 24 hours, rotating the entire heavens once every day about a fixed central Earth. Each of the inner spheres was imagined to have a single planet attached to its equator. The outermost stellar sphere carried the inner planetary spheres with it as it rotated, so the planets as well as the stars all went around the Earth once every day. However, each of the inner planetary spheres had its own separate rotation axis, so that the planets all moved relative to the rotating stellar background. The rotational axes of the planetary spheres were all parallel to the axis of the stellar sphere, which meant that the planets followed parallel paths across the sky.

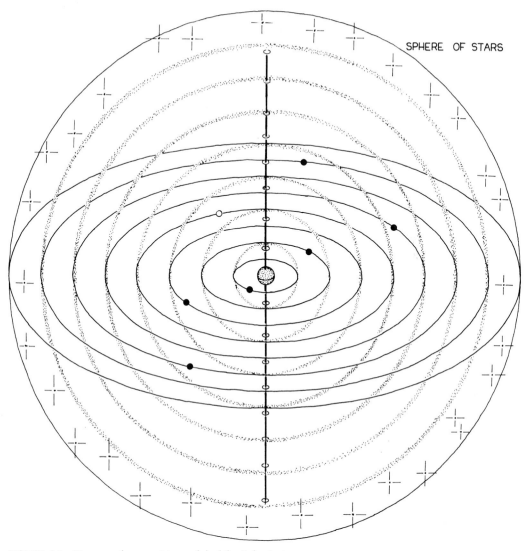

SPHERE OF STARS

FIGURE 2.1 Diagram of geocentric model of the Solar System.

The nearest sphere was that of the Moon. The lunar sphere rotated once every 27.3 days with respect to the stars. The Sun rotated once a year with respect to the stellar sphere, traveling from west to east along a path among the stars known as the *ecliptic*. During its journey across the sky, the Sun passes through 12 different constellations of stars (known as the *zodiac*), spending approximately a month in each one. Since the Moon and the other planets travel in approximately parallel paths, they too stay close to the ecliptic and are always found in one or the other of the 12 constellations of the zodiac.

This orderly motion of the planets among the stars gave rise to the pseudoscience known as *astrology*. In the prescientific era, it was believed by almost everyone that the motions of the planets had a profound influence on the lives of people and on the destinies of nations. Even today, many otherwise well-educated people still believe that the positions of the planets in the sky at the times of their birth have an effect on their careers, their finances, their marriages, and virtually every other aspect of their lives.

As time passed, this simple picture of the structure of the universe had to be modified to account for several annoying facts. First, each and every one of the planets (with the obvious exceptions of the Sun and the Moon) changes in brightness as it moves across the sky. This must mean that these planets do not always stay the same distance from Earth. Second, Mercury and Venus never wander very far from the Sun. These planets are never seen in the western sky at sunrise nor in the eastern sky at sunset. Third, when Mars is at its brightest in the nighttime sky, the planet suddenly begins to slow down in its passage among the stars. It then stops dead in its tracks and starts moving in the opposite direction. This backward (or *retrograde*) travel lasts for a few days, after which Mars comes to a halt in the stars once again. It then reverses direction and resumes its forward direction of travel.

These problems were addressed by the extremely clever artifice of the *epicycle* (Figure 2.2). The Egyptian astronomer Claudius Ptolemy (100–170 A.D.) is commonly given credit for the introduction of epicycles, although there is evidence that they had been used in astronomical theories as much as 450 years earlier. In the Ptolemaic universe, the Sun and the Moon still moved as they did in the Anaximander model, but the other planets were now pictured as

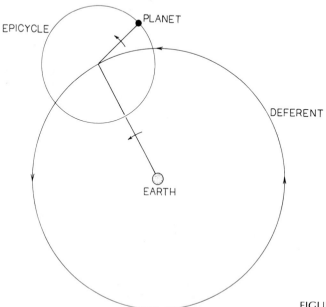

FIGURE 2.2 Planetary epicycle.

being attached to smaller invisible subspheres called epicycles. The center of each epicycle travels along a perfectly circular path (known as a *deferent*) at constant speed, but the epicycle itself rotates about its own separate axis. The planet therefore travels around the Earth along a complicated path in space.

Since the planets are not traveling around the Earth along perfectly circular orbits, they are constantly changing their distances from us, and their relative brightnesses will vary from one night to the next. Since Mercury and Venus always stay close to the Sun in the sky, these two planets must be attached to epicycles whose centers always remain on a line connecting the Sun to the Earth (Figure 2.3). The Martian epicycle was imagined to move in such a way that this planet actually does a loop-the-loop in the sky at the time when it is nearest the Earth (Figure 2.4).

It is easy to criticize the Ptolemaic picture of the universe from the viewpoint of modern knowledge, but the model did provide a description of planetary motions accurate enough to be of use in deep-water oceanic navigation. The Ptolemaic model gradually became deeply embedded in Christian theology and was to dominate astronomical thought in Europe for over 1000 years.

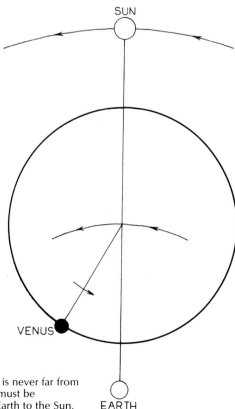

FIGURE 2.3 Epicycle for Venus. Since Venus is never far from the Sun in the sky, the center of the epicycle must be constrained to lie on a line that connects the Earth to the Sun.

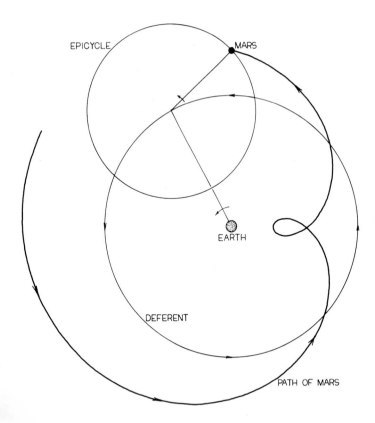

EPICYCLE

MARS

EARTH

DEFERENT

PATH OF MARS

FIGURE 2.4 Epicycle for Mars. The planet moves in such a way that it travels backward in the sky at the time that it is closest to the Earth.

THE COPERNICAN REVOLUTION

In 1543 a Polish mathematician and astronomer by the name of Nicolaus Copernicus (1473 – 1543) published a treatise that proposed a radically new model for the structure of the universe. The Earth was removed from the center of the universe and replaced by the Sun. The planets, Earth included, were pictured as traveling around the Sun along perfectly circular orbits (Figure 2.5). The planetary orbits were all nearly coplanar with each other, and all the planets traveled around the Sun in the same direction. The Copernican model of the Solar System is said to be *heliocentric,* as it places the Sun rather than the Earth at the center of the universe. The Moon, however, was still pictured as revolving about the Earth, just as it did in the Ptolemaic model. The Earth was assumed to rotate about an axis that pointed toward the North Star, so that the entire celestial sphere appeared to complete one full turn every day. It was not the sky that was turning, it was the Earth.

Mercury was placed closest to the Sun, followed at increasing distances by Venus, Earth, Mars, Jupiter, and Saturn. The Copernican model readily explains the variable brightness of the planets, since they maintain a fixed distance from

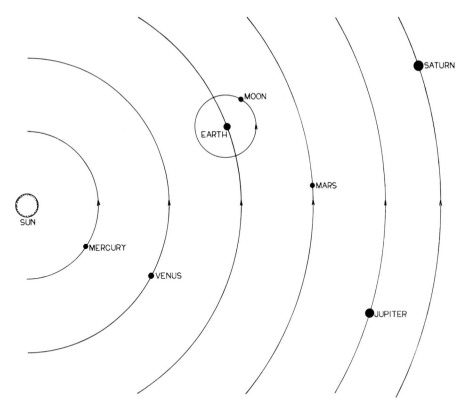

FIGURE 2.5 Diagram of the Copernican model of the Solar System.

the Sun, not the Earth. It also provides a convenient explanation of the fact that Mercury and Venus always appear near the Sun in the sky (Figure 2.6). These two planets occupy orbits with smaller radii than Earth's. The Copernican model also provides an explanation of the retrograde motion of Mars, provided one assumes that Mars travels more slowly along its orbit than the Earth does (Figure 2.7). At the time that the Earth catches up to Mars in its orbit, the red planet appears to us to be momentarily traveling backward with respect to the stellar background, for the same reason that a slower moving car appears to the occupants of a faster car to be traveling backward with respect to a distant background at the instant of passing.

There were problems with the Copernican model, and by no means did it meet with instant acceptance. There were significant departures from perfectly uniform circular motion (especially in the case of Mars), and later theorists were forced to add epicycles to the motion in order to get a better fit. The most severe problem was caused by the moving Earth. Not only was this offensive to the prevailing religions at the time, but it also did violence to the conventionally understood notions of mechanics that dated from the time of Aristotle (384 – 322 B.C.). It stood to reason that if the Earth were really moving, an object

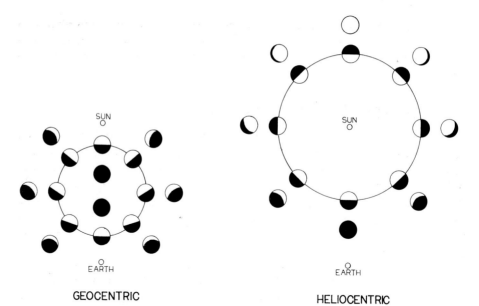

GEOCENTRIC   HELIOCENTRIC

FIGURE 2.6 A comparison of geocentric and heliocentric descriptions of the phases of Venus. In the geocentric model it is impossible to see a full Venus from Earth.

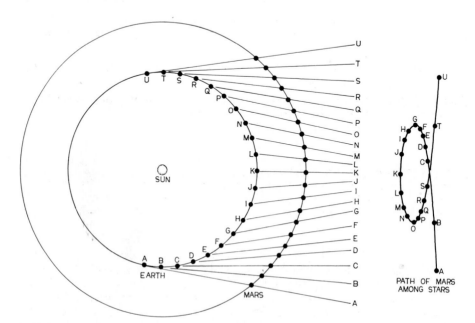

FIGURE 2.7 Heliocentric view of motion of Mars.

dropped from a height would fall far to the west of its release point rather than straight down. For these reasons, the Copernican model of the universe was actively suppressed by both the Catholic and Protestant branches of the Christian religion.

THE KEPLERIAN MODIFICATION

In the late sixteenth century, the Danish astronomer Tycho Brahe (1546–1601) built an observatory on an island in the Baltic Sea that was able to make amazingly accurate measurements of the motions of the planets. This facility was built before the advent of the telescope, but it was nevertheless able to provide planetary position measurements accurate to within a half-minute of arc. Upon Tycho's death, this vast accumulation of astronomical data fell into the hands of his associate, Johannes Kepler (1571–1630). Kepler used Tycho's data as the basis of three laws which govern the motion of the planets (Table 2.1).

Kepler adopted the basic assumptions of the Copernican universe (with the Sun rather than the Earth at the center), but he made important additions and changes. He found that a planet does not travel in a perfectly circular orbit around the Sun. It follows an orbit that is an *ellipse,* with the Sun located at one focus. The mathematical definition of an elliptical orbit is given in Figures 2.8 and 2.9. As a planet moves along its elliptical orbit, it is continually changing its distance from the Sun. The point of a planet's closest approach to the Sun is called the *perihelion;* the point of maximum departure is called the *aphelion.* The amount that the shape of an elliptical orbit departs from that of a perfect circle is measured by a quantity known as the *eccentricity.* The larger the eccentricity, the greater the departure from a circular shape. A perfectly circular orbit can be considered as a special case of an elliptical orbit—one that has an eccentricity of 0.

Kepler also discovered that a planet does not move at a constant speed along its orbit. It travels faster when it is at perihelion than when it is at aphelion (Figure 2.10). In addition, he found that planets located far from the Sun move at a much slower rate than do those near the Sun.

TABLE 2.1 KEPLER'S LAWS OF PLANETARY MOTION

1. Planets do not move in perfect circles. They travel instead in elliptical paths, with the Sun at one of the foci. The distance of a planet from the Sun will therefore change as it moves along its orbit.
2. A line drawn from a planet to the Sun will sweep out equal areas in equal times as the planet moves along its orbit. At the time when a planet comes closest to the Sun (the perihelion) it is moving faster than when it is at its most distant point (the aphelion).
3. The square of the orbital period of a planet is proportional to the cube of the average distance of that planet from the Sun. The farther the planet from the Sun, the longer its period.

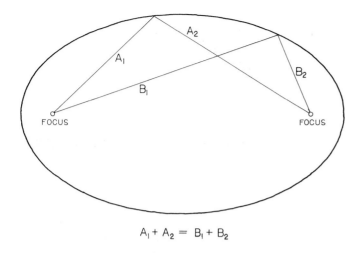

$$A_I + A_2 = B_I + B_2$$

FIGURE 2.8 Mathematical definition of an ellipse. The ellipse is defined in terms of two foci. A straight line is drawn from one of the foci to a point on the ellipse. Another straight line connects that point on the ellipse to the other focus. The sum of the lengths of the two straight lines is the same for all points on the ellipse.

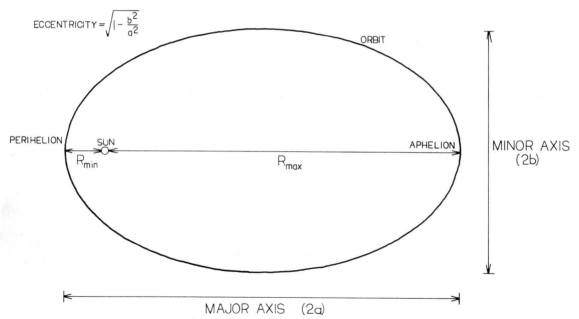

FIGURE 2.9 Definition of parameters used to describe elliptical orbits.

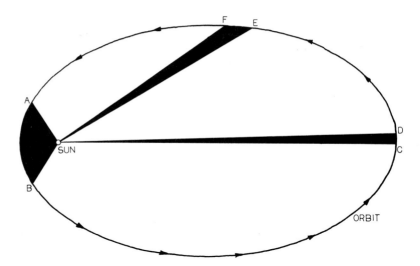

FIGURE 2.10 A planet moves along an elliptical orbit in such a way that it sweeps out equal areas in equal times. The time intervals between A and B, between C and D, and E and F are all equal, and the three black triangles all have the same area.

The properties of the orbits of the planets and moons in the Solar System are summarized in Appendixes D and E. A schematic diagram of the elliptical orbits of the nine planets is shown in Figure 2.11 *a* and *b*. The diagram is given in two parts, as it is impractical to represent all nine orbits on the same scale. The view is that which would be seen by an observer looking down on the Solar System from a point far above the Earth's north pole. Relative to this observer, the planets all travel around the Sun in a counterclockwise direction. This defines the *direct* (or *prograde*) motion. Any object orbiting the Sun in the other direction is said to have a *retrograde* motion.

Kepler's abandonment of the perfectly circular orbit as a requirement for planetary motion was perhaps the most radical change introduced by him, since even Copernicus felt that God would certainly have constrained the planets to move in the most "perfect" paths imaginable. Our biases and preconceptions die hard.

At approximately the same time that Kepler was working on his theories of planetary motion, another development occurred that was to provide striking evidence in favor of the Copernican Sun-centered universe. In Italy, Galileo Galilei (1564–1642) had heard rumors that Dutch opticians had invented a marvelous new device known as a telescope, which was capable of magnifying distant objects. Galileo gleaned enough information from descriptions of these new instruments to be able to construct his own. In the summer of 1609, Galileo turned his telescope toward the sky. He discovered that the planet Venus goes through phases just as the Moon does. Whenever Venus was closest to Earth, the planet was always in gibbous phase. When farthest from Earth, Venus was always in full phase. If the Ptolemaic model were correct, it should never be possible to see Venus in full phase. Galileo also discovered that the planet Jupiter has four tiny satellites, a sort of miniature Solar System, with four smaller objects orbiting about a larger central body.

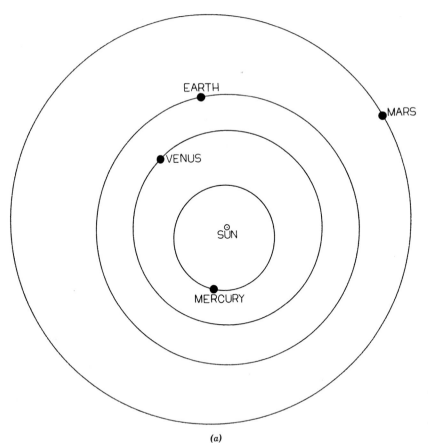

FIGURE 2.11 Orbits of the nine planets. (a) The orbits of the four inner planets. (b) The orbits of the outer five. The relative positions of the planets on January 1, 1987 are indicated.

(a)

Even though evidence gradually accumulated in favor of the Copernican model of the universe, acceptance was slow in coming because of the vigorous opposition within organized religion. Galileo was forced on pain of death to retract his views of the heliocentric nature of the Solar System. Some people, like the philosopher Giordano Bruno (1548 – 1600), were even burned at the stake for their views. Gradually, the Copernican view gained ascendancy, not so much as a result of any striking new telescopic observations or discoveries, but rather as a result of the inherent simplicity and beauty of the heliocentric universe as opposed to the Ptolemaic universe with its complicated system of epicycles.

THE NEWTONIAN MECHANICAL UNIVERSE

The final confirmation of the heliocentric picture of the Solar System had to await the advent of Isaac Newton (1642 – 1727), who lived over 100 years after the time of Copernicus. Kepler had expressed his three laws as statements of

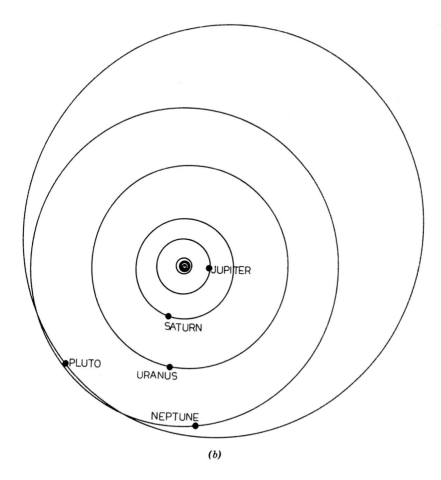

(b)

experimental fact. They were simply mathematical descriptions of how the planets moved and did not attempt to uncover the root cause of the motion. This was accomplished by Isaac Newton in his development of the science of mechanics and the theory of universal gravitation. Newtonian mechanics is summarized in his famous three laws of motion, which are listed in Table 2.2.

TABLE 2.2 NEWTON'S LAWS OF MOTION

1. All bodies at rest will remain at rest, and all bodies moving at a constant velocity will remain moving at that same velocity, unless they are compelled to change their motion by an external force.
2. If a net unbalanced force acts on a body, the body will accelerate in the direction of the force. The magnitude of the acceleration is directly proportional to the force and inversely proportional to the mass of the body.
3. If body A exerts a force on body B, then body B exerts a force on body A that is equal in magnitude and opposite in direction. (An isolated force is an impossibility; you cannot lift yourself by your own bootstraps.)

Although these laws seem almost self-evident to us now, they were by no means obvious to those living in the seventeenth century. Most "common-sense" notions of mechanics from the time of the ancient Greeks onward assumed that a body at rest was intrinsically different from one that was moving. A motionless object was imagined to be in its most "natural" state. Some sort of force was required to keep the body in motion. If the force is removed, the body will stop. Galileo was the first to cast serious doubt on this notion. He performed a series of simple experiments with bowling balls rolling along level surfaces. He noted that the smoother the surface happened to be, the farther a ball would travel before it stopped. Galileo correctly reasoned that the bowling balls slowed and stopped not because they were moving toward a more "natural" state, but because of the presence of frictional forces.

The elimination of the distinction between rest and motion at a constant velocity removed one of the strongest objections to the moving Earth. Objects sitting on the Earth's surface are not violently disturbed by our planet's rapid motion through the heavens because the law of inertia compels them to possess the same velocity as the Earth does.

GRAVITY—THE GLUE OF THE SOLAR SYSTEM

After Newton had published his three laws of motion, he looked toward the heavens to see if his laws could be used to describe the motion of the planets. He first considered the motion of the Moon, which travels in a roughly circular orbit around the Earth. Because of its circular motion, the Moon is actually undergoing an acceleration toward the Earth. According to Newton's second law, there must then be a force acting on the Moon that pulls it toward the Earth. If this force were to be mysteriously removed, the Moon would simply fly away from the Earth along a straight-line path tangential to its original orbit. But there is no apparent physical connection between the Earth and the Moon to provide such a force. Newton concluded that there must exist a different kind of force, one that acts by means of a *field* rather than by actual physical contact. Newton called this new force *gravity*.

Newton proposed that each body in the universe exerts a gravitational pull on every other body in the universe. He described the gravitational force in terms of a law of universal gravitation, which is formally stated in Table 2.3. The

TABLE 2.3 NEWTON'S LAW OF UNIVERSAL GRAVITATION

Every body in the universe exerts an attractive force on every other body in the universe. The gravitational force between any two bodies is proportional to the product of the masses of the two bodies and is inversely proportional to the square of their separation. In all cases, the gravitational force is attractive and acts along the line between the two masses.

gravitational force that holds the Moon in its orbit is exactly the same force that makes apples fall to the ground when they drop from a tree.

The Sun is far more massive than any of the planets, and its intense gravitational pull holds the planets in their orbits. Since the gravitational force on a planet is stronger the closer it is to the Sun, the planets nearest the Sun travel along their orbits at a faster rate than do those farther away. For this reason, Mercury takes only 88 Earth days to travel once around the Sun, whereas distant Pluto takes 250 Earth years to make a full circuit.

Not only do Newton's laws provide a convenient explanation of why planets travel as they do, but the laws can also be used to give a detailed quantitative description of planetary motion so precise that it can correctly predict their positions in the sky many millions of years in the future. Since Newton's laws provided such an accurate quantitative description of the motion of the planets, few educated people from the time of Newton onward could seriously doubt the truth of the heliocentric nature of the Solar System.

Until quite recently, people could only study the planets by looking at them from the surface of the Earth, using telescopes of ever-increasing power and sophistication. However, with the advent of space travel, it is now possible to send instrumented probes to explore these planets at close range, to photograph their surfaces, to study their atmospheres, and to measure their masses and densities. Detailed close-up studies have now been made of all the planets and moons out to the Uranian system. Spacecraft have landed on the surfaces of Venus and Mars and have sent pictures of their landscapes back to Earth. Samples of the lunar surface have been returned to Earth for detailed study. In 1969 humans walked on the surface of the Moon for the first time. We have gained more knowledge of our Solar System in the last 15 years than we have acquired in all our previous history. The history of space missions to the Moon and the planets is summarized in Appendix A.

The amount of knowledge that has been gathered by these exploratory craft is truly vast. It has come so rapidly that many of the data still await analysis and interpretation. For the first time in human history, we are beginning to acquire a meaningful picture of our own world within the context of the entire Solar System and are starting to appreciate the fragility of our existence when viewed in an interplanetary context. In particular, we have found that life and intelligence have appeared on no other world in the Solar System besides the Earth.

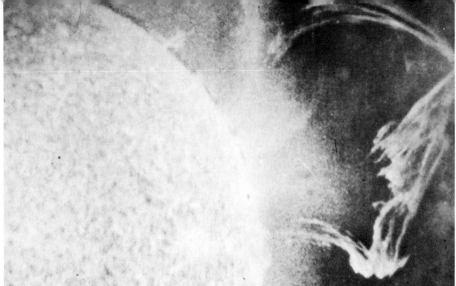

3
Our Star, the Sun

The Sun is the most prominent object that ever appears in the sky. It is the ultimate source of nearly all of the energy used on Earth. Without its presence, all life would soon perish. Impressed with its obvious power and importance, many ancient societies worshiped the Sun as a deity. The ancient Sumerians thought of the Sun as a god, to which they gave the name Uto. Uto evolved into Marduk — the chief deity of the Babylonians. The Aztecs worshiped a particularly bloodthirsty Sun god, one which demanded constant human sacrifice. The ancient Incas thought of themselves as Children of the Sun. The pharaoh Akenaten (1379 – 1362 B.C.) introduced Sun worship into Egypt in the form of the god Aten. Akenaten's new religion did not last very long, but his cult of Sun worship was the first to introduce the concept of only one god rather than many.

A SPHERE OF HYDROGEN AND HELIUM

The Sun (Figure 3.1) is by far the largest and most massive single object in the Solar System. It contains 99.9 percent of the matter in the Solar System. In numbers, this is 1.99×10^{30} kg, 1000 times more massive than the largest planet, Jupiter, and 333,400 times more massive than the Earth. The diameter of the Sun is 1.39×10^6 km, about 109 times larger than the Earth. This gives the Sun an average density of 1.4 g/cm³, about one-quarter of that of Earth and actually

Erupting prominences on the Sun taken by an ultraviolet camera carried aboard the Skylab orbiting space station. (Photograph courtesy of NASA)

22

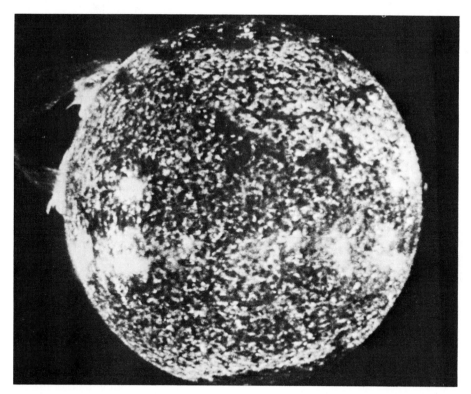

FIGURE 3.1 Photograph of the Sun taken in ultraviolet light of ionized helium. This photograph was taken in 1973 by the instrument of the Naval Research Laboratory aboard the *Skylab* space station. A single prominence, 40 times larger than the Earth, is shown at the limb.

only slightly higher than that of water. Even though the Sun is physically much more massive than the Earth, it must be made of a material that is far less dense.

The Sun is so hot that all the matter in its interior must be in the plasma state, consisting of ionized species and free electrons. It is possible to determine the elemental composition of the Sun by examining the spectrum of the light emitted from the visible surface. These studies indicate that 94 percent of the atoms in the Sun are hydrogen atoms, with most of the rest being helium. Other elements in the Sun are present in much lower abundances. For every 10,000 atoms of hydrogen, there are 7 atoms of oxygen, 4 of carbon, 1 of nitrogen, and so on. Elements of higher atomic weight are even more scarce; the relative atomic abundance generally decreases with increasing atomic weight. Only 2.5 percent of the solar mass is made up of "heavy metals," which is an astronomer's shorthand for any element with an atomic weight greater than that of helium.

The relative abundances of the elements found in the Sun are listed in Appendix B. Properties of the Sun are summarized in Appendix C.

THE COSMIC POWER SOURCE

The rate at which energy is given off by the Sun is known as the *luminosity* and is given in units of watts. The solar luminosity if 3.9×10^{26} W, which seems to be constant from one year to the next to within 0.1 percent. The average power emitted per unit area on the solar surface is 6.44×10^7 W/m^2, which corresponds to a body at an *effective temperature* of 5500°C. The solar spectrum is shown in Figure 3.2. It quite nicely matches that of an ideal *blackbody* radiator at a temperature of 5500°C. Such a body emits its maximum light intensity at a wavelength of about 5000 Å (1 angstrom = 10^{-10}m). This wavelength happens to fall in the yellow region of the visible light spectrum, accounting for the Sun's characteristic color.

Superimposed on the continuous solar spectrum are thousands of narrow absorption lines. These are produced by the presence of elements in the outer-most layers of the Sun and are called *Fraunhofer lines*. They are named in honor of Joseph von Fraunhofer (1787–1826), a German physicist who spent much of his professional life studying sunlight. Each element has its own individual set of Fraunhofer lines. Consequently, a detailed examination of the positions and intensities of the Fraunhofer lines can be used to measure the relative abundances of the elements that are present in the Sun. It is interesting to note that the element helium was first detected on the Sun before it was discovered on Earth!

How has the Sun kept on shining so steadily for such a long time? This question was a point of great controversy for a long time. During the nineteenth century, the only known source of radiant energy was chemical combustion, the process responsible for the light emitted by fires. A calculation first performed by the famous English physicist Lord Rayleigh (1842–1919) indicated that the energy yield from typical chemical combustion reactions is so low that the Sun should burn itself out in only a few thousand years. However, the abundant geological and fossil evidence being uncovered for the first time during this century seemed to indicate that the Earth and Sun had been around for a lot longer than that, millions of years or even greater. This paradox was the source of much acrimonious debate between physicists, geologists, and biologists throughout most of the late nineteenth century. The physicists argued that the fossil record had to be wrong; the geologists and biologists held that the Sun must shine by some as yet undiscovered new and exotic energy conversion process.

It turned out that the biologists were right. The first clue came when Henri Becquerel (1820–1891) discovered the phenomenon of radioactivity. Some atomic nuclei are unstable and will give off energy in the form of radiation. Sometimes this radiation is in the form of high-energy particles such as electrons, neutrons, or alpha particles (^{4}He nuclei). Sometimes it takes the form of high-energy electromagnetic radiation (either X rays or gamma rays). A second clue came when Albert Einstein announced his special theory of relativity in 1905. This theory predicted that it is in principle possible to convert matter directly into energy. This conversion would produce far more energy than could

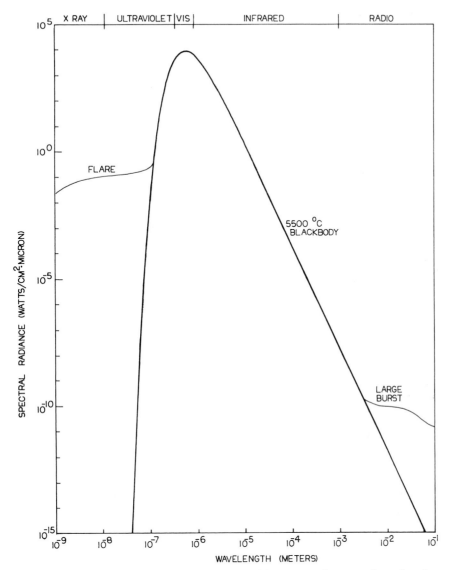

FIGURE 3.2 The spectral radiance of the surface of the Sun as a function of wavelength. The peak emission lies within the narrow band of visible light. For wavelengths lying in the range between the near ultraviolet and the far infrared the spectrum of the Sun closely matches that of an ideal "blackbody" radiator at a temperature of 5500°C. At other wavelengths the intensity of the light from the Sun is highly variable and depends to a large extent on the level of solar activity. At radio wavelengths there is a component that changes in intensity with the 11-year solar cycle. At far-ultraviolet and X-ray wavelengths there is a strong component that is enhanced at the time of a solar flare.

be obtained by ordinary chemical combustion. The English physicist A. S. Eddington was perhaps the first to propose that the conversion of matter into energy was the mechanism that makes the Sun shine.

In 1939, Hans Bethe proposed that thermonuclear fusion reactions between the nuclei deep in the solar interior is the primary solar energy source. The mass of the Sun is so great that gravitational compression in the interior produces extremely high densities and temperatures, far greater than any existing on Earth, although recent laser-induced fusion experiments have come close to reproducing the conditions of the solar interior in the laboratory. The material at the center of the Sun is so hot that all of the atoms located there are completely stripped of their electrons, leaving a dense plasma of protons, alpha particles, other atomic nuclei, and free electrons, all flying around at tremendous velocities. At such extremely high velocities, some of the nuclei can fuse together when they collide, releasing energy.

The chain of *fusion reactions* believed responsible for the Sun's heat is listed in Table 3.1. In the first step, two hydrogen nuclei react to form a deuteron, a positron, and a neutrino. The positron reacts with an electron to produce a pair of gamma rays. The deuteron reacts with another proton to produce a ^{3}He (helium 3) nucleus and a gamma ray. Finally, the ^{3}He nucleus reacts with another ^{3}He nucleus to produce a stable ^{4}He nucleus, a pair of protons, plus still another gamma ray. The net result of the chain of reactions is the disappearance of four protons and two electrons, and the production of one ^{4}He nucleus, eight gamma rays, plus two neutrinos. The hydrogen in the core is gradually depleted as the fusion reactions proceed, and the concentration of helium steadily builds up.

The energy given off in these fusion reactions is released in the form of a flood of neutrinos and gamma rays. Compared to conventional chemical combustion, the energy released is enormous. There is enough energy given off by the fusion of 1 g of hydrogen to lift a weight of 10,000 tons to a height of 25,000 feet above the Earth. Neutrinos are electrically neutral particles, generally thought to be massless. They react so weakly with other matter that virtually all of these particles escape from the Sun and fly away into space. On the other hand, the gamma rays produced by the fusion reactions undergo numerous collisions with the electrons and ions in the interior, gradually giving up their

TABLE 3.1. FUSION REACTIONS IN THE SUN[a]

^{1}H	+	^{1}H$\longrightarrow ^2$H	+	e^+	+	ν	
e^+	+	$e^-\longrightarrow \gamma$	+	γ			
^{2}H	+	^{1}H$\longrightarrow ^3$He	+	γ			
^{3}He	+	^{3}He$\longrightarrow ^4$He	+	$2\,^1$H	+	γ	
net: 4 ^{1}H	+	2 $e^-\longrightarrow ^4$He	+	8γ	+	2ν	

[a]Abbreviations and symbols: ^{1}H, proton (hydrogen nucleus); ^{2}H, deuteron (heavy hydrogen nucleus); ^{4}He, alpha particle (helium nucleus); ^{3}He, helium-3 nucleus; e^-, electron; e^+, positron; ν, neutrino; γ, gamma ray.

energy in the form of heat. It takes several million years for a typical gamma ray to work its way out to the solar surface. By the time that it does, it has lost most of its energy and has been reduced to radiation of much longer wavelength.

THE INTERIOR OF THE SOLAR FURNACE

Even though the interior of the Sun is inaccessible to direct observation, theoretical physicists have been able to produce mathematical models that describe some of its more prominent structural features. A schematic diagram of the solar interior is shown in Figure 3.3.

The solar *core*, where all of the thermonuclear reactions take place, is estimated to have a temperature of $1.5 \times 10^7 °C$ and a density of $150\,g/cm^3$. Half of the solar mass is concentrated in a central sphere that occupies only 2 percent of the total solar volume. Both the temperature and the density steadily decrease with distance from the center. By the time that one is halfway to the surface, the temperature has dropped to $6 \times 10^6 °C$ and the density is down to 1 g/cm^3. At the surface the density is only $2 \times 10^{-8}\,g/cm^3$, far less than that of air. The temperature there is only $4230°C$.

The rest of the solar interior is divided into two distinct regions, a *radiative interior* and a *convective outer shell*. The inner 590,000 km of the Sun (which includes the dense energy-producing core) consists of completely ionized gases, which are highly transparent to the electromagnetic radiation streaming outward from the center. At a distance of $0.85R\odot$ (solar radius) from the center, the temperature is low enough that some electrons can bind to nuclei, forming partially ionized atoms. This causes the gases of the Sun to become much more

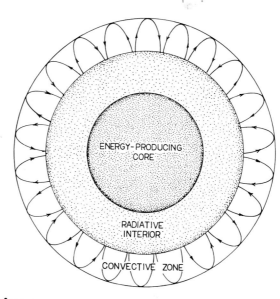

FIGURE 3.3 The interior of the Sun.

opaque to the radiation coming from within. This increased opacity causes a large amount of the solar energy streaming outward from the interior to be strongly absorbed. The onset of strong energy absorption at his particular level sets up a convective instability that produces turbulent currents of gas, which rapidly carry the solar energy the rest of the way outward to the surface.

Theoretical calculations indicate that the Sun has been converting hydrogen into helium for approximately 4.7 billion years. If the Sun continues to consume hydrogen at the same rate, it will take about 100 billion years for it to exhaust all its fuel. The actual life expectancy of the Sun is probably a good deal shorter than this because only the hydrogen in the dense core ever gets hot enough to undergo fusion. The initial composition of the Sun (by weight) was probably quite similar to that which is currently found in its outermost surface; that is, it was 78 percent hydrogen, 20 percent helium, and 2 percent heavier elements. Over the years, a large fraction of the hydrogen in the core has been converted into helium. It is estimated that the solar core currently contains about 35 percent hydrogen and 63 percent helium. There is probably enough hydrogen left in the core to keep the Sun burning at more or less the same rate for another 5 billion years. The Sun is, therefore, now at about the halfway point in its life.

THE SUN'S VISIBLE SURFACE: THE PHOTOSPHERE

The visible surface of the Sun is called the *photosphere*. It is sufficiently transparent so that one can see into the Sun a distance of about 150 km. The temperature at the outside of the photosphere is 4200°C; at the inside, it is 8200°C. The effective temperature of the Sun, as calculated from the total energy emitted from the surface, is 5500°C.

The visible surface of the Sun has a granular texture (Figure 3.4). A typical *granule* is 1000 km across and consists of a brighter region surrounded by a

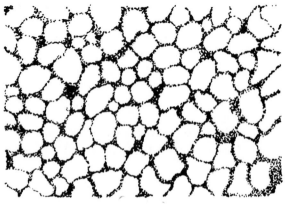

FIGURE 3.4   The Sun's visible surface.

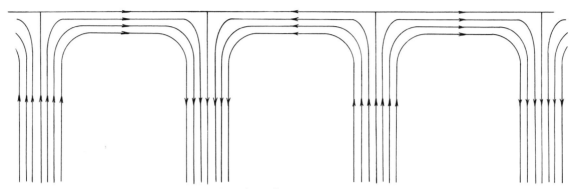

FIGURE 3.5   Convective heat currents near the solar surface.

darker border. Granules are produced by convective heat currents flowing outward from the interior. The bright center of a granule is approximately 100°C hotter than its darker surroundings. The lighter regions of granules are hotter rising gases, whereas their darker borders are produced by cooler descending gases (Figure 3.5). The speeds of the gases involved in these turbulent vertical motions can be as high as a few tenths of a kilometer per second. The granular structure of the photosphere is constantly changing. A typical granule will last a few minutes before it fades out and is replaced by another. Granules tend to organize themselves into cells of 300 or so. These cells, termed *supergranules*, typically last about a day before they disperse. The surface of the Sun is constantly seething and bubbling, much like a pot of boiling water on a stove.

THE ROTATION OF THE SUN

Like each of the planets, the Sun rotates about an axis. Some surface detail is visible on the Sun, and crude estimates of the solar rotation can be made by watching these features move across the solar disk. However, the Sun does not have a solid surface like the Earth, and easily visible features generally do not survive for very long. It is better to study the rotation of the Sun by measuring the *Doppler shifts* of the light emitted from the various parts of the solar disk (Figure 3.6).

When a source of light approaches an observer (or vice versa), the wavelengths of the absorption lines are all decreased. All of the lines are moved toward the blue end of the spectrum. The lines are said to be *blue-shifted*. When the light source is moving away from the observer, the wavelengths of the absorption lines are all increased. They are moved as a unit toward the red end of the spectrum. They are said to be *red-shifted*. One can determine the rate at which the Sun is rotating by pointing a spectroscope toward different places on the solar limb and measuring the differing amounts of Doppler shifts that are present in the absorption lines.

The measurements indicate that the solar rotation is in the direct sense, that is, in the same direction that the Earth travels around the Sun. The equator

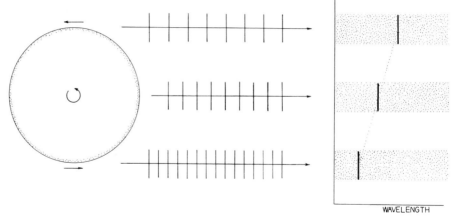

FIGURE 3.6 Doppler shift measurement of the solar rotation rate.

of the Sun is approximately parallel to the ecliptic, so the polar regions of the Sun can be seen only at extremely oblique angles from Earth. A point on the solar equator completes one turn every 24 days, 16 hours, which corresponds to a tangential velocity of 2.0 km/sec.

However, the Sun appears to be rotating at a significantly slower rate at higher latitudes (Figure 3.7). At 30°C from the equator, the period is 26 days. At 75°, it is 33 days. Near the poles, the rotational period may be as long as 35 days. The *differential rotation* of the Sun was first reported by the English astronomer Richard Carrington (1826–1875) in 1861.

SUNSPOTS

The most prominent features on the solar photosphere are the *sunspots* (Figure 3.8). They are dark, roughly circular regions that occasionally appear on the otherwise rather featureless surface. Sunspots seem to have first been reported by Galileo and independently by Johann Fabricus (1564–1617) in the year 1611, although they are apparently described in some ancient Greek writings. The Chinese had known of them for at least 1000 years before Galileo.

Sunspots can range in size from a 1000 km across up to as much as 150,000 km, although the average size is about 10,000 km. The relative darkness of sunspots is a contrast effect; they are localized cool regions on the solar surface that are roughly 2000°C lower in temperature than the surrounding photosphere. The dark central portion of a sunspot is usually surrounded by a somewhat brighter outer region that is marked by radially oriented patterns of filaments. The darker central region of a sunspot is called the *umbra*; the brighter outer part is called the *penumbra*.

The eighteenth-century English astronomer William Herschel (1738–1822) thought that sunspots were "holes" in a glowing solar atmosphere, permitting a cooler (and presumably habitable) surface to be seen! Others imagined

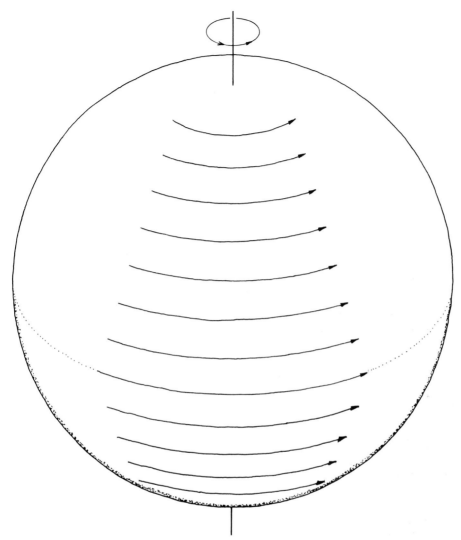

FIGURE 3.7 The differential rotation of the Sun. The equatorial regions of the Sun rotate at a slightly faster rate than do the polar regions.

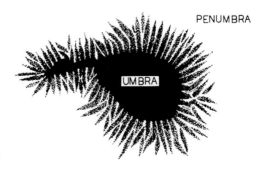

PENUMBRA

UMBRA

FIGURE 3.8 Drawing of a sunspot.

that sunspots were solar hurricanes or other such violent phenomena. However, the cool interior of a spot is relatively quiet in comparison to the simmering, turbulent surface of the outside photosphere.

The cause of sunspots remained mysterious until 1908. In that year, the American astronomer George Ellery Hale discovered that sunspots have intense magnetic fields associated with them. This discovery was made possible by prior knowledge of the *Zeeman effect,* a phenomenon that takes place when atoms are placed in a strong magnetic field. The characteristic optical absorption lines of the atoms are split into several separate, closely spaced components by the field, the amount of splitting being directly proportional to the magnitude of the field. By studying the polarization of the light that is absorbed by the atoms, it is even possible to determine the magnetic field's direction!

The magnetic fields in the vicinity of sunspots can be up to 10,000 gauss (=1 tesla) in intensity, many thousands of times stronger than Earth's magnetic field and equivalent to the strongest magnetic fields produced in laboratories on Earth. The highest magnetic field intensities are found in the cool central umbras of sunspots, with weaker fields existing in the warmer outer penumbras. Superimposed on the intense sunspot fields is a far weaker global magnetic field of a few gauss in intensity. It is complicated in structure and variable in intensity and does not always have a north – south polar orientation as the terrestrial magnetic field does.

Are strong magnetic fields the cause of sunspots or are they an effect? It is generally thought that the strong magnetic fields associated with sunspots are responsible for their low temperature although the details are poorly understood. High-energy charged particles cannot readily enter regions where intense magnetic fields are present, so convective currents from the interior tend to be suppressed. Consequently, sunspots are relatively "cold" regions on the hot solar surface. The heat leaking upward from the interior has to go somewhere; some of it is diverted to the edges of sunspots, which are a bit hotter than the surrounding photosphere.

THE 11-YEAR SUNSPOT CYCLE

A typical sunspot comes and goes within a week; although some of the larger ones can last a month or more. Throughout the eighteenth century, it was generally recognized that the number of spots on the Sun at any time did not remain the same from one year to the next but varied in some sort of cyclical pattern. During the early nineteenth century, the German amateur astronomer Heinrich Schwabe (1789–1875) spent 20 years observing the Sun. In 1843 he announced that sunspots appeared and decayed in an 11-year cycle (Figure 3.9).

The sunspot cycle is not perfectly regular; successive maxima can be as close as 7.5 years and as far as 16 years apart. At the minimum in the cycle, there are only three or four spots seen on the entire solar disk. Sometimes there are none at all. At the maximum, there can be 100 spots or more. At the beginning of a cycle, the first spots start to break out at latitudes approximately 40° from the

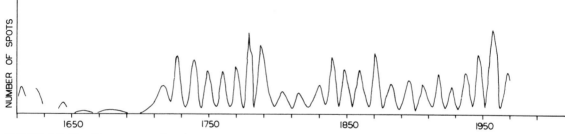

FIGURE 3.9 The 11-year sunspot cycle.

solar equator. As these initial spots fade and die, they are replaced by newer and more numerous spots at progressively lower latitudes, closer to the equator. The maximum occurs about 4 years after the beginning of the cycle, at which time most spots are concentrated about 15° from the equator. The number of spots begins to decline thereafter but the equatorial migration continues. By the time that the last few spots finally vanish, they have approached within 8° of the equator. Oddly enough, sunspots are never found above 45° latitude and they tend to avoid a narrow belt around the equator.

Sunspots usually appear in pairs, with magnetic field lines leaving one of the spots and entering the other (Figure 3.10). One of the spots is a magnetic north pole — the other, a south pole. The relative magnetic polarities of the individual spots in a pair exhibit an extremely interesting pattern. Every one of the spots that is leading in the direction of solar rotation in the northern hemisphere has the same polarity! A similar leading – trailing pattern is also seen in the southern hemisphere but the polarity is reversed! When the sunspot cycle is repeated 11 years later, the same sort of pattern is exhibited, but the polarity is reversed in both hemispheres! This behavior is shown in Figure 3.11. Because of the polarity change at the start of each cycle, one should strictly speak of a 22-year sunspot cycle rather than an 11-year cycle.

Although the cause of the 11-year sunspot cycle is very poorly understood, it is generally suspected that the differential rotation of the Sun is responsible (Figure 3.12). The equatorial regions of the Sun are rotating at a slightly

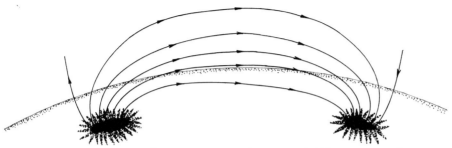

FIGURE 3.10 Sunspots generally occur in pairs, with magnetic field lines emerging from one spot and entering the other.

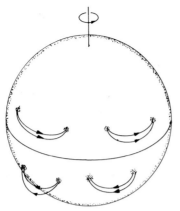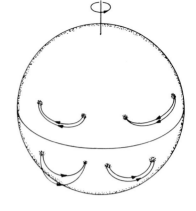

FIGURE 3.11 Magnetic fields and sunspot cycles.

faster rate than the polar regions. This differential rotation results in a strong electrical current in the gaseous ions just below the surface, which in turn produces an intense subsurface magnetic field. As the differential rotation continues, the subsurface magnetic field becomes stronger and stronger and the field lines get wound tighter and tighter together. More and more rotational kinetic energy is converted into magnetic energy. The buildup of magnetic energy cannot continue forever. Eventually, the subsurface magnetic field becomes so strong that it bursts through the photosphere in isolated areas. The places where the interior solar magnetic field penetrates the photosphere are

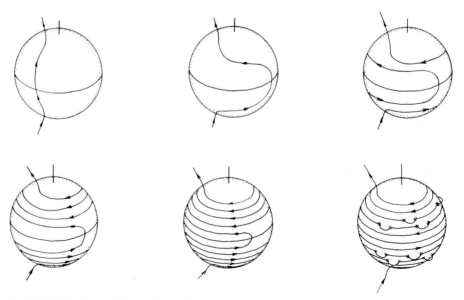

FIGURE 3.12 Differential rotation and the sunspot cycle.

seen as dark spots on the surface. The formation of sunspots releases the magnetic energy stored in the solar interior, leaving the Sun free to begin another round of the cycle.

In 1894 E. Walter Maunder, the superintendent of the Solar Division of the Royal Greenwich Observatory in London, reported that his examination of old astronomical records revealed that there had been a 70-year period in the seventeenth and early eighteenth centuries during which almost no spots were seen on the Sun (Figure 3.9). Two maxima took place in the 30 years following the discovery of sunspots. However, the spots declined in number to a low level in 1645 and then vanished almost completely until the year 1715. The sunspot cycle we know today started back up again in 1715 and has continued ever since. The cause of the *Maunder minimum* is uncertain; apparently there are longer-period variations superimposed on the 11-year cycle.

THE SOLAR ATMOSPHERE

During a total solar eclipse, the Moon completely covers the solar photosphere. In the instant just before totality is reached, a bright red flash will appear on the solar limb just outside the photosphere. This red flash is caused by the light emitted by the *chromosphere* — the lowest level of the solar atmosphere (Figure 3.13).

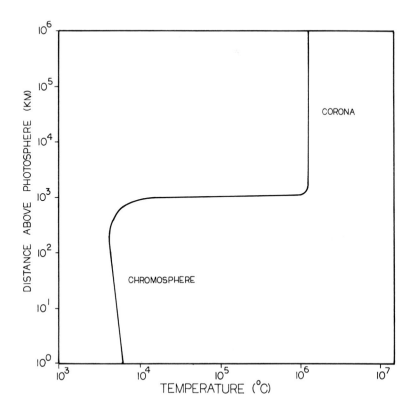

FIGURE 3.13 Vertical profile of the solar atmosphere.

The chromosphere consists of a layer of hot gases approximately 2000 to 3000 km thick. The chromosphere is variable in extent and can reach out as far as 4000 km from the visible surface. The chromosphere is much hotter than the photosphere and is rich in ionized gases (that is, composed primarily of protons and electrons). The density of gas in the chromosphere steadily decreases with altitude; at the very top of the chromosphere, the density is 100,000 times less than it is at the surface of the photosphere.

The cooler lower layers just above the photosphere are at 4200°C, cool enough to exhibit absorption lines attributable to the presence of partially ionized calcium, sodium, iron, nickel, magnesium, aluminum, titanium, and chromium atoms. The temperature of the chromosphere rapidly rises with altitude; it can get as hot as 50,000°C in its very uppermost levels. At such elevated temperatures, the gases of the chromosphere are hot enough to produce strong emission lines, the most intense of which is due to hydrogen. The strongest hydrogen emission line is the bright red Hα line and is responsible for giving the chromosphere its characteristic red color. In addition, the high temperature of the upper chromosphere results in a strong ultraviolet and X-ray emission. Because such radiation is strongly absorbed by the Earth's atmosphere, it can only be studied from outer space.

The temperature of the chromosphere steadily rises with increasing altitude above the solar photosphere. The energy source responsible for the high temperature of the upper chromosphere is mysterious. Most solar physicists believe that it is heated by the turbulence at the surface of the photosphere that is generated by the convective currents coming upward from the interior.

At an altitude of 5000 to 10,000 km above the photosphere, the characteristic red color of the chromosphere disappears. It is replaced by a continuous white background light that extends outward as far as a couple of million kilometers from the Sun. This is the *corona*. It can be seen from Earth in particularly spectacular fashion at the time of a total solar eclipse.

The corona is distinguished by its enormously high temperature. Values as high as many millions of degrees Celsius have been observed. The transition between the chromosphere and the corona is quite abrupt, the temperature jumping from 50,000°C to several million degrees Celsius in only a few tens of kilometers. The origin of the high temperature of the solar corona is very poorly understood, but it is generally thought that it is caused by the dissipation of frictional energy generated by convective motions at the surface of the photosphere.

The characteristic white color of the solar corona is caused by light that is scattered from the highly energetic free electrons that are present in the hot coronal plasma. These hot electrons scatter some of the light emitted from the surface so that the visible-light portion of the corona spectrum closely matches that of the solar photosphere. In addition, the high temperature of the corona produces an intense ultraviolet and X-ray emission.

Superimposed upon the continuous white-light spectrum of the corona is a series of emission lines. Their discovery caused a sensation since they did not match the emission spectra of any known element on Earth. At first it was thought that they must be produced by some as-yet-undiscovered element present on the Sun. This mystery element was given the name *coronium*. In 1940 the Danish scientist Bengt Edlen showed that these anomalous lines were not produced by any new element but were instead emission lines produced by familiar elements that were in highly ionized states. The temperature of the corona is so high that atoms can be stripped of as many as a dozen or more electrons, leaving highly ionized species behind. The most prominent coronal emission line is an intense green line produced by iron atoms that have lost 13 electrons.

Millions of tiny spikes extend outward from the upper chromosphere and protrude into the corona. These spikes, called *spicules,* are 300 to 1000 km wide and are several thousand kilometers high. They are the visible signs of massive vertical currents of hot gases. The vertical velocities of these gases can reach speeds as high as tens of kilometers per second. In the telescope, spicules look like blades of grass or tongues of fire. They typically last only a few minutes. The rapid motions of the gases in the spicules may have something to do with the high temperature of the corona.

The shape and extent of the solar corona are largely dictated by the pattern of the solar magnetic field, in particular by the intense magnetic field lines that penetrate the photosphere in the regions of the sunspots. The highly energetic hot ions of the corona are constrained to travel along the magnetic field lines that originate farther down in the solar interior.

The intense magnetic field lines that connect the members of a sunspot pair to each other are strong enough to trap large numbers of protons, electrons, and other ions. Once entrapped, these ions rapidly cool and recombine, creating a long, thin "tube" of cool, dense, un-ionized gas (primarily hydrogen) that loops through the intensely hot corona. The gas inside the magnetic flux tube is actually cooler than the photosphere. When it is seen up against the visible surface of the Sun, it shows up as a thin, dark line rich in prominent hydrogen absorption lines. Such features are known as *filaments*. They can last as long as several months before they disperse. When flux tubes appear in the limb of the Sun, they show up as long, thin loops of glowing gas that protrude many thousands of kilometers from the solar disk. When they are seen up against the limb in this fashion, these tubes are called prominences. On occasion, prominences can reach out as far as 10^6 km from the surface.

THE WIND FROM THE SUN

The solar corona is somewhat leaky and many of its particles have enough energy to escape from the Sun altogether and flow into outer space. Most of these particles are protons and electrons, although some heavier ions manage to escape as well. This stream of particles flowing away from the Sun is known as the *solar wind*. The existence of the solar wind was first confirmed by the early Soviet and American spaceprobes launched during the late 1950s, although scientists had long suspected that it was present. Typical particle speeds are 300 to 800 km/sec, and the average particle flux at the Earth is about $1 \times 10^8/m^2/sec$.

Most of the solar wind particles appear to come out of "holes" in the solar corona which are free of magnetic field lines. There are recurrent "gusts" in the solar wind as these coronal "holes" change their position with respect to the Earth. The solar wind coming from the poles of the Sun cannot be directly observed, but it can nevertheless be indirectly studied by examining the way it affects radio signals from distant celestial radio sources as they pass through it. The solar wind coming from the poles is relatively steady and is 50 percent faster than the solar wind in the plane of the solar equator. The polar regions appear to be permanent coronal "holes."

During the period of maximum sunspot activity, the size of the solar corona is larger than normal, and the solar wind is much more intense. This enhanced solar activity is produced by the release of the stored-up magnetic energy generated by the differential rotation of the Sun. Some of these solar wind particles impinge upon the Earth. Most are prevented from striking the atmosphere by the terrestrial magnetic field. However, a few become trapped in the magnetic field and spiral along the magnetic field lines. They enter the atmosphere over the north and south poles, ionizing air molecules as they travel. These ionized air molecules in turn emit visible light, producing the stunningly beautiful *auroras* often seen in extreme northern and southern latitudes at times of high solar activity.

EXPLOSIONS ON THE SUN

Perhaps the most spectacular thing that ever happens on the Sun is the *solar flare*. The existence of solar flares was first reported by the English astronomer Richard Carrington in 1859. Solar flares are intense discharges of energy that originate in the chromosphere. They typically last for a few minutes, but some of the larger flares can last as long as 1 hour. During a flare, large amounts of matter are violently ejected from the Sun, some particles reaching velocities as great as one-third the speed of light (10^8m/sec). A solar flare also releases an intense flood of X rays and ultraviolet light as well as a powerful surge of radio noise. A single large flare can give off enough energy to supply all the power needs of the United States for the next 60,000 years!

The cause of solar flares is unknown, but flares do tend to occur more frequently during periods of high sunspot activity. At the maximum in the sunspot cycle, a flare will occur about once an hour, and a particularly large flare will occur about once a month. Solar flares are of more than strictly academic interest, as they can have profound effects on the Earth. Some of the flood of matter ejected from the Sun during a flare will strike the Earth and disrupt the terrestrial magnetic field. Such an event is called a *magnetic storm*. During magnetic storms, all sorts of strange and unusual things can happen. Magnetic compasses are often upset, making ocean-going navigation hazardous. The large number of particles injected into the terrestrial magnetic field causes spectacular auroral displays. The flood of X rays emitted during a solar flare greatly increases the number of charged particles in the ionosphere. This causes the level of the ionosphere to shift, altering the pattern of long-range radio communication. On some occasions, the ionosphere becomes dense enough to absorb the radio waves impinging on it from below, shutting off intercontinental radio communication altogether. Last, but perhaps far from least, the intense radiation emitted during flares can be a real hazard for astronauts in outer space. One should therefore never plan a long-duration space mission during times of peak solar activity.

FLIGHTS TO THE SUN

The closest approach yet made to the Sun by a spacecraft is 4×10^7 km, reached by the German–American solar probe *Helios 2*. The European Space Agency (ESA) and the American National Aeronautics and Space Agency (NASA) had planned to collaborate on a highly sophisticated solar exploration mission in the late 1980s. A pair of spacecraft were to be placed in solar orbits highly inclined with respect to the ecliptic plane. This would give humanity its first clear view of the polar regions of the Sun. One craft was to be built in Europe, the other in the United States. Both craft were to be initially aimed toward Jupiter. Upon encounter with Jupiter, the craft would be deflected into highly elliptical solar orbits, which would bring them out of the ecliptic plane and pass them over the solar poles. Unfortunately, budgetary problems caused the American probe to be canceled in 1981. The European craft survived. It has been given the name *Ulysses*.

Ulysses was originally scheduled to be sent into space aboard the Space Shuttle *Challenger* in May of 1986. Once in Earth orbit, *Ulysses* was to be dispatched toward Jupiter by the use of a special upper stage (known as Centaur) that would burn high-energy liquid-oxygen/liquid-hydrogen fuel. However, the January 1986 destruction of the *Challenger* and the loss of its seven crew members forced an indefinite delay in the American space program, and the launch of *Ulysses* was postponed. The subsequent reappraisal of the space program resulting from the *Challenger* disaster led NASA officials to conclude that it was too dangerous to carry the Centaur upper stage aboard the shuttle. The Centaur project was canceled. A different upper stage must be developed before *Ulysses* can be flown. Alternatively, it might be possible to switch *Ulysses* to an expendable booster, most likely a Titan 34D7. In either event, the launch of *Ulysses* will probably be delayed until at least 1989.

In 1978 the Jet Propulsion Laboratory of Pasadena, California, proposed a close-encounter solar mission for the 1990s. The project is named *Starprobe*. The craft is to carry a conical ablative heat shield that will permit it to approach to within 2.5×10^6 km of the solar photosphere. During the encounter, the craft will study the solar wind and the magnetic fields within the corona. As in the case of the ESA solar polar mission, the close approach to the Sun will be made possible by an encounter with Jupiter. As yet, no funding has been provided.

PLANETS OF ROCK:
The inner solar system

PART TWO PHOTO:
Viking orbiter photograph of the southern hemisphere of Mars. The bright area near the bottom is the south polar cap. The Argyre basin is at the center. (Photograph courtesy of NASA)

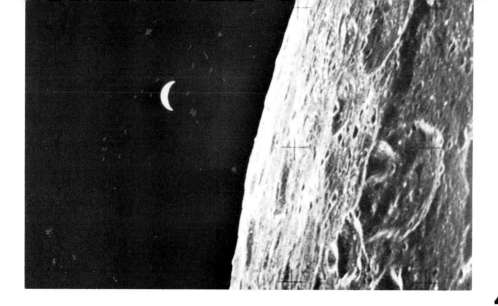

4

The Earth from a Planetary Perspective

It can perhaps be said that the most aesthetically pleasing products that have been produced by the space age are the color photographs of the Earth that have been taken from outer space (Figure 4.1). In these pictures, the Earth appears as a blue sphere interspersed with brownish areas, hauntingly beautiful up against the inky blackness of outer space. Apart from its beauty, there is not much to distinguish the Earth from the other planets in the Solar System. Some of the planets are much larger and some are a good deal smaller. What makes the Earth truly unique is that it is probably the only planet in the Solar System that has life.

Since we are so intimately familiar with the Earth, it is perhaps only natural for us to dismiss our planet as somehow trivial and uninteresting. However, the advent of space travel has given us a lot of new knowledge about the other planets in our Solar System and has forced us to look at our world in a new perspective. In particular, we have discovered that the existence of human life and intelligence on Earth is the happy result of a long series of extremely fortunate circumstances. If any one of them had been seriously altered, we would not be here.

Photograph of Earth taken from the Apollo 11 command module in orbit around the Moon. (Photograph courtesy of NASA)

FIGURE 4.1 Photograph of Earth taken by *Apollo 11*. The African continent is at the center left. Saudi Arabia is at the top. (Photograph courtesy of NASA)

THE EARTH'S ORBIT

The Earth's orbit around the Sun is approximately circular (with an eccentricity of only 0.017). The average Earth–Sun distance is 1.49×10^8 km (93 million miles). This distance defines the *astronomical unit* (or AU), a scale often used in discussing distances within the Solar System. The eccentricity of the Earth's orbit causes our planet to be about 5×10^6 km closer to the Sun at the time of perihelion than it is at the time of aphelion. Consequently, sunlight on Earth varies in intensity by about 6 percent over the course of a full year. Perihelion occurs on January 4, near the beginning of winter in the northern hemisphere.

Earth's elliptical orbit around the Sun defines a fixed plane in the sky, one which maintains a constant orientation relative to the stars (Figure 4.2). Astronomers refer to this plane as the *ecliptic plane;* it is a convenient reference for describing the positions of the distant stars. A hypothetical observer stationed in space above the north pole will see the earth travel around the sun in a counterclockwise direction. This defines the direct (or prograde) sense (Figure 4.3). An object traveling around the Sun in the opposite direction is said to have a retrograde orbit. The other eight planets in the Solar System also have prograde orbits; they all travel around the Sun in the same direction as the Earth does.

The time taken for the Earth to make one complete orbit of the Sun with respect to the fixed stars is called the *sidereal year.* This is 365.24 days long. During its revolution around the Sun, the Earth travels at an average speed of 30 km/sec.

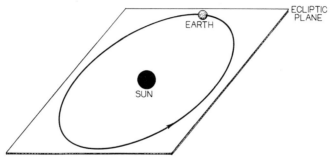

FIGURE 4.2 The ecliptic plane — the plane of Earth's orbit around the Sun.

THE ROTATION OF THE EARTH

The Earth rotates once a day on its axis, bringing alternate periods of light and dark to most points on the globe. The rotation is in the direct sense, that is, counterclockwise as viewed from the north pole (Figure 4.4). As seen from the Earth, the Sun travels across the sky in a clockwise direction, rising in the east and setting in the west. The *solar day* is defined as the time interval between the

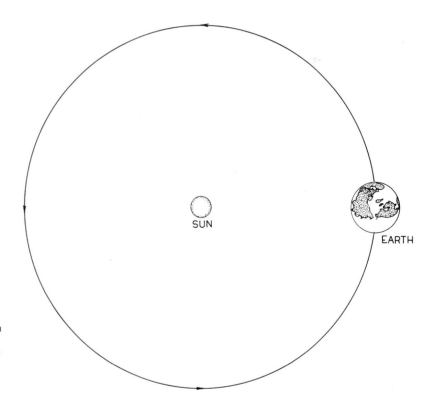

FIGURE 4.3 Earth's orbit around the Sun. If one is looking down on the Solar System from a point above the Earth's north pole, then the Earth travels in a counterclockwise direction about the Sun.

FIGURE 4.4 Definition of direct rotation. To an observer located above the north pole the Earth appears to be rotating in a counterclockwise direction about its axis.

instant that the Sun reaches the highest point in the sky and the same instant the next day. Because the Earth's orbit is not a perfect circle, the length of the solar day depends on the time of year. The solar day is slightly longer when Earth is at aphelion than when Earth is at perihelion. The *mean solar day* is defined as the length of the solar day averaged over the entire year. It is arbitrarily taken to be exactly 24 hours long.

Sometimes, another type of day is discussed. This is the *sidereal day,* which is defined as the time taken for the Earth to make one complete turn with respect to the fixed stars. Because of the Earth's revolution around the Sun, the sidereal day is shorter than the mean solar day by 1 part in 366 (about 4 minutes) (Figure 4.5).

Over human historical time scales the sidereal day is constant in length. However, over geological times there is a slow but steady change. Accurate time measurements indicate that the length of the day is increasing at an average rate of about 20 sec every 1 million years. The slow braking of the Earth's rotation is primarily due to tidal coupling with the nearby Moon. The rotational history of the Earth can be traced back to ancient times by studying the growth patterns of fossil shells and corals. These studies tell us that 1.5 billion years ago the day was only 10 to 12 hours long!

THE FOUR SEASONS

Perhaps the most readily noticed periodic phenomenon that takes place in our lives is the coming and going of the *seasons.* Contrary to general opinion, the eccentricity of the Earth's orbit has nothing to do with the passage of the seasons. Winter in the northern hemisphere actually comes when the Earth is at

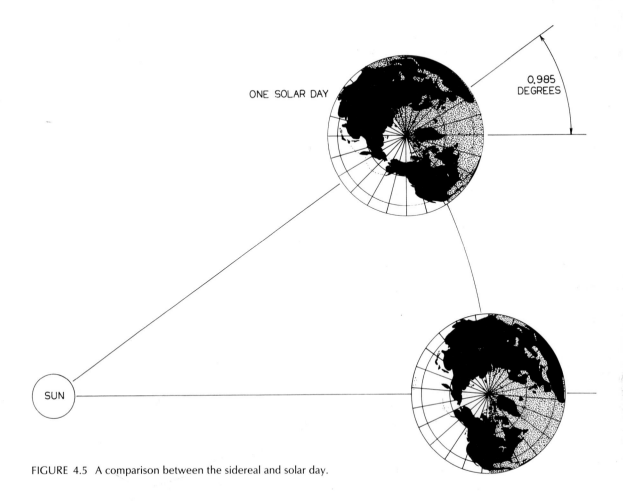

FIGURE 4.5 A comparison between the sidereal and solar day.

perihelion, closest to the Sun. The seasons are caused by the fact that the Earth's equator is not parallel to the ecliptic plane. The angle between them is approximately 23.5°. As the Earth travels around the Sun, the Earth's rotational axis remains pointing toward the same spot in the sky (toward the star Polaris in the constellation Ursa Minor). Consequently, the maximum apparent height of the Sun in the sky at any one location on the Earth will be different at different times of the year. The higher the average height of the Sun in the sky, the greater the average amount of solar energy that falls on a unit area of the surface and the higher the average temperature (Figure 4.6).

In the northern hemisphere the Sun reaches its highest point in the sky on June 21, the first day of summer. On that day the noonday Sun appears directly overhead at points lying along the 23.5°N latitude line (the so-called Tropic of Cancer). This is a line parallel to the equator, passing just south of the tip of Florida. This particular event is called the *summer solstice* and was considered a special instant in time by many ancient cultures, so special, in fact, as to have mystical or religious significance. Stonehenge, an ancient megalithic monument located near Salisbury in southern Britain, was built in such a way that its major axis pointed toward the spot on the local horizon where the Sun rose at the time of the summer solstice (Figure 4.7).

The Sun is at its lowest point in the northern sky on December 21, the first day of winter. On that day, the noonday Sun is directly overhead at all points along the 23.5°S latitude, the Tropic of Capricorn. When the northern hemisphere is in winter, the southern hemisphere is in summer.

The noonday Sun is directly overhead at the equator at two instants during the year, on March 21 (the *vernal equinox*) and on September 21 (the *autumn equinox*). The word "equinox" means "equal night" — at the time of the equinoxes, the daylight and nighttime periods are equal in duration. At the time of the vernal equinox the Sun appears to be located in the constellation Pisces. The line between the Earth and the Sun's location in the sky at the time of the vernal equinox changes only slowly in orientation from year to year and is often used as a baseline from which astronomical measurements of the positions of the stars are made.

Near the poles the axial inclination can cause some rather bizarre effects that appear odd to those people who are used to living at lower latitudes. The best-known of these effects is the so-called midnight sun. The summer sun will appear in the sky all day long for at least one day out of the year at all points northward of latitude 66.5°N (the Arctic Circle), and at all points southward of latitude 66.5°S (the Antarctic Circle). During the winter in these polar regions, there is at least one day out of the year in which the Sun never rises in the sky at all.

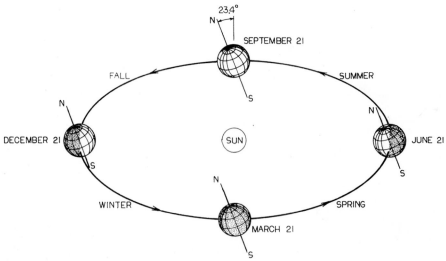

FIGURE 4.6 Origin of the seasons. The Earth's equator makes an angle of 23.5° with respect to the ecliptic plane. During the northern hemisphere's summer, the north pole is tilted toward the Sun. During the northern hemisphere's winter, the north pole is tilted away from the Sun.

FIGURE 4.7 Stonehenge, a megalithic monument in Britian. This is a view taken from the interior of the monument, looking east. At the time of the summer solstice, the Sun rises directly above the stone seen at the center of the middle arch.

VARIATIONS IN THE EARTH'S MOTION

The Earth's orbital eccentricity seems to be constant over time scales of years or centuries, but over periods of many millenia there are definite changes. There is a slow periodic drift in orbital eccentricity, between a minimum of virtually 0.00 and a maximum of about 0.06. This oscillation in eccentricity has a period of approximately 100,000 years. Earth's eccentricity is currently near the lower limit of these extremes, but is slowly advancing toward larger values. The changes of eccentricity seem to be driven by periodically varying gravitational forces exerted on Earth by the giant planet Jupiter.

Because of its rapid rotation the Earth is slightly bulged at the equator; the polar diameter is slightly less than the equatorial diameter. As a consequence of the Earth's equatorial bulge, both the Sun and the Moon can exert small but significant gravitational torques on our planet. These torques cause the rotational axis of the Earth to *precess* slowly in space in a conical fashion (Figure 4.8). Relative to terrestrial observers, the location of the Earth's north pole appears to traverse a circular path in the stars, taking 25,800 years to make a full circuit. At present, the pole star is Polaris. However, at the time the Pyramids of Egypt were built, Alpha Draconis was the pole star. Thirteen thousand years from now, Vega will be.

This slow axial precession causes the Sun's apparent position in the sky at the time of the vernal equinox to advance slowly along the celestial equator,

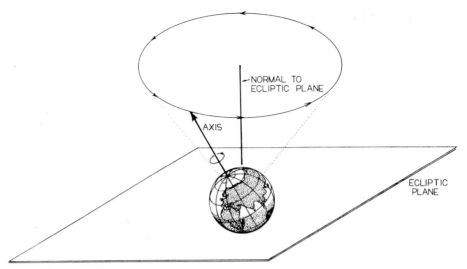

FIGURE 4.8 Precession of the equinoxes. The Earth's axis slowly precesses in a conical fashion about a line perpendicular to the ecliptic.

moving in and out of the 12 constellations of the zodiac. Consequently, this phenomenon has come to be known as the *precession of the equinoxes.*

The precession of the equinoxes causes the changing seasons to get out of phase with the sidereal year. The *tropical year* is defined as the length of time that it takes for the Sun to return to the vernal equinox. It is 20 minutes longer than the sidereal year.

The precession of the equinoxes takes place at such a slow rate that only precise astronomical measurements can detect any changes in the sky that take place over time intervals as short as a few years. Nevertheless, the phenomenon was known in ancient times. The Greek astronomer Hipparchus (190 – 120 B.C.) was the first to report its existence. The vernal equinox is now located in Pisces, but is slowly moving along the ecliptic toward the constellation Aquarius. The precession of the equinoxes is the source of many cyclical theories of history and has been incorporated into astrology. Since we now know that the precession of the equinoxes is caused by the Earth's response to perfectly natural forces, much of the mysticism and superstition surrounding the phenomenon is removed.

Superimposed upon the precession of the equinoxes is a periodic change of the angle between the Earth's equator and the ecliptic plane. This angle bobs back and forth between a minimum of 22.1° and a maximum of 24.5°, with a period of approximately 41,000 years. The last maximum occurred about 9000 years ago, and the current trend is toward a smaller tilt. Like the precession of the equinoxes, the gradually changing axial inclination is driven primarily by gravitational torques exerted on the Earth by the Sun and the Moon.

THE EARTH'S CHANGING ORBIT—
THE CAUSE OF THE ICE AGES

There is abundant geological evidence that over the past million years the Earth has been subjected to a succession of ice ages, which have periodically covered much of its surface with massive glaciers. At the beginning of an ice age cycle, the average overall temperature of the entire Earth begins to decrease. This decline causes the amount of ice and snow near the poles to increase. As the ice and snow coverage at the poles steadily rises, the temperature drops still further. The falling temperatures cause the sheets of ice to creep farther and farther toward the equator. At the height of the last ice age, areas as far south as the central United States became encased in massive sheets of ice. Eventually, the temperature decline ceases, and the temperature begins to rise once again. The warmer temperatures melt the outer edge of the ice pack, forcing the ice to recede toward the poles. We are currently in an interim period, recovering from the previous ice age and awaiting the next.

What is the cause of the ice ages? A possible clue may be found in the geological record of the past million years. A careful examination of this record seems to indicate that the waxing and waning of the ice has followed a remarkably regular cyclic pattern. The time interval between successive ice ages has been approximately 100,000 years. The remarkably consistent periodicity of the ice ages has prompted several workers to look for extraterrestrial causes. In 1941 the Serbian meteorologist Milutin Milankovich suggested that the periodic waxing and waning of ice ages was caused by the slowly changing orbital eccentricity of the Earth.

Why should the small variation in the eccentricity of the Earth's orbit affect the weather so drastically? Because of the eccentricity of the Earth's orbit, the intensity of sunlight that strikes our planet varies over the course of a year. Sunlight is slightly weaker when our planet is at aphelion than when it is at perihelion. Since the Earth moves more slowly along its orbit at aphelion than it does at perihelion, our planet spends more of its time distant from the Sun than it does closer to the Sun. When the eccentricity of the Earth's orbit increases, this effect is exaggerated, causing the average annual solar flux to decline by a slight amount. Acting over the course of many years, even a relatively tiny decrease in the solar energy input to the Earth will cause the average temperature of the planet to decline.

While the orbital eccentricity of the Earth is increasing, there is a progressive decline in the average temperature of the hemisphere that is in summer at the time of aphelion. Cooler summers mean less snow melting and a slow but steady growth of the polar ice caps. The growth of glaciers near the poles increases the overall reflectivity of the Earth, decreasing the amount of solar energy absorbed and lowering the average temperature of the entire planet. The declining temperature in turn causes the ice pack to advance even farther toward the equator. Over the course of many years, enough ice can accumulate to cover areas quite distant from the poles with massive glaciers as much as 1 mile thick. The drop in average temperature required to trigger a major ice age is rather small; at the height of the last ice age (about 18,000 years ago) the average temperature was only 5°C cooler than it is now. The glaciation process reverses itself when the orbital eccentricity begins to decrease once again, bringing warmer summers to the hemisphere tilted toward the Sun at aphelion and forcing the ice pack back toward the poles.

The dominant effect seems to be the growth of the north polar ice caps rather than the southern ones, since the southern hemisphere has no large landmasses between 40°S and 70°S upon which large glaciers can grow. The last ice age ended approximately 11,000 years ago; the next one will come 90,000 years from now.

THE EARTH'S INTERIOR—JOURNEY TO THE CENTER OF THE EARTH

Contrary to popular understanding, most of the ancients were perfectly aware that the Earth was not flat. The first ancient philosopher known to have taught that the Earth is curved was Anaximander of Miletus. Anaximander, however, pictured the Earth as a cylinder, not a sphere. The Greek mathematician and philosopher Pythagoras (546–480 B.C.) seems to have been the first to suggest that the Earth has a spherical shape. Reasonably accurate measurements of its diameter were made as early as 250 B.C. by the Greek astronomer Erastosthenes (276–196 B.C.).

The modern value for the mean radius of the Earth is 6378 km. The Earth is, however, slightly flattened at the poles, being about 50 km smaller in diameter there than it is at the equator. The fractional difference between the polar and equatorial diameters is called the *oblateness* and is about 4 parts in 1000. It is a consequence of the Earth's spin—the matter near the equator being pulled away from the rotational axis by the centrifugal force produced by the rapid rotation.

The presence of the Moon enables the mass of the Earth to be measured. It is 5.98×10^{24} kg, which gives the Earth an average density of 5.5 g/cm³. The density can be used to get some idea of the internal composition of the Earth. The soil and rocks near the Earth's surface have an average density of only 3.0 g/cm³ so the interior of the Earth must be made of significantly denser material.

However, the interior of the Earth is under extremely high pressure, and the matter located there is under heavy compression. Theoretical calculations indicate that if it were magically possible to remove this high pressure, the average density of the Earth would be only 4.04 g/cm³.

Even though the interior of the Earth is inaccessible to direct examination, it can nevertheless be studied by sending seismic waves through it. These investigations have established that the Earth's interior can be divided into three distinct regions: a central *core,* an inner *mantle,* and an outer *crust* (Figure 4.9).

At the center of the Earth lies the core. The Earth's core is characterized by an abrupt change in the seismic response at a depth of 3000 km below the surface. The material below this depth will not transmit seismic shear waves.

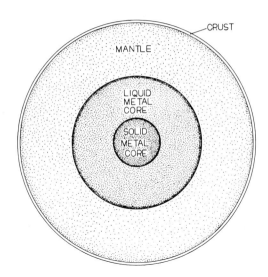

FIGURE 4.9 The interior of the Earth.

This must mean that the matter at the center of the Earth is in the liquid rather than the solid state.

The radius of the central core is 3480 km. The outer boundary of the core is quite sharp, and the density at the interface abruptly jumps from 6.0 to 9.9 g/cm³. The temperature within the core is estimated to be 2700 to 3000°C, high enough to keep all of the material located there in the liquid state. The pressure in the core is an enormous 1.5×10^6 to 3.0×10^6 atm. The molten core occupies only 16 percent of the Earth's volume but comprises 31 percent of its mass.

The only known liquid substance with a density as high as 10 g/cm³ is pressurized molten metal. Of all the metallic elements, only iron and nickel are thought to be sufficiently abundant in the universe to have made up so massive a core. A 4 percent solution of nickel in iron (perhaps diluted with 8–10 percent sulfur) is thought to be the approximate composition, since such a mixture will satisfy all of the density and melting point requirements. The high concentrations of iron, nickel, and sulfur in the core may explain why these elements are significantly depleted over their cosmic abundances in the outer crust of the Earth.

There is some evidence for the presence of a small, solid kernel at the very center of the Earth. This inner core is probably made almost entirely of solid crystalline iron and is relatively poor in sulfur. Its density is 13 g/cm³, the pressure is 3.6×10^6 atm, and the temperature is 3000°C. The pressure is so high that it prevents the metal from melting.

Outside the core lies the *mantle*. The mantle accounts for nearly half the Earth's radius, 83 percent of its volume, and 67 percent of its mass. Seismic studies indicate that the upper mantle has an average density of 3.3 g/cm, but there is an increase of density with depth, reaching a value as high as 6.0 g/cm³ at a depth of 3000 km just outside the core. At that depth the pressure is 1.5×10^6 atm and the temperature is 2500°C.

Although we have never been able to drill a hole deep enough to reach the mantle, we nevertheless have access to samples of rock that were once located there. Numerous volcanic lavas contain certain coarse-grained dark rocks that are classified as *peridotites*. They are rich in a mineral called *olivine* [which has the chemical formula $(Mg,Fe)_2 SiO_4$]. Laboratory experiments show that the olivine in these rocks must have solidified at pressures at least as high as 10,000 to 20,000 atm and at temperatures of the order of 1000 to 1400°C. These conditions are believed to exist at depths of 30 to 60 km, in the very uppermost parts of the mantle. These high-pressure minerals were stabilized when the rocks containing them came to the surface and rapidly cooled.

There is another type of rock that may originate from even farther down in the mantle. This is *kimberlite,* a magnesium- and potassium-rich volcanic rock with a high water content. It is of special economic importance because it is the primary source of diamonds. Diamond is an exceedingly rare crystalline modification of carbon, one that can form only at pressures of 25,000 to 50,000 atm and at temperatures of 800 to 1400°C. Such conditions are believed to exist at depths of 150 to 300 km, well into the upper reaches of the mantle.

It is reasonable to presume that much of the mantle is composed of peridotite minerals. Rocks such as peridotite that are rich in magnesium and iron are termed *ultramafic* (*ma-,* magnesium; and *-fic,* ferric or iron). More than 90 percent of the weight of the mantle is believed to be composed of ultramafic rock, with the remainder being primarily oxides of elements such as aluminum, calcium, and sodium. As much as 0.1 percent of the weight of the mantle may be composed of materials such as water and carbon dioxide.

Despite the high temperatures throughout the lower mantle, we know it is nonetheless almost entirely solid, since it will transmit seismic shear waves. However, the very uppermost layers of the mantle directly beneath the crust are at a temperature just below the melting point, and are able to flow (or "creep") under the application of stress. Some regions of the upper mantle are actually molten, giving rise to volcanism when this material is forced upward to the surface.

The outermost crust is, of course, the part of the Earth that is most familiar to us. It has been the subject of intensive study by geologists over the last 200 years. It seems to be chemically and seismically distinct from the upper mantle. The crust is approximately 30 km thick and has a density ranging from 2.7 to 3.0 g/cm³. It is rich in silicate minerals, with aluminum, iron, calcium, sodium, potassium, and magnesium atoms being abundant.

The rocks and minerals found in the crust fall into three general classes: *igneous, sedimentary,* and *metamorphic.* Igneous rocks have a volcanic origin; they were produced when molten material (known as *magma*), originating from deep in the Earth, came to the surface, cooled, and solidified. Sedimentary rocks, on the other hand, have an oceanic origin. They are the products of the gradual accumulation of minerals that slowly settled to the ocean bottoms. Metamorphic rocks are a composite variety. They have a sedimentary origin, but they show evidence of having been subjected to extended periods of extreme heat and high pressure subsequent to their formation.

The crust is laterally subdivided into two distinctly different types of regions, the lower-lying *oceanic floors* and the elevated *continental landmasses*. They are chemically and morphologically quite different from each other. At the present time, about 70 percent of the Earth's surface is covered by oceanic floors, leaving only 30 percent for the continental landmasses.

Oceanic crust is generally rather thin, with an average thickness of only 6 km. The average density of the material in the oceanic crust is about 2.9 g/cm³. The rocks on the ocean floors are primarily made up of silicate chemical compounds that are rich in magnesium, iron, calcium, and aluminum. Plagioclase and pyroxene are the dominant minerals present. The oceanic crustal rocks are of a type classified by geologists as *basalts*. Basalts are hard, dense, darkly colored rocks that have a rather fine-grained crystalline structure. Basalts have a volcanic origin; their fine-grained structure is evidence that these rocks have rapidly cooled and solidified from an initially molten state. They were formed when hot lava coming from the interior of the Earth encountered the cold waters of the ocean depths.

The continental landmasses have an average density of 2.7 g/cm³, slightly less than that of typical oceanic crust. Continental crust is generally much deeper than oceanic crust; crustal thicknesses as high as 70 km have been measured. Continents are usually situated at much higher elevations than oceanic crust, the average difference in height between the two being approximately 5 km. This difference in elevation is maintained by the geological phenomenon known as *isostasy,* in which dense, low-lying oceanic crust hydraulically supports an elevated, lighter continental landmass.

The rocks and minerals that make up the continents are classified by geologists as *granites*. Granitic rocks are somewhat different in composition and structure from the rocks that make up the ocean floors. Granitic rocks are richer in sodium, potassium, and silicon than basaltic rocks, but they are poorer in iron, magnesium, and calcium. Granites are much lighter in color than basalts and have a coarser-grained crystalline structure. The coarse-grained crystalline structures of granitic rocks indicate that they must have cooled at a much slower rate than did the basaltic rocks of the ocean floors.

THE ATOMIC CLOCK AND THE AGE OF THE EARTH

How old are the rocks, mountains, and soil of the Earth? Are they only a few thousand years old, or have they been there forever? Fortunately, there is an extremely precise means by which one can determine the answer to this question. This is the method of *radioactive dating,* which measures the length of time that a rock has existed in the solid state. At the time of the initial formation of a solid rock from a cooling liquid melt, various amounts of radioactive materials (in particular, rubidium 87, thorium 232, uranium 238, uranium 235, potassium 40, samarium 147, and rhenium 187) are trapped in the crystalline minerals that form the rock. These radioactive elements decay at known rates to produce stable "daughter" elements that remain trapped in the rock. By measuring the relative amounts of parent and daughter isotopes present in the interior of a rock, the amount of time that has elapsed since its initial solidification can be determined.

Rocks vary widely in their radioactive ages. Some rocks are relatively young, with ages measured in millions of years or less. Other rocks are incredibly old, having ages measured in billions of years. However, it is quite rare to find any rocks that are much older than 2.8 billion years. For a long time, the oldest known rocks were those found in the Isua formation located in Greenland. This formation consists of greenstone rocks and associated sediments and has been dated at 3.8 billion years. In 1983 a few grains of zircon minerals 4.1 to 4.2 billion years in age were found in metamorphosed rocks on Mt. Narryer in Western Australia. The oldest rocks tend to be found near the centers of large continental landmasses, with progressively younger and younger rocks being found closer to their edges. Rocks found at the bottoms of oceanic basins are typically much younger than those coming from continental landmasses, with no deep oceanic crust being any older than 200 million years. The widely varying ages of crustal rocks suggest a long and complex history of solidification, melting, and reformation, a process that continues to the present day.

Is there any way that the age of the Earth itself can be determined? Radioactive dating can unfortunately measure only the time elapsed since the last solidification of a rock. It can tell us little about the prior time spent by the rocky material in the molten state, because parent and daughter elements are likely to be hopelessly intermixed. The complete melting of a rock and its subsequent resolidification "resets" the radioactive clock back to zero. There is, fortunately, an extremely clever method by which this problem can be circumvented. This involves a study of the stable isotopes of lead. The lead currently found in terrestrial rocks is primarily of four different isotopes, ^{206}Pb, ^{207}Pb, ^{208}Pb, and ^{204}Pb. The first three of these are radiogenic; they can be produced only by the radioactive decay of heavier elements. They are, in fact, daughter products of the decomposition of the radioactive isotopes ^{238}U, ^{235}U, and ^{232}Th, respectively. Lead 204 is primordial, which means that it cannot be created by the radioactive decay of any heavier element. Any primordial lead currently found in terrestrial rocks must have been present here on this planet ever since the Earth itself was formed. As the Earth has aged, more and more radiogenic lead has been created, whereas the amount of primordial lead has remained constant. Every time that a rock melted and reformed, its primordial lead was progressively diluted with more and more radiogenic lead. The older the rock, the larger the primordial/radiogenic lead ratio. Some lead-rich minerals have managed to crystallize with no significant amounts of uranium or thorium parent isotopes. Their lead isotopic abundances must currently be exactly the same as they were when they first solidified. By measuring the relative amounts of the lead isotopes present in these minerals, the "lead evolution" time that elapsed before their final crystallization can be determined.

The results seem to indicate that the Earth probably first formed as a solid, coherent body about 4.7 billion years ago. The fact that the radioactive age of every single rock on the Earth's surface is very much less than 4.7 billion years is indicative of the extensive surface-altering geological activity that has taken place in the past.

THE EARTH'S DRIFTING CONTINENTS — THE RING OF FIRE

In the early part of the twentieth century, it was suggested that the continents and oceans of the Earth were slowly but constantly changing shape and sliding about on the surface. The most obvious clue was the close match of the western shore of the Africa with the eastern shore of South America, almost as if these two landmasses had originally been part of one larger continent but had subsequently split apart. The first systematic theory of *continental drift* was proposed by the German meteorlogist Alfred Wegener in 1929, but it was not until the 1960s that measurements of the amount of magnetism in deep oceanic rocks were able to verify the model experimentally.

The outermost several kilometers of the crust is quite rigid and brittle. It is known as the *lithosphere*. However, there is a layer of partially molten material just below it. This layer (known as the *asthenosphere*) is approximately 150 km thick and can plastically flow and deform under the application of stress.

The crust and mantle down to a depth of 100 km are divided laterally into 10 major *continental plates* that cover the entire Earth. Plate boundaries follow the long and narrow regions where earthquakes and volcanoes are frequent. These plates "float" on top of the asthenosphere and can migrate laterally over the surface. The rate of plate motion is of the order of a few centimeters per year, so the map of the Earth changes significantly over a time period as short as 0.5 million years (Figure 4.10).

The force that drives continental drift is the convective upward flow of molten lava that takes place along *rift zones*. Rift zones are usually found at the bottoms of deep oceans and can be many thousands of kilometers in length. The heat and high pressure produced by the molten rock flowing upward along a rift zone forces the adjacent plates apart. Once the lava produced by this submarine volcanic activity reaches the surface, it cools and solidifies to form fresh new oceanic crust (Figure 4.11).

It is estimated that there are at least 20 eruptions of these submarine rift volcanoes every year. Most occur at such depths in the ocean that they are scarcely noticed on land. In many cases, enough volcanic lava accumulates along an active rift zone to form a chain of high mountains. Such a submarine mountain chain is known as an *oceanic ridge*. Oceanic ridges typically lie about 3 km above the average level of the surrounding ocean floor, so they would certainly be counted as major mountain ranges if they were on land. The mid-Atlantic ridge is perhaps the best known of these submarine mountain ranges. It stretches virtually from pole to pole, passing down the middle of the Atlantic Ocean. The ridge neatly marks the boundary between the diverging plates carrying the African, European, and American continents. The Americas moved west, while Europe and Africa moved east.

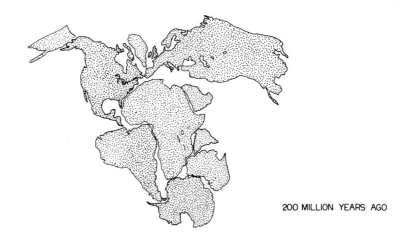

200 MILLION YEARS AGO

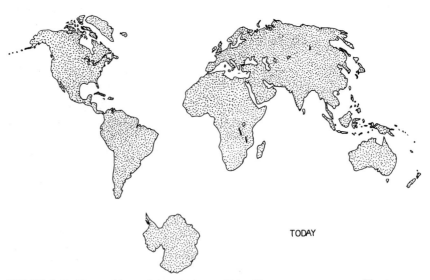

TODAY

FIGURE 4.10 The positions of the continents 200 million years ago, as compared to their positions today. Two hundred million years ago all landmasses were assembled into a supercontinent known as *Pangaea*. Individual continental landmasses subsequently split off from Pangaea and have been moving apart ever since.

Since plates are being forced apart in some regions, it naturally follows that there must be other places in which plate collisions are taking place. In some locations where crustal plates are currently in collision, new mountains are being created. The most obvious example is the Himalayan mountain chain, located along the line where the Indian continental plate is being forced up

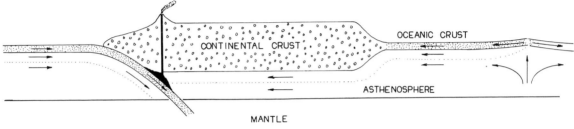

FIGURE 4.11 A slice through the Earth's crust.

against the Asian landmass. In other regions, one continental landmass is actually forced to slide up over the other as they collide. The lower plate is pushed down into the asthenosphere, where it is remelted. This process is known as *subduction*. It creates deep *oceanic trenches,* such as the one near the Marianas Islands in the Pacific Ocean.

Intense volcanism is often found in the vicinity of the subducting crust. An erupting volcano is one of the most awe-inspiring and frightening events to take place in nature. Hot lava is driven upward from the interior of the Earth along cracks in the crust. Once it reaches the surface, it flows out and covers the surrounding terrain. The steady accumulation of this lava often forms a high mountain, usually with a circular opening (or *caldera*) at the top. In addition to the usual massive flood of molten lava, an erupting volcano will generally also eject a towering cloud of ash, dust, and superheated steam. In addition to water vapor, poisonous gases such as carbon dioxide (CO_2), sulfur dioxide (SO_2), hydrogen sulfide (H_2S), and hydrogen chloride (HCl) are usually also released during an eruption. A particularly violent volcanic eruption can release as much energy as the explosion of a nuclear weapon. Heavy loss of life and vast property damage are the usual results whenever a volcanic eruption takes place near an inhabited area.

Volcanoes found along subduction zone boundaries are called *subduction volcanoes* by geologists. Most subduction volcanoes are located on the overlying plate, usually along a line where the underlying (or descending) plate is 100 to 200 km below the surface. It is not clear whether the lava, gas, dust, and ash that is ejected from these volcanoes is produced by heat generated by mechanical friction between the sliding continental plates, or whether it is a result of deep fractures formed in the overlying plate that allow molten material from greater depths in the mantle to reach the surface.

Most of the volcanoes caused by crustal subduction are found on land and are hence quite readily noticed by humans. It is estimated that Earth has approximately 500 to 600 of these volcanoes, most of them lying along the island arcs of the western Pacific (Aleutians, Kuriles, Japan, Marianas) and along the western North and South American coasts. This is the so-called ring of fire, which encircles the entire Pacific basin. Japan alone has 50 active volcanoes along parts of its four island arcs. Another line of subduction volcanoes extends from the Mediterranean Sea through Asia all the way to Indonesia.

Driven by the combined mechanisms of ocean floor spreading and crustal subduction, there is a continuous recycling of rocky material back and forth between the solid lithosphere and the partially molten asthenosphere upon which it floats. It is estimated that roughly one-third of the mantle (corresponding to the outer 700 km) has been recycled back and forth between the surface and lower depths over the lifetime of the Earth.

THE EARTH'S MAGNETIC FIELD

The Earth possesses an intrinsic magnetic field of maximum intensity approximately 0.5 gauss at the surface. Although its true origin is largely unknown, this field can be visualized as being produced by a giant bar magnet located about 400 km from the Earth's center. This imaginary magnet makes an angle of about 24° with the rotational axis (Figure 4.12). Although there is still no adequate theory of the origin of Earth's magnetic field, most models assume that the field is maintained by a pattern of persistent electric currents existing deep within the molten core. The electric currents are driven by the combined effects of heat convection and planetary rotation.

The *magnetic poles* are defined as the two places on the Earth's surface where the magnetic field lines are entirely vertical. The north magnetic pole is located at 75°N, 100°W, at a point in the Viscount Melville Sound in northern Canada. The south magnetic pole is at 67°S, 142°E, at a point near the coast of Antarctica.

The Earth's magnetic field seems to be constant from one year to the next, but over longer periods of time there are significant changes. The magnetic poles migrate around the geographical poles in a roughly circular path, spanning about 20° in longitude every 500 years. The intrinsic magnetism in old Roman pottery, as compared to that in identical modern replicas, indicates that the terrestrial magnetic field was 50 percent stronger at the time of the Caesars than it is now.

Information about terrestrial magnetism in the more distant past can be acquired via a study of magnetic rocks, particularly those on the ocean floors. As molten iron-rich rocks cool and solidify, they acquire a net magnetization in response to the magnetic field present at the time of their solidification. Some of the oldest rocks yet found exhibit a magnetization, indicating that the Earth has had a magnetic field for at least 3.5 billion years. Examination of remnant magnetization within more recently formed volcanic rocks shows definite evidence that the Earth's magnetic field has reversed its direction at irregular intervals in the past, perhaps at least 10 times in the past 5 million years. The effect of this field reversal on terrestrial life is a matter of conjecture.

THE EARTH'S MAGNETOSPHERE

The terrestrial magnetic field deflects most of the solar wind particles impinging upon the Earth, preventing them from striking the upper atmosphere and disrupting the delicate ozone layer that protects the surface against harmful

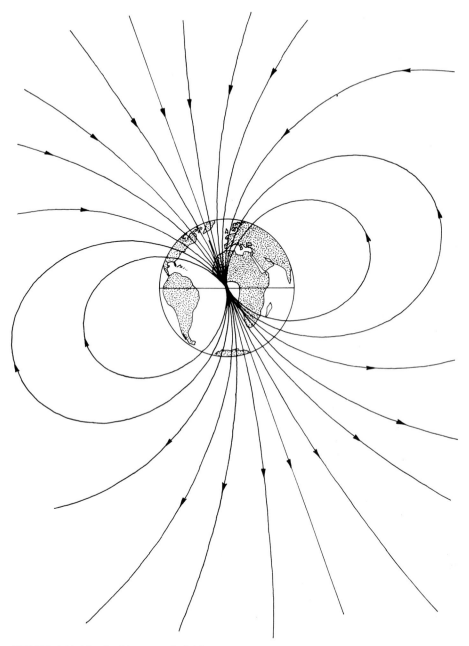

FIGURE 4.12 The Earth's magnetic field.

ultraviolet light. Since the flow of the solar wind past the Earth is supersonic, a *shock wavefront* is set up on the sunward side of the Earth. The steady pressure of the solar wind compresses the magnetic field lines on the sunward side of the Earth, so that these lines do not extend outward any farther toward the Sun than approximately 100,000 km [$15R_e$ (Earth radii)], well within the Moon's orbit. This maximum extent of the field lines is called the *magnetopause,* and its precise location changes with time in response to the variable pressure exerted by the solar wind. On the antisolar side of Earth, the field lines extend outward perhaps as far as 6×10^6 km from the Earth (Figure 4.13).

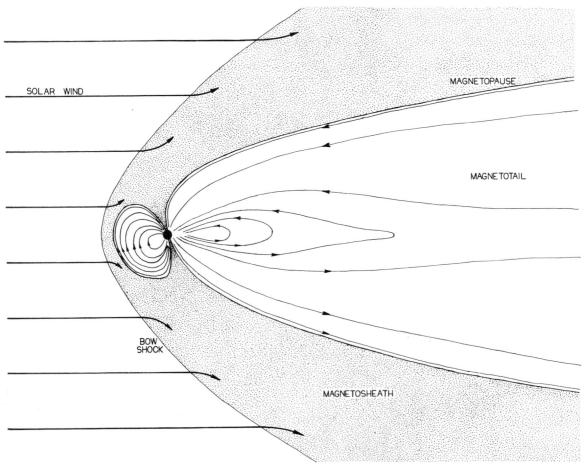

FIGURE 4.13 Diagram of the Earth's magnetosphere. The solar wind is deflected by the Earth's magnetic field, so that most of these particles do not strike the Earth. Since the solar wind flow is supersonic, a *shock wave* is set up in the Sun-facing direction. At the shock front the solar wind particles are abruptly slowed down from 400 to 250 km/sec. In the Sun-facing direction, the magnetic field lines are compressed by the pressure exerted by the solar wind. In the antisolar direction, the interaction with the solar wind pulls the field lines out into a long *magnetotail,* which extends well beyond the orbit of the Moon. The *magnetopause* is the outer boundary of the magnetosphere across which the solar wind particles cannot readily pass. Between the shock front and magnetopause is the *magnetosheath,* a transition region characterized by jumbled and chaotic magnetic field lines.

There is dense region of particle radiation just above the Earth's atmosphere that is a direct result of the terrestrial magnetic field. This radiation pattern was discovered by the first American earth satellite, *Explorer I* (launched in January 1958). Its existence was suspected as far back as 1907 as a probable cause of auroral phenomena. The radiation consists of a plasma of high-energy protons, electrons, helium, oxygen, and nitrogen ions. Most of this plasma is confined to a region lying roughly in the plane of the Earth's magnetic equator. The entire pattern of magnetic field lines, along with the associated radiation, is called the *magnetosphere.*

The particles that are trapped within these radiation belts appear to come from two different sources. One source is the upper ionosphere. Some of the ions created by the ultraviolet destruction of upper atmospheric gas molecules are ejected into outer space, where they become trapped in the magnetic field, spiraling along the magnetic lines of force. The other source appears to be the solar wind. About 1 in 1000 solar wind particles impinging upon the Earth's magnetosphere ends up trapped by the magnetic field, primarily over the north and south magnetic poles.

There are two toroidal-shaped regions where the radiation is particularly intense, one at an altitude of 2000 to 5000 km, and the other at an altitude of 13,000 to 19,000 km (Figure 4.14). These regions are called *Van Allen belts* in honor of James A. Van Allen, who played a leading role in their discovery and interpretation at the dawn of the space age in the late 1950s. The inner toroidal region of the Van Allen belts consists mainly of high-energy protons and electrons. The intensity of the radiation there is fairly constant, averaging 2×10^4 protons/cm²/sec and 2×10^9 electrons/cm²/sec. The outer zone is almost exclusively populated with fast-moving electrons. This zone is characterized by a highly variable radiation intensity that changes with the level of solar activity. During peak periods of solar activity the radiation in the outer zone can give a dose as high as 10 roentgens/hour, enough to be a health hazard to astronauts flying in this region of space. It is therefore a good idea to pass through the outer regions of the Van Allen belts as quickly as possible.

When large numbers of charged particles impinge upon the upper atmosphere at the poles, they ionize some of the molecules in the air. These ionized molecules give off light, creating the stunning phenomena known as the *aurora borealis* (north pole) and *aurora australis* (south pole), which can be seen in extreme latitudes. The aurora is especially prominent at times when the Sun is active. It is estimated that the average amount of power that is dumped by the Earth's magnetosphere into the northern and southern polar regions of the atmosphere is about a 0.1 tw (1 terawatt = 10^{12} watts), approximately the amount of power that is generated by all of the electrical power plants on Earth.

THE HYDROSPHERE

One of the most abundant chemical compounds on the Earth is water (H_2O). Approximately 2 parts in 10,000 (by mass) of the entire planet is made up of water. Part of the Earth's supply of water exists in the vapor phase as a compo-

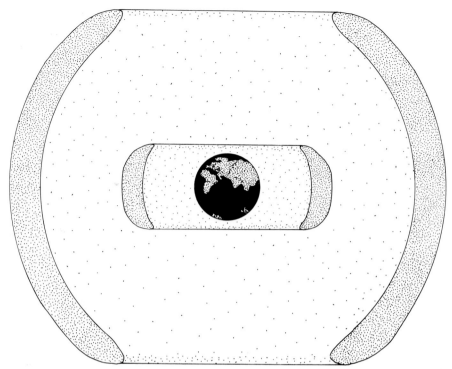

FIGURE 4.14 The Van Allen radiation belts of the Earth.

nent of the atmosphere. Other water is chemically combined with the rocks and minerals of the crust, forming compounds known as *hydrates*. However, the vast majority of the Earth's inventory of water is present in the liquid state at the surface, in the form of extensive oceans, rivers, and lakes. The presence of vast amounts of liquid water at the surface of the Earth is perhaps the single most important condition that has made it possible for life to survive on this planet and on no other in the Solar System.

Earth also has a great deal of frozen water ice present on its surface. Approximately 7 percent of the Earth's surface is covered by solid ice, mainly at the cold northern and southern poles. The south polar cap contains about twice as much ice as the northern one does. About 10 percent of the Earth's water is tied up as solid ice in the polar caps at any one time. If some catastrophe were to cause all the polar ice suddenly to melt, the oceanic water levels would rise by about 200 m, flooding most of the coastal regions on the planet.

THE ATMOSPHERE

A particularly significant factor, as far as life is concerned, is the Earth's dense atmosphere. Seventy-eight percent (by volume) is nitrogen (N_2), 21 percent is oxygen (O_2), and about 1 percent is argon (Ar). Water vapor (H_2O) is also present in the air, the exact amount of which varies with surface location and the season of the year. The water content in the atmosphere ranges from virtually 0 to as high as 4 percent. There is a constant exchange of water between the oceans and the atmosphere via the cycle of evaporation, condensation, and precipitation.

There are many other gases present in the atmosphere, but their relative abundances are much smaller. The atmosphere has about 330 parts per million (ppm) carbon dioxide (CO_2), 70 ppm helium (He), 18 ppm neon (Ne), 1.5 ppm methane (CH_4), and 1 ppm krypton (Kr). Gases such as hydrogen (H_2), nitrous oxide (N_2O), carbon monoxide (CO), ammonia (NH_3), nitrogen dioxide (NO_2), sulfur dioxide (SO_2), hydrogen sulfide (H_2S), and ozone (O_3) are found only in trace amounts of 0.5 ppm or less.

Air pressure is usually measured in units of *atmospheres* (atm). The average pressure of the air at sea level is arbitrarily defined to be exactly 1 atm. One atmosphere of pressure is numerically equal to $101,500 \text{ N/m}^2$ in metric units. In English units, 1 atm $= 14.7 \text{ lb/in.}^2$.

As anyone who has ever climbed a mountain or flown in an airplane can testify, the pressure of the atmosphere declines steadily with altitude. The pressure drops by approximately a factor of 2 for every 5.5-km increase in altitude. The change in average temperature of the atmosphere with height is much more complicated and depends to a large extent on the nature of the energy inputs to the atmosphere at different heights above the ground. There are five distinct vertical subdivisions, termed the *troposphere, stratosphere, mesosphere, ionosphere,* and *thermosphere* (Figure 4.15).

The troposphere is the lowest layer of the atmosphere. It extends upward from the surface to an altitude of approximately 10 km. Without exception, all living land creatures reside within this thin layer of air. The average temperature at the surface is between 10 and 20°C, but there is a uniform decline of temperature with height. The rate of temperature decline (the so-called *lapse rate*) is of the order of 5 to 10°C for each kilometer of altitude, the exact value being a function of the local humidity of the air. At an altitude of 10 km, the average temperature is down to a frigid −50°C and the pressure is 0.25 atm.

The energy source responsible for the relatively high temperature of the troposphere is the absorption of incident solar energy at the Earth's surface. Since the atmosphere itself is largely transparent to visible sunlight, the average temperature of the troposphere decreases as one gets farther and farther above the warm surface.

At a height of 10 km, the average air temperature reaches a minimum value. This level of minimum temperature is called the tropopause. Above the tropopause, the temperature begins to rise with ascent. This region of rising air temperature defines the stratosphere. The temperature reaches a value nearly that of sea level at an altitude of 50 km. However, the pressure at that altitude is only 0.001 atm.

The rising temperature of the stratosphere must mean that there is some sort of energy source located there. The mechanism responsible for the heating of the stratosphere is *oxygen photochemistry.* Oxygen molecules in the upper

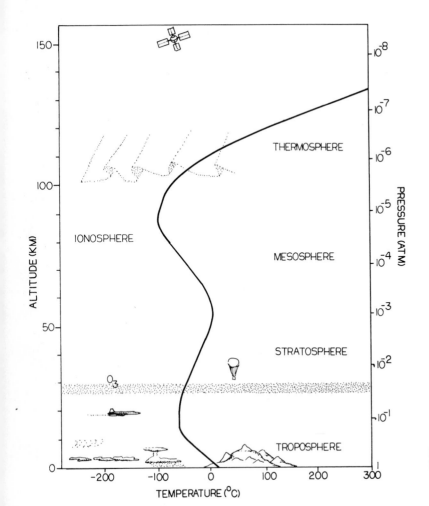

FIGURE 4.15 Verticle profile of the Earth's atmosphere.

stratosphere absorb low-wavelength ultraviolet sunlight. The energy carried by this ultraviolet light splits the oxygen molecule (O_2) into its constituent oxygen atoms. These energetic oxygen atoms then bump into other molecules in the upper atmosphere, transferring their excess energy into heat. A few of the oxygen atoms are able to combine with oxygen molecules, forming ozone (O_3). Most of the stratospheric ozone occupies a relatively thin layer at an altitude of 30 km. Within the *ozone layer*, the relative ozone concentration is as high as 12 ppm. This trace of ozone acts in turn to block out ultraviolet light of even longer wavelength. The net result of oxygen photochemistry is to prevent ultraviolet light of wavelength shorter than 3500 Å from reaching the ground. This shielding is all-important for life on Earth; were this ultraviolet light to reach the ground undiminished, unprotected organisms would be killed in a matter of minutes.

At a height of 50 km the temperature reaches a maximum. Above this point the atmospheric temperature begins to decline with height once again, and the *mesosphere* begins. At an altitude of 80 km the temperature is down to −100°C and the pressure is only 0.00004 atm. The air throughout the mesosphere is so thin that ultraviolet absorption is not a significant heating factor.

At an altitude of 80 km, there is a highly variable layer of charged particles. This is the ionosphere, which is sustained by the absorption of ultrashort-wavelength ultraviolet light and X rays emitted by the Sun. The ionosphere moves irregularly up and down, and the density of charged particles there shows a strong correlation with solar activity. The ionosphere is best known to most people for the role that it plays in reflecting long-wavelength radio waves, permitting radio stations to broadcast over intercontinental distances.

Above the ionosphere, the air temperature begins to rise once again, reaching a temperature as high as 700°C at an altitude of 130 km. This is the thermosphere, which is heated by impact with highly energetic charged particles coming from the Sun and from cosmic space. The height of the thermosphere rises and falls in response to heating produced by solar activity, particularly during solar flares. The high temperature of the thermosphere is not a hazard to spacecraft passing through it, since the pressure there is only 1×10^{-7} atm. By the time that an altitude of 150 km is reached, the atmosphere has merged into the near-vacuum of outer space.

LIFE AND AIR—AN ETERNAL SYMBIOSIS

Unlike other planetary atmospheres, the Earth's protective air blanket has been largely formed and shaped by biological activity. To some extent, the activities of humanity itself have even played a significant role (Figure 4.16).

The best-known of these interactions is, of course, the *oxygen/carbon dioxide* cycle. Photosynthetic organisms take some of the carbon dioxide in the air and combine it with water in the presence of light to create sugar for use as food. During the photosynthetic process, oxygen is released into the air as a by-product. Other creatures use this oxygen for respiration, releasing carbon dioxide back into the air in the process. After the death of an animal or plant, some of the organic carbon in its dead body gets oxidized during the decay process and is returned to the atmosphere as carbon dioxide. Virtually all of the oxygen currently in the atmosphere was originally placed there by the action of photosynthetic organisms, primarily by microscopic algae living in the oceans. The Earth seems to be the only planet in the Solar System with an oxygen-rich atmosphere, created by the presence of abundant life.

Carbon dioxide is also cycled back and forth between the atmosphere, the oceans, and the outer crust. Carbon dioxide is highly soluble in water. When in solution it forms carbonate or bicarbonate ions. The solubility is so high that the oceans currently contain 60 times more CO_2 than the atmosphere does. The oceans also contain large amounts of dissolved calcium ions. The carbonate ions in the water will react with this calcium to form the insoluble compound calcium carbonate. This calcium carbonate slowly settles to the bottom of the oceans, where it ultimately forms into limestone minerals and other sedimentary rocks. Over the years, these carbonate minerals have become an important compo-

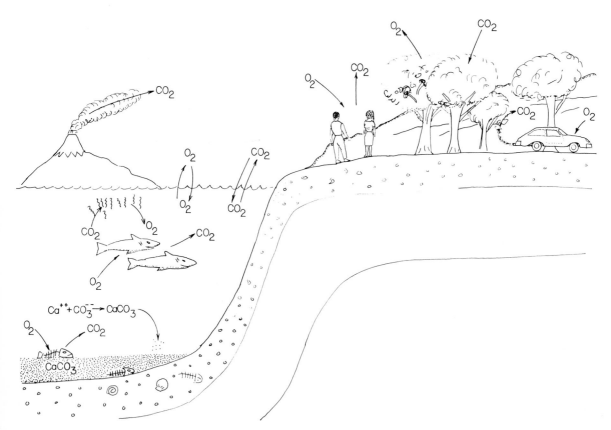

FIGURE 4.16 Diagram of oxygen/carbon dioxide cycle.

nent of the Earth's crust. They play an important role in geological processes. When subduction of continental plates takes place, carbonates are driven deep down into the upper mantle. When these subducted carbonates are melted by the extreme heat and pressure, some of them are converted back into carbon dioxide and are returned to the air via volcanic activity. In other locations crustal carbonate minerals are lifted out of the water by collisions between continental plates. Once out of the water these minerals are slowly weathered, the carbon in the rocks being oxidized and returned to the atmosphere as CO_2. Most of Earth's supply of carbon dioxide is currently chemically combined within the rocks and minerals of the outer crust. It is estimated that if all of the carbon dioxide currently "locked" into the rocks as carbonate minerals were somehow to be forced out into the atmosphere, the air pressure would rise to 25 atm.

The dominant gas in the atmosphere is nitrogen. Nitrogen is rather chemically inert but is nevertheless essential to life. In order to employ the nitrogen in the air for biological purposes, Nature has evolved efficient means for its conversion into the more useful nitrate or nitrite form. This can be done on a relatively small scale by the action of lightning; but far more important are the nitrogen-fixing microorganisms that perform this process by use of complex biochemical systems. Some of the nitrates and nitrites that are synthesized in this manner enter the bodies of living organisms, whereas others end up enriching sedimentary rocks in nitrogenous minerals. Other microorganisms have evolved that convert nitrates and nitrites back into N_2, returning the nitrogen to the atmosphere from whence it came.

Many of the trace gases currently in the atmosphere also have a biological origin, particularly methane and ammonia. The primary source of the trace of methane currently found in the atmosphere is the anaerobic decomposition of organic matter. This occurs whenever dead plants become buried in sediments devoid of oxygen. The methane thus generated is released to the atmosphere when geological action brings this material back to the surface. The concentration of methane in the air stays at a very low level because it is very prone to decomposition by reaction with the active chemical species that are created by ultraviolet photochemistry of oxygen and other gases in the upper atmosphere.

Ammonia gas is released into the atmosphere when the nitrogenous organic compounds in dead animals and plants are decomposed by the actions of microorganisms. The ammonia concentration in the air is so slight because ammonia is highly soluble in water. Ammonia injected into the atmosphere "rains out" within a matter of days. It ultimately ends up in the oceans, where it reacts with acidic impurities to form nitrogen-rich sedimentary deposits that slowly settle to the bottom. Ammonia is also destroyed by participation in a complex sequence of photochemical reactions in the stratosphere. It is decomposed into various nitrogen oxides and into molecular nitrogen.

THE GREENHOUSE EFFECT

The surface temperature averaged over an entire year over the entire surface of the Earth is $+13°C$ (55°F). This temperature is hot enough to keep most of the Earth's water from freezing into solid ice. However, it is not so warm that all the water in the Earth's oceans is driven out into the atmosphere as hot steam. Either alternative would have destroyed all life on the planet. Over the past 4 billion years of Earth's history, a temperature balance has been maintained, delicately teetering between a perpetual ice age on the one hand, and death in a bath of superheated steam on the other.

The temperature balance raises an interesting problem. The effective temperature of the Earth (as determined by the energy balance between the sunlight that is absorbed by the planet and that which is reflected back into space) is only $-20°C$, far below the freezing point of water. The surface of the Earth is at least 30°C warmer than it would be if it were heated simply by absorbing the solar energy incident upon it. If sunlight were the sole source of heat for the lower atmosphere, all of the Earth's water would have frozen into solid ice many years ago!

The cause appears to be a *greenhouse effect* resulting from the presence of the water vapor and carbon dioxide in the atmosphere. Both of these gases are completely transparent to visible light. However, they are both strong absorbers of infrared light. When visible light from the Sun strikes the surface of the Earth, part of the solar energy is absorbed and the surface is warmed. The rest of the sunlight is reflected back into space. The warm surface of the Earth will tend to reradiate its excess heat energy back into space. However, this heat radiation takes place primarily in the infrared region of the spectrum rather than in the visible. The infrared radiation emitted by the ground is absorbed by the carbon dioxide and water vapor in the atmosphere above the surface. The heat energy is trapped near the ground and does not escape into outer space — hence the life-preserving warmth.

If all of the carbon dioxide and water vapor were suddenly to vanish from the air, the greenhouse heating of the Earth's lower atmosphere would cease. The average surface temperature would soon drop below the freezing point of water. More water would freeze into solid ice. The presence of more ice would produce a brighter surface, causing a larger fraction of the incident sunlight to be reflected back into space. This would cause the net solar energy input to the planet to decrease, causing the overall temperature to decline still further and even more water to freeze. Eventually all of the water in the Earth's oceans would freeze into solid ice. Such a process is called *runaway glaciation.* If this were to happen, it would be an unparalleled disaster, resulting in the destruction of all living things on the planet. The Earth would freeze to death.

A significant increase in the amount of water vapor and carbon dioxide in the atmosphere would be just as disastrous. Larger amounts of carbon dioxide or water vapor in the atmosphere would enhance the greenhouse effect and would cause the average surface temperature to become warmer. Warmer temperatures would in turn evaporate some ocean water, raising the atmospheric humidity still higher. Higher temperatures would also force some of the carbon dioxide gas dissolved in the oceans back into the atmosphere. The increased water vapor and carbon dioxide content in the air would in turn trap still more heat and would produce even higher temperatures. The temperature could rise so high that all the water in the rivers, lakes, and oceans of the Earth would vaporize and all of the carbon dioxide would be baked out of the rocks, producing a dense, crushing atmosphere of superheated gases. Such a rapid, uncontrolled rise in temperature has come to be known as a *runaway greenhouse effect.* This too would be a disaster, completely destroying all forms of life. The Earth would smother to death.

Earth has been exceedingly fortunate in having precisely the right amount of heat-trapping gases in its atmosphere. There is enough to keep the oceans from freezing over, but not so much that the oceans all boil away. This fortunate state of affairs has been made possible by a complex equilibrium between the carbon dioxide stored in the atmosphere, dissolved in the oceans, bound inside living organisms, or locked inside carbonate crustal minerals. If this equilibrium were to be seriously disrupted, catastrophe would result.

THE PRIMITIVE EARTH

Although there are considerable disagreements about the particular details, it appears that the Earth initially formed in a more or less "cold" state out of material that condensed from a vast nebula of gas and dust that occupied our region of space about 4.7 billion years ago. The original proto-Earth that accumulated from this collection of debris was undoubtedly far more homogeneous in composition than it is now, with metallic and rocky debris intermixed more or less uniformly throughout. The interior of the planet was also much cooler.

However, many radioactive isotopes of thorium, uranium, rubidium, and potassium had been incorporated into the body of the new planet. The early Earth must have been five times more radioactive than it is now, and the thermal energy released by the decay of these radioactive elements was great enough to force the interior of the planet to become very hot. There was probably sufficient internal radioactive heating to melt the entire Earth. While the Earth was in the molten state, the denser metallic (largely nickel and iron) material sank to the interior to form the core, leaving the lighter silicate material "floating" on top to form the mantle and crust. At this time, the heat was so intense that the entire surface of the Earth must have been covered by an "ocean" of red-hot molten lava.

Soon thereafter the internal radioactive heating abated and the molten surface of the Earth began to cool. Eventually the surface temperature became low enough for a solid crust to form. The first solid crust may have appeared as early as 4.2 to 4.3 billion years ago. The interior still remained entirely molten, and intense volcanic activity and massive lava flows must have been prevalent almost everywhere on the surface of the young Earth. Because of the intense level of geological activity, no permanent record of this early period has survived. Geologists refer to this hidden early period of Earth's history as the *Hadean epoch*.

By 3.8 billion years ago the outer surface of the Earth had cooled sufficiently for a permanent solid outer crust to form. This marks the beginning of the *Archean* epoch. The early Archean crust was quite thin and fragmentary, and no large continental land masses like those present today existed. The surface of the Earth was broken up into many thousands of tiny plates, each of which melted, solidified, and then melted again many, many times. The volcanic activity at this time was so intense that relatively few rock formations dating from the Archean epoch survive today.

Approximately 2.5 billion years ago, the first permanent granitic continents appeared. This event marks the end of the Archean and the beginning of the *Proterozoic* (the era of "primitive life"). The early granitic continents that formed at this time are known as *cratons*. Many of these cratons are still present today. They are generally found at the centers of large continental landmasses. The cratons seem to have acted as nucleation centers for the subsequent growth of younger continental crust at their periphery. The continental landmasses have been steadily growing in size ever since. Once they are formed, continental landmasses are rarely destroyed. Oceanic crust, on the other hand, has been formed, remelted, and then reformed a number of times over the long history of the Earth.

The various subdivisions of Earth's geological history are listed in Table 4.1. The evolutionary history of Earth's crust is summarized in Figure 4.17.

EARTH'S FIRST ATMOSPHERE

The origin of the Earth's atmosphere and its subsequent evolution are subjects of great controversy within the geological and astronomical communities at present. The Earth probably acquired some sort of atmosphere almost immediately after it formed. This early atmosphere was undoubtedly gathered up by the young Earth from the ample amounts of uncondensed gases still left in the solar nebula. This first atmosphere must have been largely composed of hydrogen and helium, since these gases are found today in abundant amounts in the atmospheres of the Jovian planets. Jupiter was massive enough to retain its

TABLE 4.1. GEOLOGICAL TIME INTERVALS

Cryptozoic eon (4.45 billion – 700 million years ago)
   Hadean era (4.45 – 3.8 billion years ago)
   Archean era (3.8 – 2.5 billion years ago)
   Proterozoic era (2.5 billion – 700 million years ago)

Phanerozoic eon (700 million years ago – present)
   Paleozoic era (700 – 230 million years ago)
      Eocambrian period (700 – 560 million years ago)
      Cambrian period (560 – 485 million years ago)
      Ordovician period (485 – 435 million years ago)
      Silurian period (435 – 410 million years ago)
      Devonian period (410 – 350 million years ago)
      Mississippian period (350 – 320 million years ago)
      Pennsylvanian period (320 – 290 million years ago)
      Permian period (290 – 230 million years ago)
   Mesozoic era (230 – 65 million years ago)
      Triassic period (230 – 192 million years ago)
      Jurassic period (192 – 135 million years ago)
      Cretaceous period (135 – 65 million years ago)
   Cenozoic era (65 million years ago – present)
      Paleogene period (65 – 25 million years ago)
         Paleocene epoch (65 – 54 million years ago)
         Eocene epoch (54 – 36 million years ago)
         Oligocene epoch (36 – 23 million years ago)
      Neogene period (25 million years ago – present)
         Miocene epoch (25 – 6 million years ago)
         Pliocene epoch (6 – 2 million years ago)
         Pleistocene epoch (2 million years ago – 10,000 years ago)
         Holocene epoch (10,000 years ago – present)

primal atmosphere and remains largely hydrogen and helium to this day. However, the young Earth was much less massive than Jupiter, and our planet quickly lost its original hydrogen – helium protoatmosphere to outer space.

Earth probably did not begin to acquire its permanent atmosphere until somewhat later, only after radioactive heating of the interior had begun. When the interior began to melt, some gases were baked out of the molten rocks by the heat and were forced to the surface during volcanic eruptions. These gases formed Earth's first permanent atmosphere. A planetary atmosphere that is acquired via outgassing from the interior rather than directly from the solar nebula is called a *secondary* atmosphere.

The primary evidence that Earth's atmosphere is secondary is derived from the relative scarcity of neon in the current atmosphere. Neon is a "noble gas"; it cannot react with other elements to form chemical compounds. Once neon is introduced into the atmosphere, it should remain there forever. Neon is actually quite common in the universe; in the Sun it is almost as abundant as nitrogen. One might then expect that any terrestrial atmosphere that was largely acquired directly from the primitive solar nebula be rich in neon. However, Earth's present atmosphere has only a very small trace of neon. The virtual absence of neon in the current atmosphere can be taken as strong evidence that Earth's atmosphere was acquired from the interior of the planet rather than from outer space.

Other evidence for the secondary nature of the terrestrial atmosphere can be derived from the argon gas in the atmosphere. Like neon, argon is a noble gas and does not form compounds with other elements. Elemental argon has three isotopes, ^{36}Ar, ^{38}Ar, and ^{40}Ar. They differ from each other only in mass. They are otherwise chemically identical. Argon 36 and argon 38 are primordial isotopes. Neither one of them can be created by the radioactive decay of any heavier element. Any ^{36}Ar or ^{38}Ar presently on Earth must have been here ever since the initial formation of our planet. Argon 40 is a radiogenic isotope. It can only be created by the decay of a radioactive isotope of potassium, ^{40}K. Most of the argon in the Sun is primordial, but the vast majority of the argon in the Earth's atmosphere is radiogenic. The argon gas in the terrestrial atmosphere is largely a product of internal geological activity. It came from the interior of the Earth, not from outer space.

The precise chemical composition of Earth's first permanent atmosphere is still a matter of hot controversy. The first systematic model of Earth's primeval atmosphere was proposed by University of Chicago chemist Harold Urey during the early 1950s. He suggested that this atmosphere had a composition dictated largely by the elemental composition that existed in the solar nebula at the time of the initial formation of the Earth. If one takes an atomic mixture of solar composition and allows it to cool and condense to form simple gaseous compounds, one obtains mostly H_2O, CH_4, H_2S, NH_3, and of course H_2. Chemists classify such a gas mixture as *reducing,* since there are many molecules rich in hydrogen atoms, but there is no free molecular oxygen (O_2).

Urey pictured the reduced gases in the primeval atmosphere as having been formed by chemical reactions between the atoms in the original solar

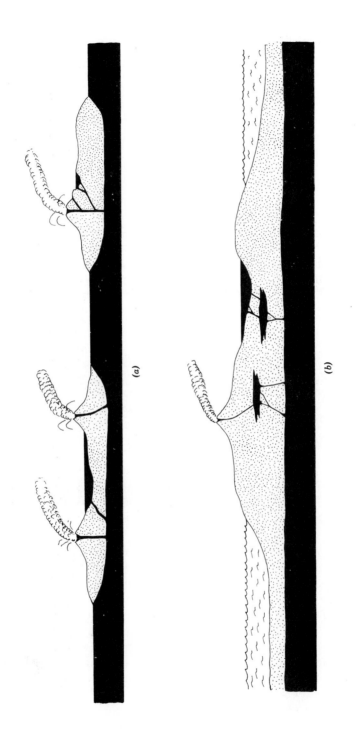

(a)

(b)

74

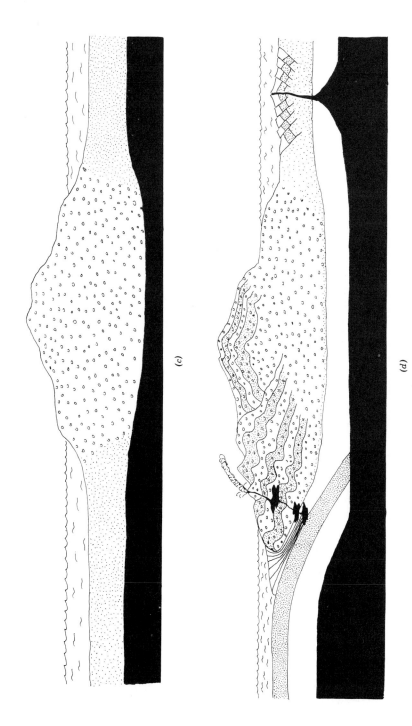

(c)

(d)

FIGURE 4.17 Schematic view of history of the Earth's crust. As seen (a) 4.2 billion years ago; (b) 3.8 billion years ago; (c) 2.5 billion years ago; (d) today.

75

nebula. As the Earth accumulated from the dust and debris of this nebula, some of these gases became trapped in the interior of the new planet. Shortly thereafter radioactive heating within the newly formed Earth forced many of these gases out of the interior and upward toward the surface. These gases formed Earth's first permanent atmosphere.

Urey's model of the primitive terrestrial atmosphere was accepted for many years as being a reasonable working hypothesis. Recently, however, a somewhat different model has gained many adherents. This model proposes that Earth's first atmosphere bore little relationship to the elemental composition of the solar nebula. Instead, it was produced by the vapors that were given off as by-products of chemical reactions in the interior that were driven by the intense radioactive heating. These vapors were subsequently forced upward to the surface. These gases may have been H_2O, CO_2, Cl_2, N_2, SO_2, and H_2S, since these vapors are the primary gaseous products of present-day volcanic eruptions. This mixture of gases is considerably less reducing than the ammonia – methane primeval atmosphere proposed by Urey.

The primary argument used in support of a weakly rather than a strongly reducing primitive atmosphere is the evidence for the presence of massive amounts of carbon dioxide in the early terrestrial atmosphere. Some of the oldest known rocks are rich in carbonate minerals, indicating the presence of large amounts of carbon dioxide at the time of their formation.

Recent mathematical models of the primitive terrestrial atmosphere indicate that even if the Earth did originally possess large amounts of ammonia and methane, these gases could not have been retained for very long. At that time, the young Sun was a very strong source of ultraviolet light. It may have given off as much as 10,000 times as much ultraviolet light as it does now. Both methane and ammonia are particularly prone to being destroyed by ultraviolet photochemistry.

The photochemistry of water probably also played a key role in the removal of methane and ammonia from the atmosphere. The temperature at this time must have been quite high, and the Earth's primeval atmosphere must have been much more humid than today's relatively dry air. Highly energetic ultraviolet light from the early Sun tore apart many of the water vapor molecules, producing oxygen and hydrogen atoms. The hydrogen escaped into space, but the oxygen remained behind to attack the methane molecules, eventually creating carbon dioxide. The carbon dioxide currently found in the atmosphere, in sedimentary rocks, and in living creatures may have originally been created by the destruction of primordial methane gas many millions of years ago. Ammonia reacts with oxygen and other intermediates to form various nitrogen oxides as well as molecular nitrogen (N_2). The decomposition of ammonia may be the major source of the nitrogen gas found in today's atmosphere. In addition, ammonia is extremely soluble in water. The small amount of ammonia that managed to escape photodestruction must have been very quickly "washed out" of the atmosphere when the first oceans formed.

THE PRIMEVAL GREENHOUSE

By the time that the Earth first acquired a solid crust, it probably already possessed a dense atmosphere of reduced gases such as carbon dioxide, nitrogen, water vapor, ammonia, and methane mixed together in uncertain proportions. Because of the intense volcanic activity present on the newly solidified Earth, the reducing atmosphere of the early Archean must have been much denser and far hotter than the relatively thin, oxygen-rich atmosphere of today. At that time, all of the Earth's water was in the vapor phase as superheated steam. The surface pressure may have been as high as several hundred atmospheres.

Gradually, the atmosphere began to cool off. At some point, the average temperature of the air dropped below the boiling point of water. When this happened the water vapor in the atmosphere began to condense and it started to rain. As the temperature continued to decline, more and more water fell from the atmosphere. There was so much water vapor in the air that it probably rained for many millions of years all over the world. The water from this million-year rainstorm formed the first oceans.

We know extensive oceans were already in existence at least 3.8 billion years ago, as some of the oldest known rocks have a sedimentary origin and large amounts of liquid water must have been present at the time that these rocks were deposited. Liquid water has been present on Earth in copious amounts ever since.

However, the survival of large amounts of liquid water at this early time presents geologists and astronomers with a serious theoretical problem. Long-established physical models of stellar evolution predict that stars such as the Sun undergo a slow but steady increase in luminosity as they age. Theoretical calculations indicate that the Sun of 3.5 to 4.0 billion years ago must have been 15 to 30 percent dimmer than it is now. This raises a troubling paradox: the early Sun was so weak that the average surface temperature of the primitive Earth should have very rapidly cooled below the freezing point of water. All of the oceans should have frozen into solid ice almost immediately after they formed!

This obviously did not happen. On the contrary, much geological evidence seems to indicate that the average temperature at the surface of the Earth during the early Archean was actually considerably warmer than the current temperature. Average water temperatures during these early times can be inferred from a measurement of deuterium/hydrogen abundance ratios in chert minerals (crystalline siliceous sedimentary rocks). These measurements indicate that cherts dated at 3.5 billion years were deposited at oceanic temperatures greater than 50°C. More recent cherts were deposited at successively lower temperatures, indicating a gradual cooling trend over the past few billion years.

What is responsible for this odd state of affairs? The cause is thought to be Earth's ancient reducing atmosphere. Ammonia, methane, carbon dioxide, and water vapor are efficient heat-trapping agents; they transmit visible light quite efficiently, but they strongly block infrared radiation. Earth's primitive reducing atmosphere provided a protective "blanket" that kept the planet's surface warm enough to prevent water from freezing, even in the face of the weak early Sun. Over the years, both ammonia and methane have been removed from the atmosphere, and most of the carbon dioxide has been incorporated into the limestone crust and into living creatures. Consequently, the greenhouse heating of the Earth has undergone a slow but steady decline, resulting in a gradual cooling trend over the past 4 billion years. This trend will continue into the indefinite future, provided human beings do nothing to disturb it.

OXYGEN AND EVOLUTION

A careful study of the geological record can trace the presence of life on Earth back as far as 3.5 billion years ago. The first living creatures were exceedingly primitive, single-celled organisms that lived exclusively in the oceans. Even today, the most complex of living beings are still largely made of water, reflecting the origin of all life in the primeval seas.

At the time of the origin of life, Earth's atmosphere was reducing, with gases such as carbon dioxide, methane, ammonia, and water vapor mixed together in uncertain proportions. There was little or no molecular oxygen present. An experiment first performed by University of Chicago researchers Stanley Miller and Harold Urey demonstrated that the molecules in such a mixture of reduced gases will readily react with each other to form complex organic substances whenever the mixture is subjected to any sort of energetic disturbance, such as an electrical spark discharge, a beam of ultraviolet light, a burst of high-energy ionizing radiation, or even excess heat. In the dense, reducing atmosphere of 4 billion years ago the lightning flashes in ancient thunderstorms and the intense ultraviolet light emitted by the Sun must have generated a steady supply of complex organic molecules, which rained down from the sky and fell into the oceans.

Numerous experiments have shown that the presence of even a trace amount of molecular oxygen in the gas mixture undergoing energetic excitation severely inhibits the synthesis of complex organic molecules. Rather than organic molecules, a dull and uninteresting inorganic "smog" is synthesized. If Earth's ancient atmosphere *did* have appreciable amounts of oxygen gas, organic molecules could never have been formed and life would never have formed on the Earth.

The newly formed oceans of the early Archean probably contained as much as 1 percent organic material. At some unknown time in the past, a few of these organic molecules happened by chance to gather together into an arrangement that was able to make a copy of itself. The first living cell was born. These first cells must have been exceedingly simple organisms, consisting of little more than a membraneous bag enclosing a "soup" rich in proteins, food molecules, and nucleic acids. Such simple one-celled organisms are classified by biologists as *procaryotic*. Cells of this type still exist today. The first living cells probably derived their food energy from the anaerobic fermentation of glucose, a process still used today by many primitive single-celled creatures that live in oxygen-poor environments.

The first procaryotes were probably entirely *heterotrophic;* they were able to pull all of the energy-producing and structure-building molecules that they needed directly from the rich organic soup in the surrounding water. However, as these primitive creatures multiplied in number, the organic nutrients dissolved in the oceans were gradually consumed. If this had continued unabated, the nutrients would have been exhausted and all life on Earth would have perished.

The destruction of life on Earth was averted by the appearance of a radically new and different means of energy conversion, a process known as *photosynthesis.* Approximately 3.3 billion years ago a creature evolved that was capable of using the energy available in sunlight to convert the carbon dioxide dissolved in the oceans into food. Life escaped from a total dependence on organic nutrients in the water; photosynthetic organisms prospered and flourished, whereas their heterotrophic cousins perished. The oceans rapidly turned green with life.

The first photosynthetic cells were probably not nearly as complex as the green plants of today. They probably manufactured glucose (chemical formula $C_6H_{12}O_6$) out of carbon dioxide and hydrogen sulfide via the reaction

$$6CO_2 + 12H_2S \xrightarrow{\text{light}} C_6H_{12}O_6 + 6H_2O + 12S \qquad (4.1)$$

A similar reaction is employed by the closest living relatives of these early photosynthetic cells, the green and purple bacteria, which live in anaerobic environments rich in sulfur. Hydrogen sulfide is now relatively rare in the terrestrial atmosphere and oceans, but it must have been much more abundant in the early reducing environment in which life originated.

Although biological production of oxygen probably did not appear until approximately 2 billion years ago, there was nevertheless a steady growth in the amount of atmospheric oxygen all throughout the Archean era. The absorption of ultraviolet light by water vapor molecules high in the atmosphere was primarily responsible. Ultraviolet light energy tears water molecules apart, splitting them into oxygen and hydrogen atoms. The vast majority of the oxygen atoms produced by this process were immediately consumed by the ammonia and methane still present in the early atmosphere, producing a steady buildup in atmospheric carbon dioxide and nitrogen. However, a few of these oxygen atoms lasted long enough to combine with each other and form molecular oxygen. This oxygen steadily accumulated in the atmosphere and in the oceans below.

At this time, the oceans contained much water-soluble reduced (ferrous) iron, produced by runoff from the early landmasses. Reduced iron reacts with molecular oxygen to form insoluble ferric oxide (Fe_2O_3). The ferric oxide produced by reaction with this primordial oxygen slowly settled to the ocean bottoms, forming geological features known as *banded iron formations* in sedimentary rocks.

Some of the oldest sedimentary rocks have banded iron formations, indicating that some sort of oxygen source must have existed at this time. However, these ancient sedimentary deposits also have a lot of reduced uranium (uranite), which is unstable in the highly oxidizing atmosphere of today. The oxygen content of the oceans and the atmosphere above must then have been quite low all throughout this early period. The precise amount of oxygen in the reducing atmosphere of the early Archean era is in dispute; estimates range from as low as 10^{-15} to as high as 10^{-3} of the present atmospheric level.

About 2 billion years ago, a new and more advanced form of photosynthesis appeared on the Earth, one that could use water rather than hydrogen sulfide as the hydrogen donor. It is represented by the reaction

$$6CO_2 + 6H_2O \xrightarrow{\text{light}} C_6H_{12}O_6 + 6O_2 \qquad (4.2)$$

The first cells capable of this type of photosynthesis were the *cyanobacteria*, which were the prototypes for the blue-green algae in the oceans as well as for all green plant photosynthesis. Instead of rare gases such as hydrogen sulfide, photosynthetic organisms could now use the abundant water molecule itself as a primary reactant. The cyanobacteria prospered and multiplied. They entirely displaced their anaerobic predecessors in most ecological niches.

This new process is called *aerobic photosynthesis,* since one of the "waste products" of the reaction is oxygen gas. The first cyanobacteria were so instantly successful and proliferated so rapidly that the geological record shows clear evidence of their appearance. The rapid multiplication of these creatures emitted a sudden burst of oxygen into the environment, which quickly and efficiently cleared all of the reduced iron from the water. This resulted in the creation of enormous, massive banded iron formations in the sedimentary rocks. These formations are the primary source of iron ore today. All of the large iron ore formations are approximately 2 billion years old and have traditionally been used to date the first appearance of the cyanobacteria. In addition, reduced uranium disappears from sedimentary deposits laid down after this time.

Once the dissolved iron was swept out of the oceans, the oxygen gas generated by aerobic photosynthesis began to enter the atmosphere. A record of a sudden infusion of oxygen into the air about 2 billion years ago may be preserved in the so-called *red beds*. These are deposits consisting of many fine sand particles coated with ferrous oxide. Since most red beds are found in geological strata of continental origin, it is thought that they were formed by exposure to oxygen in the atmosphere rather than underwater.

The sudden transition from a reducing to an oxidizing atmosphere also had profound effects on the Earth's climate. For the first 2 billion years of the Earth's history, the surface was so warm that solid ice could not form anywhere on the planet. The geological record shows that there was a rather sudden and abrupt worldwide cooling trend about 2 billion years ago. Major glaciation appeared at the north and south poles for the first time. What happened? It seems that the oxygen-initiated destruction of the reducing atmosphere was responsible. The burst of photosynthetic oxygen emitted into the air destroyed the remaining traces of ammonia and methane in the atmosphere within only a few million years. The removal of these gases resulted in a severe diminution in the greenhouse heating of the atmosphere, and the Earth turned sharply cooler.

The oxygen content of the air has continued to rise ever since, and more and more carbon dioxide has been incorporated into sedimentary rocks and into living creatures. Consequently, the average temperature of the surface of the Earth has slowly but steadily declined over the last 2 billion years of the planet's history.

The increased availability of oxygen made possible another watershed in the history of life — the appearance of the first *eucaryotic* cell. Eucaryotic cells are larger and more complex than procaryotic cells. The most important innovation introduced by the eucaryotic cell is its ability to use molecular oxygen to provide energy. This process is known as *respiration* and involves the use of oxygen to consume glucose by the reaction

$$C_6H_{12}O_6 + 6O_2 \longrightarrow 6CO_2 + 6H_2O \tag{4.3}$$

This process is the "inverse" of aerobic photosynthesis, since glucose is degraded back to carbon dioxide and water. Respiration yields back 18 times as much energy per glucose molecule consumed as does anaerobic fermentation. By breaking down glucose into water and carbon dioxide, virtually all of the biologically useful energy can be recovered. The evolution of aerobic respiration made it possible to utilize the energy stored in glucose much more efficiently, and eucaryotic cells flourished and multiplied. They were so successful that they entirely displaced their procaryotic predecessors in most environments. All multicellular organisms from amoebae to human beings are made up of eucaryotic cells and all require oxygen for survival and growth.

With the advent of oxidative respiration, the oxygen/carbon dioxide cycle we know today became established. Photosynthetic cells and respiratory cells became symbiotically dependent on each other; the success of one led to the proliferation of the other. The advent of the oxygen/carbon dioxide cycle led to a massive proliferation of life in the oceans and to an unprecedented, rapid rise in the oxygen content of the air.

For the first 3 billion years of evolutionary history, life consisted exclusively of single-celled organisms. Roughly 1 billion years ago, the first complex multicellular creatures appeared. This important evolutionary step was made possible by the rapidly rising concentration of oxygen in the environment. These creatures rapidly multiplied and diversified. The sudden appearance of multicellular creatures marks the beginning of the *Phanerozoic eon* (the eon of "abundant life").

The rapid expansion of life in the oceans produced a continual increase in the oxygen content of the atmosphere. By approximately 400 million years ago, enough oxygen had accumulated in the atmosphere to permit life to spread onto dry land. Since then, life has multiplied, diversified, and spread over the entire Earth to occupy every available niche. All of the beauty and diversity that we see in the natural world today is a consequence of a steadily rising oxygen content in the atmosphere, made possible by a complex sequence of events that began nearly 4 billion years ago.

5

The Moon — Earth's Nearest Neighbor

Earth's single moon has for a long time been the object of intense human interest. People must have speculated about the Moon for about as long as the human race has been on this planet (Figures 5.1 and 5.2). They wondered what it was like on the Moon and, in particular, if there were any beings living up there. Up until the year 1609 people could do little more than idly speculate about the Moon. In that year Galileo turned his telescope toward Earth's neighbor. He described a world covered with mountains, hills, valleys, seas, and large land-masses, a world that might not be very different from the Earth. As telescopes improved in quality over the years, much more about the Moon was learned. It was gradually transformed in human imagination from an Earth-like world, per-haps rich with life, into an airless and desolate world completely devoid of living creatures.

All observations of the Moon up to the mid–twentieth century were, of necessity, made from the surface of the Earth. Within the last 20 years, the advent of space travel has made it possible to observe the lunar surface from close up. The Moon is the only other body in the entire Solar System upon

Photograph of the crater Copernicus on the Moon taken by the probe *Lunar Orbiter 2*. Copernicus is a relatively young lunar crater, located just south of Mare Imbrium. At the time that the photograph was taken, the craft was located 250 km south of the crater, flying 45 km above the lunar surface. (Photograph courtesy of NASA)

FIGURE 5.1 Telescopic photo of the Moon. (Photograph courtesy of Lick Observatory)

FIGURE 5.2 Relative sizes of the Earth and Moon.

whose surface humans have walked, and it will probably remain so for the rest of this century.

THE PHASES OF THE MOON

The most readily noticeable aspect of the Moon is that it periodically changes its appearance in the sky. At the start of a cycle the Moon appears as a bright circle in the sky — the *full moon*. A shadow then starts to creep over the Moon, starting from the eastern limb. The shadow gets larger and larger until only one-half of the Moon is visible. This is the *third quarter*. The shadow continues to grow until only a thin crescent on the western limb is visible. The Moon then briefly disappears (the *new moon*). The Moon begins to reappear shortly thereafter, first as a thin crescent on the eastern limb, then increasing to exactly one-half the face, at *first quarter*. The Moon continues to grow until its full face is visible once again, and the cycle is repeated (Figure 5.3).

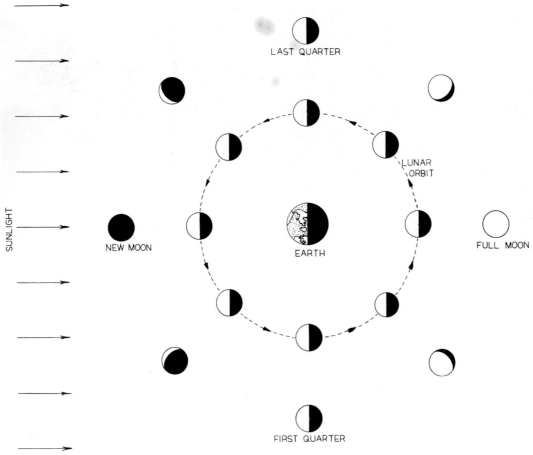

FIGURE 5.3 The phases of the Moon.

Primitive people thought that the periodic waxing and waning of the Moon was the work of a deity or supernatural force. Today, we know that it is caused by the fact that the Moon has no intrinsic luminosity of its own. It can shine only by reflecting light from the Sun. Only one-half of the surface of the Moon can ever be illuminated at any one time. This fact seems to have been first recognized by Thales of Miletus (624 – 546 B.C.). The Moon changes its appearance in the sky because it revolves around the Earth. When the Moon and Sun are on opposite sides of the Earth, the fully illuminated face of the Moon is turned toward us, and we see a full moon. When the Moon passes between Earth and the Sun, the dark side of the Moon faces us, and we see a new moon.

THE ODD LUNAR ORBIT

The Moon travels around the Earth in an elliptical orbit, moving in the direct sense. The Moon therefore travels around the Earth in the same direction that the Earth revolves around the Sun (Figure 5.4). It also travels around the Earth in the same direction that the Earth rotates about its axis.

The time taken for the Moon to complete one full revolution of the Earth is called the *month*. There are two types of month that are commonly discussed, neither of which is equal to the calendar month. The first of these is the *sidereal month*, which is the time taken for the Moon to return to the same position in the sky relative to the fixed stars. This is 27.3217 days in length. The other is the *synodic month,* which is the time interval between successive full moons. This is 29.5306 days, 2.2 days longer than the sidereal month. The reason for the difference is the Earth's motion along its orbit around the Sun as the Moon revolves (Figure 5.5).

The distance between the Earth and Moon was known in ancient times. It was first estimated with some degree of accuracy by Aristarchus of Samos (310 – 230 B.C.) in the third century B.C. Modern measurements give a mean lunar distance of 384,400 km. However, the lunar orbit is actually rather eccentric. The distance of closest approach to Earth (the *perigee*) is 363,300 km. The Moon's maximum distance from the Earth (the *apogee*) is 405,510 km. Because of gravitational forces exerted by the Sun, the eccentricity can vary between 0.04 and 0.06; the average eccentricity of the lunar orbit is 0.0549.

The plane of the Moon's orbit is not parallel to the Earth's equator. It is, in fact, much more closely parallel to the ecliptic plane (making an angle of 5.63°) than it is to the terrestrial equator. The Moon therefore appears to travel along the constellations of the zodiac in a path close to that followed by the Sun and the other planets. The two points in the sky where the path of the Moon crosses the ecliptic plane are called *nodes*. The *ascending node* is the point where the Moon crosses the ecliptic plane going south to north; the *descending node* is the crossing point going from north to south. The imaginary line connecting the two nodal points to each other is called the *line of nodes* (Figure 5.6).

If it were not for the Sun, the nodal points would be fixed in space. However, the Sun also exerts a gravitational pull on the Moon, forcing the nodes to undergo a slow motion along the ecliptic in the retrograde (east-to-west) direc-

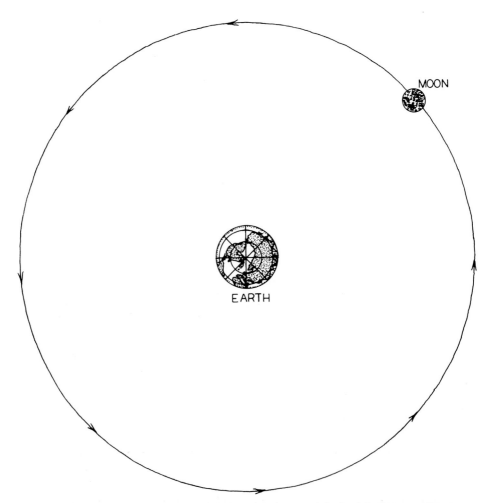

FIGURE 5.4 The orbit of the Moon. The Moon travels around the Earth in the same direction that the Earth rotates on its axis.

tion. The plane of the lunar orbit slowly turns about an imaginary axis oriented perpendicular to the ecliptic plane. This motion is called the *regression of nodes*. It takes 18.6 years for the line of nodes to make a full circuit of the sky (Figure 5.7). The slow rotation of the plane of the lunar orbit is sometimes measured by reference to yet another type of month, the so-called *draconic* (or *draconitic*) *month*. The draconic month is the time interval between two successive passages of the Moon through one of its nodes. It is 27.2122 days long.

The *line of apsides* is an imaginary line drawn across the widest part of the Moon's orbit, from apogee to perigee (Figure 5.8). The Sun's gravity forces the lunar perigee to move slowly eastward (in the prograde direction) among the stars, making a full circuit of the sky once every 8.85 years. This phenomenon is called the *advance of the line of apsides* (Figure 5.9). The length of time that it

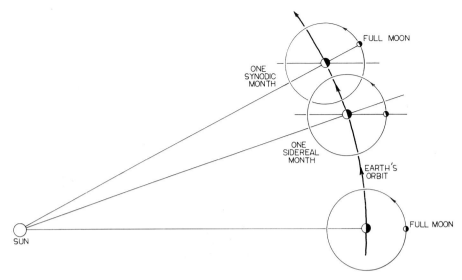

FIGURE 5.5 Difference between synodic and sidereal month. Because of the Earth's revolution around the Sun, the interval between successive full moons is always longer than the time taken for the Moon to return to the same position in the sky with respect to the fixed stars.

takes for the Moon to pass from one perigee to the next is called the *anomalistic month* and is equal to 27.5546 days. It is 5.5 hours longer than the sidereal month.

SOLAR AND LUNAR ECLIPSES

When seen from the Earth, the relative sizes of the Moon and the Sun are approximately the same. This coincidence is the origin of one of the most beautiful and spectacular events ever to take place in the entire natural world —the *solar eclipse,* which happens when the Moon passes in front of the Sun (Figure 5.10).

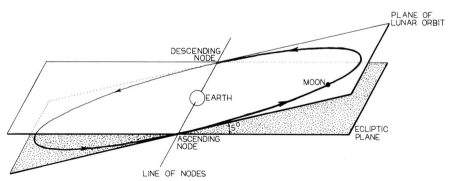

FIGURE 5.6 Line of nodes, the imaginary line joining the two points in the sky where the lunar orbit crosses the ecliptic plane.

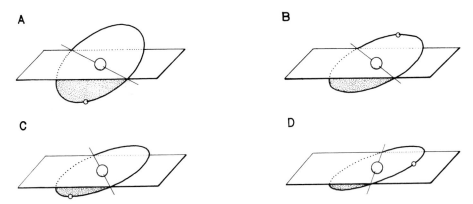

FIGURE 5.7 Regression of nodes. Gravitational forces exerted by the Sun cause the nodal points to move slowly along the ecliptic in a retrograde direction.

The prospect of the Sun momentarily "going out" was a frightening prospect for ancient peoples, and solar eclipses were objects of superstitious dread. As more and more of the Sun started to disappear, it was imagined that it was being devoured by some sort of invisible monster or demon. Special magical ceremonies were hastily commenced to induce the Sun's return. Oddly enough, they worked every time. The name of the first person to figure out what is actually going on during a solar eclipse has been lost to history.

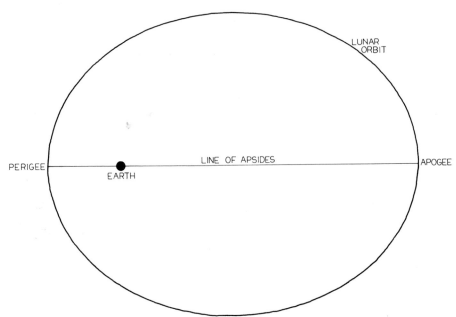

FIGURE 5.8 Line of apsides, the imaginary line drawn across the length of the lunar orbit, connecting perigee and apogee.

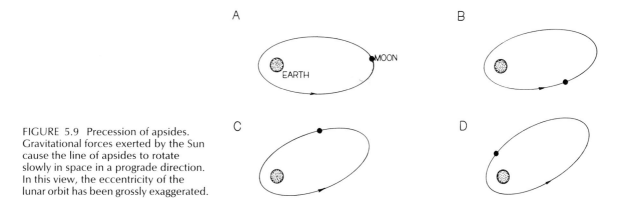

FIGURE 5.9 Precession of apsides. Gravitational forces exerted by the Sun cause the line of apsides to rotate slowly in space in a prograde direction. In this view, the eccentricity of the lunar orbit has been grossly exaggerated.

There are two types of solar eclipses, partial and total. A *partial* eclipse occurs when the Moon blocks only part of the solar disk. A *total* eclipse occurs when the Moon covers the entire solar disk (Figure 5.11). During a total solar eclipse, a momentary "night" takes place and the spectacular solar corona becomes visible as a dazzling white light surrounding the darkened solar disk. The type of eclipse that you observe depends on where you are on the Earth at the time of the eclipse. Because the Sun is physically so much larger than the Moon, the shadow of the Moon consists of a long, thin cone that just barely reaches the Earth. This cone is called the *umbra*. Only those people inside the umbra can see a total eclipse. There is a much larger region of partial shadow outside the umbra. It is known as the *penumbra*. People inside the penumbra see only a partial eclipse. Because the Moon's shadow on the Earth is so narrow, only a very few observers get a chance to see a total eclipse. On the average, a total eclipse will be visible at any given point on the Earth only once every 360 years! Many people (including the author) have never seen one.

FIGURE 5.10 Origin of solar eclipse. A solar eclipse occurs when the Moon passes in front of the Sun, casting a shadow on the Earth.

The Moon — Earth's Nearest Neighbor **89**

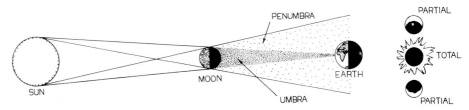

FIGURE 5.11 Umbra and penumbra. The shadow cast by the Moon has two components: a region of full shadow (the umbra) in which all of the light from the Sun is blocked, and a region of partial shadow (the penumbra) in which only part of the light from the Sun is blocked. Observers located within the umbra see a total eclipse, those within the penumbra see a partial eclipse.

Because of the eccentricity of the lunar orbit, total solar eclipses can vary in duration and type. The longest lasting total eclipses occur when the Moon's shadow on the Earth is the largest. This will happen when the Moon is at perigee (closest to Earth) at the time the Earth is at aphelion (farthest from Sun). Under those conditions, the shadow that the Moon casts on the Earth is 270 km in diameter, and totality will last 7.5 minutes. The shortest total eclipses occur when the Moon is farthest from Earth. In such cases the cone of the Moon's shadow does not quite reach the Earth, and the total eclipse is annular. In annular eclipses the apparent size of the Moon is slightly smaller than that of the Sun, and a ring of light coming from the limb of the Sun is still visible even at totality (Figure 5.12).

The Moon can also be eclipsed. Since the Moon shines only by reflected

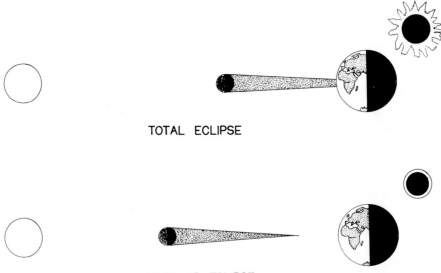

TOTAL ECLIPSE

ANNULAR ECLIPSE

FIGURE 5.12 The difference between total and annular solar eclipses. In a total eclipse the full face of the Sun is obscured by the Moon. In an annular eclipse the Moon is sufficiently far from the Earth that it blocks only the center of the Sun's face, leaving points on the solar limb still visible.

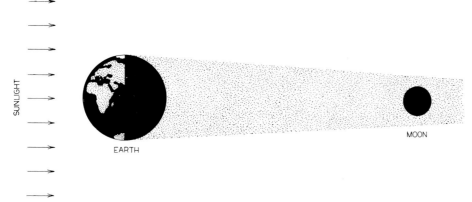

FIGURE 5.13 Origin of lunar eclipse. A lunar eclipse occurs when the Moon passes through the shadow cast by the Earth.

light, it can be eclipsed when it passes inside the shadow cast by the Earth (Figure 5.13). Unlike the case of the solar eclipse, everyone in the night hemisphere of Earth sees exactly the same thing at the time of a lunar eclipse. Because they can be seen by everyone on half the planet when they occur, lunar eclipses are not nearly as fascinating as solar eclipses. Lunar eclipses can last for several hours, and totality can last for an hour or more.

Even at totality the eclipsed Moon is usually not completely invisible. More often than not the eclipsed Moon is faintly illuminated by sunlight that is refracted from the upper atmosphere of Earth. The color of the dimly lit eclipsed Moon is generally reddish, due to preferential scattering of blue light by the Earth's atmosphere.

THE SAROS CYCLE

Solar eclipses can occur only at the time of the new moon, and lunar eclipses, only at the time of the full moon. If the lunar orbit were exactly parallel to the ecliptic plane, one solar and one lunar eclipse would occur every month. Eclipses do not occur nearly this often because the plane of the lunar orbit is tilted with respect to the ecliptic plane. Most of the time, the Moon passes significantly above or below the Sun (or the Earth's shadow) and no eclipse is seen (Figure 5.14).

Only when the Sun happens to lie along the line of nodes is there a possibility of an eclipse. A solar eclipse takes place when the Moon is passing through a node at the time of the new moon; a lunar eclipse occurs when the Moon passes through a node at the time of full moon (Figure 5.15). The Sun falls along the line of nodes twice every year. The time interval between two successive passages of the Sun through a node is called the *eclipse year*. It is 346.6 days long, 18.6 days shorter than the sidereal year. The eclipse year differs from the sidereal year because of the regression of nodes. An *eclipse season* is the time

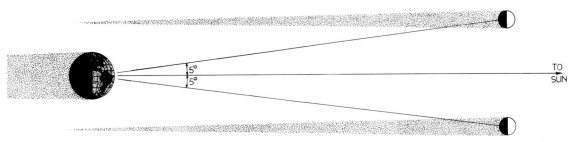

FIGURE 5.14 Most of the time the Moon passes significantly above or below the line connecting the Sun and Earth and no eclipse is seen.

interval during which the Sun is close enough to a node so that an eclipse would occur if the Moon should happen to pass through that node. Some of these eclipses will be lunar, others will be solar. There can be as many as seven eclipses in a year, including at least two and as many as five solar eclipses.

If eclipses could have been predicted, some of the terror that they aroused could have been alleviated. In ancient cultures, a great deal of economic and political power could be gained by anyone able to do this. It is not clear who was the first to accomplish this feat. Thales of Miletus is commonly given credit for the first correct prediction of an eclipse (585 B.C.), although Babylonian astronomers may have had some success at least a couple of centuries earlier.

In the ancient world it was generally recognized that eclipses occurred in a cyclic pattern. If an eclipse happened to occur at a certain location during a given year, one could be sure that another eclipse of a similar type would occur at the same location exactly 18 years, 11⅓ days later. This interval is known as the *saros*.

The origin of the saros cycle lies in the phenomenon of the regression of nodes. By a peculiar mathematical coincidence, 223 synodic months is almost

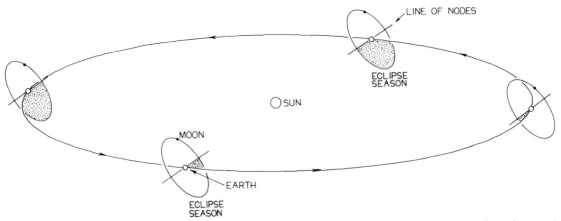

FIGURE 5.15 The plane of the lunar orbit remains approximately fixed in space as the Earth revolves about the Sun. A lunar or solar eclipse can occur only when the line of nodes of the lunar orbit happens to be pointed toward the Sun.

exactly equal to 242 draconic months. This time interval is equal to 18 years, 11⅓ days of calendar time (plus or minus 1 day, depending on how many leap years fall in the interval). After one saros interval, the Sun and Moon will have returned to almost exactly their same relative positions in the sky, and the same pattern of eclipses will be repeated.

Since many eclipses take place in each saros interval, there are many overlapping saros cycles simultaneously taking place. The extra one-third of a day in the saros means that the Earth will be 120° out of phase when solar eclipses in a given saros cycle return. Successive solar eclipses in a cycle will be displaced around the Earth 120° westward in longitude. For example, a total solar eclipse visible over Europe will reappear over North America exactly one saros interval later. After three saros intervals have passed (54 years, 34 days of calendar time), the extra one-third of a day is used up and solar eclipses will once again appear over the same spot.

LUNAR ROTATION—THE TIDAL COUPLE

The Moon rotates on its axis in the direct sense, with a period exactly equal to the sidereal month. The Moon's rotation is said to be *synchronous* with its revolution. We therefore see only one side of the Moon; the other face is perpetually hidden from view (Figure 5.16). The hidden face of the Moon is sometimes colloquially referred to as the "dark side," which is a misnomer in the sense that both sides receive the same average amount of sunlight. The satellites of many other planets also exhibit synchronous rotation, indicating that the effect is a fairly common phenomenon throughout the Solar System.

The fact that the lunar day and the sidereal month happen to coincide

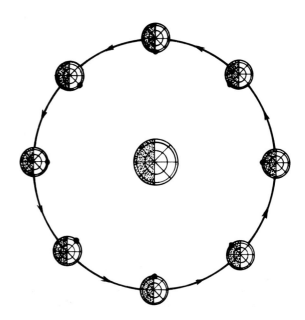

FIGURE 5.16 Resonant rotation of the Moon. The black dot shows the location of a feature on the surface of the Moon. The rate at which the Moon rotates on its axis exactly matches the rate at which it revolves about the Earth. Consequently, the same face of the Moon is always presented to the Earth.

cannot be accidental. It can be explained by the presence of a strong *tidal coupling* between Earth and Moon. The Earth's gravitational field attracts the near side of the Moon more strongly than the far side, stretching the Moon slightly out of spherical shape. The early Moon probably rotated on its axis at a much more rapid rate than it does now. As the Moon rotated, the tidal bulging produced internal frictional forces that gradually slowed down the lunar rotation rate. This braking continued until the lunar rotation rate finally matched the sidereal month.

Even though the sidereal month precisely matches the rotation rate, the eccentricity of the lunar orbit permits us to see slightly different parts of the lunar disk at different times of the month. The lunar rotation rate is constant, but the Moon is moving along its orbit at a faster rate when it is at perigee than when it is at apogee. This causes the lunar rotation momentarily to get out of phase with its revolution. Consequently, slightly different faces of the Moon are presented to us at different parts of its orbit (Figure 5.17). In addition, the lunar equator is not precisely parallel to its orbital plane. The angle between them is 6.5°, permitting us to see over the lunar poles at the time that the lunar rotational axis is canted toward Earth (Figure 5.18). These effects (known as *librations*) permit us to see as much as 59 percent of the lunar surface at various different times.

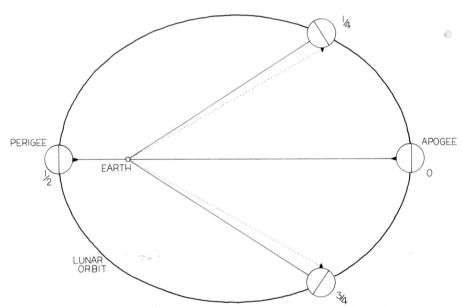

FIGURE 5.17 Librations in longitude. Because the Moon's orbit around the Earth is not a perfect circle, slightly different aspects of the lunar face are presented to Earth at different times in the month. The black triangle represents the location of a prominent feature on the Moon, perhaps a mountain or a large crater. In this diagram, the eccentricity of the lunar orbit has been grossly exaggerated in order to show the effect more clearly.

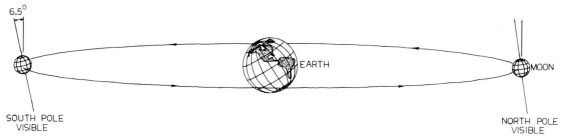

6.5°

SOUTH POLE
VISIBLE

EARTH

MOON

NORTH POLE
VISIBLE

FIGURE 5.18 Librations in latitude. Because the Moon's equator is not precisely parallel to the plane of its orbit, it is possible to see "over" the lunar poles by a slight amount.

THE EARTH'S OCEANIC TIDES

The Moon also exerts tidal forces on the Earth, but the effect is not nearly so dramatic. The primary effect on Earth is the creation of strong oceanic tides. The lunar gravitational field causes the water in the oceans to bulge outward a distance of 1 m or so. Because the lunar gravitational force varies inversely with the square of the distance to the Moon, there are two water bulges — one in the lunar direction and the other in the antilunar direction. As the Earth rotates with respect to the Moon, the water level alternately rises and falls as these bulges of water are swept around the world. Since there are two bulges, there are two high tides per day (Figure 5.19). Because of the Moon's motion along its orbit, the time of high tide at any one place on Earth will come approximately 1 hour later each day.

The Sun also exerts tidal forces on the Earth, but the effect is only about one-half as great. When the Sun, Moon, and Earth lie along a straight line (at the time of the new or full moon), the tidal forces exerted by the Sun and Moon act in the same direction, producing especially high high tides and low low tides. These are called *spring tides*. At the time of the first and last quarters, the lunar and solar tidal forces are perpendicular to each other, producing lower high tides and higher low tides. These are called *neap tides*. The most intense tides of all occur when the Moon is at perigee at the time of the new or full moon (Figure 5.20).

One of the subtle side effects of the oceanic tides raised on Earth by lunar gravitational forces is the generation of friction between the oceans and the continental landmasses as the water level rises and falls. Acting over many years, this tidal friction is slowing down the rotation of the Earth. Eventually the Earth's rotation will be slowed sufficiently so that it exactly matches the sidereal month. When this happens, the Moon will always appear to be above the same spot on

MOON

FIGURE 5.19 The origin of oceanic tides on Earth.

EARTH

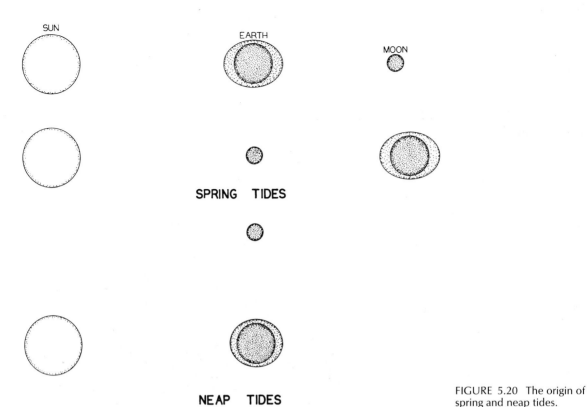

SUN

EARTH

MOON

SPRING TIDES

NEAP TIDES

FIGURE 5.20 The origin of spring and neap tides.

the surface of the Earth. People in one hemisphere would see a perpetual Moon in the sky, whereas people in the other hemisphere would never see the Moon at all. For the near future there is no reason for alarm, since the length of Earth's day is currently increasing by only 1.8 sec every 100,000 years. It will take many billions of years before Earth's day matches the sidereal month.

The tidal coupling between Earth and Moon is not only increasing the length of Earth's day; it is also forcing the Earth and Moon farther apart. Because of the tidal friction between water and land, the Earth's tidal bulge does not point precisely toward the Moon. Since the Earth rotates faster than the Moon revolves, the tidal bulge is pulled slightly ahead of the Moon. As a consequence, there is a small gravitational force exerted by the Earth on the Moon in a direction tangent to the lunar orbit. Since this force is in the same direction as the Moon's motion, the Moon is slowly being nudged into a higher orbit. The Moon is currently receding from the Earth at a rate of 3.2 cm/year, certainly a negligible process as far as the history of human existence upon the Earth is concerned, but significant over the long history of the Solar System. At the time of the initial formation of the Solar System 4.7 billion years ago, the Moon must have been considerably closer to the Earth than it is now. It has been slowly moving away ever since. Since the Moon was so much closer to the Earth in ancient times than it is now, the tidal forces exerted on the Earth must have been very much stronger and may have played an important role in shaping Earth's surface and atmosphere. They may have even played a role in the appearance and subsequent evolution of life.

LUNAR EXPLORATION—THE COLD WAR IN SPACE

The thought of actually traveling to the Moon to look at the landscape in person has intrigued people for thousands of years. It was not until the advent of rocketry and space travel that it became possible. The history of lunar exploration closely matches the history of international rivalry between the United States and the Soviet Union. Both nations became locked into a competitive program of lunar exploration during the 1950s and 1960s, perhaps more for reasons of national prestige or perceived military advantage than for any pressing scientific necessity.

The Soviet Union took an early lead. They were first to impact a spacecraft on the surface of the Moon. This was the probe *Luna 2,* which crashed on the lunar surface in September of 1959. In October of 1959, the Soviets placed *Luna 3* in a highly elliptical Earth orbit that took it around the Moon to photograph the hidden side for the first time. The Soviets were also the first with a soft lunar landing—*Luna 9,* which touched down on the Moon in February of 1966 and sent back a series of surface photographs. They were first yet again with a lunar satellite, *Luna 10,* which entered lunar orbit in April 1966.

The Soviet lead in lunar exploration seemed insurmountable at the time. Early in the 1960s, President Kennedy committed the United States to the landing of humans on the Moon. The American lunar exploration program got off to a shaky start, with a long string of embarrassing failures. After several early mishaps, the last three probes in the Ranger series were able to send back close-up television photographs of the lunar surface in the moments just before they crashed onto the surface. In late 1966 the United States placed the first of several spacecraft in lunar orbit to make detailed maps of both the near and far sides of the Moon. In that same year, the United States placed the first of its Surveyor landers on the Moon to provide a detailed picture of the surface environment in preparation for later landings.

The race for the Moon ended in July 1969, with the landing of the *Apollo 11* astronauts. Six Apollo landings were made, the last of which took place in December of 1972. A dazzling battery of scientific tests and experiments were performed on the lunar surface by the Apollo astronauts. Scientific instruments left behind on the Moon sent a continual stream of useful data back to the Earth for several years thereafter. Nearly 1000 kg of lunar surface material was brought back to Earth for study.

Although the Soviet Union had also hoped to land people on the surface of the Moon, insoluble problems with their large rocket booster caused them to abandon their plans during the early 1970s. Instead, they used automated spacecraft, which were remotely controlled from Earth. Three spacecraft in the

Luna series succeeded in landing on the lunar surface and returning capsules containing small amounts of lunar soil safely to Earth. Two other Luna landers deployed automatic rovers onto the surface. These carried out extensive maneuvers and transmitted television pictures and other valuable data back to Earth. Whether or not the Soviet Union will ever try again to send people to the Moon is a matter of conjecture.

THE LEGACY OF APOLLO

The initial phase of lunar exploration is apparently over, no spacecraft having been sent toward the Moon since 1976. When it shall resume again is uncertain. Nevertheless, the harvest of useful lunar data was a rich one.

THE INTERIOR

Spacecraft exploration gave a reliable value for the lunar mass and made it possible to construct a realistic model of the lunar interior. The mass of the Moon is about 1.2 percent that of Earth, and the radius is 1738 km. There is no noticeable oblateness. The average density of the Moon was found to be 3.34 g/cm³, appreciably less than Earth's. The Moon and Earth must therefore be structurally quite different. The lunar surface materials brought back to Earth for study have an average density of about 2.96 g/cm³, only slightly less than the overall density of the entire Moon. This must mean that there is very much less differentiation of material with depth inside the Moon than there is inside the Earth.

Seismographic tests performed by instruments left behind by the Apollo landing teams indicate that there is probably a very small central core 400 to 700 km in diameter. It probably takes up no more than 5 percent of the total lunar mass. This core may still be partially molten, although the temperature at the center is probably not much above 1500°C. The composition of the core is uncertain, but it may be rich in iron and sulfur in combination. The pressure at the center is about 10,000 atm, not enough to cause any significant amount of compression. Surrounding the core is a 1000-km-thick mantle. The mantle is completely rigid down to a depth of at least 600 km below the surface, although it could still have some fluid properties at greater depths near the core. At the surface there is a crust 60 to 100 km thick (Figure 5.21).

The automatic seismic detectors left behind by the Apollo landings did detect a "moonquake" from time to time, but these are much less frequent than terrestrial quakes and are far less intense. They originate much deeper in the interior than do terrestrial quakes, and they appear to fall into two distinct classes. Moonquakes of the first type appear to originate from centers located 600 to 800 km below the surface, and their occurrence appears to be correlated with the motion of the Moon around the Earth. Forty centers of seismic activity are known, all but two of which lie on the near side. Half of the centers are active when the Moon is at perigee, the other half when the Moon is at apogee, so this type of moonquake is a response of the lunar interior to the periodic flexing exerted by the Earth. The other type of seismic disturbance appears to be

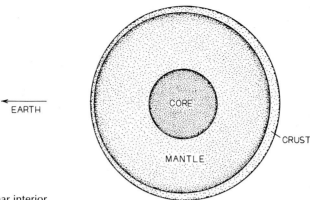

FIGURE 5.21 Diagram of lunar interior.

unrelated to the Earth. This type originates below a depth of 1000 km, near the boundary of the core. The origin of these is uncertain.

The early lunar flyby and orbital spacecraft found that the Moon has no significant magnetic field. Because of the lack of a magnetic field the Moon has no Van Allen radiation belts, and the surface is exposed to the constant erosive forces of the solar wind. However, it was found to everyone's surprise that the lunar rocks brought back to Earth exhibited a residual magnetism. This may indicate that the Moon at one time had a sizable magnetic field of up to 1 gauss in intensity. Older rocks tend to have larger residual magnetism than younger ones, indicating a magnetic field that rapidly died out over time.

The *Explorer 35* lunar satellite discovered that there are small regions on the surface that have relatively high magnetic fields (of the order of 0.0001 gauss). These regions are coherent over many tens of kilometers, and are termed *magcons*. The most intense magcons are found near the Van de Graaf crater on the far side of the Moon. However, there is no simple moonwide field distribution and no unique north–south polar orientation.

THE BARREN SURFACE

Chemical analyses of the lunar surface rocks that were returned to Earth demonstrated that they were rather similar to terrestrial rocks, with silicates of such metals as iron, aluminum, titanium, and magnesium being abundant. There are, however, important differences. The more *volatile* (that is, low-melting-point) elements such as zinc, lead, gold, silver, indium, thallium, and bismuth are relatively less abundant on the Moon than they are on Earth. The Moon is correspondingly richer in high-melting-point *refractory* elements such as magnesium, sulfur, and titanium. The relatively large abundance of refractory elements probably means that the Moon was subjected to high temperatures early in its history. Because of the low lunar gravity many of the more volatile elements were evaporated away to space by the intense heat. The oxidized compounds on the surface are generally in their lowest states of oxidation, which indicates that the Moon never had any significant oxidizing atmosphere at any time in its past. In fact, some of the iron in the Moon rocks appears in the form of metallic crystalline occlusions rather than as an oxide.

Rocks found on the Earth usually have significant amounts of water chemically combined with the minerals in their interiors. Such water is called *water of hydration*. Terrestrial minerals typically contain 1 to 2 percent (by weight) water of hydration. However, Moon rocks have absolutely no water of hydration at all. There are no rocks on the Moon that bear any similarity to terrestrial shales or limestones. Such rocks are formed by sedimentary deposition under water, and their total absence on the Moon must mean that the lunar surface never had any appreciable amount of liquid water at any time in the past. Any water initially present on the Moon must have quickly evaporated away into space under the influence of the intense heating that was present early in its history. The lunar soil has only a few parts per million of organic (carbon-containing) chemical compounds, and there are no signs of any life forms ever having been present.

Surface drilling experiments and Earth-based radar studies indicate that the crust consists of highly fractured material and loose rubble down to depths of at least a few meters. The outer zone of fine soil and loose rocks is known as the *regolith*. The rubble zone on the lunar surface is a very poor conductor of heat. The low thermal conductivity means that the surface cools off very rapidly with the coming of night and heats up rapidly with the coming of sunrise. The top meter or so of the lunar crust is "plowed over" once every 100 million years by the continual impact of micrometeors. The Apollo and Luna landers have sampled lunar rocks down to a depth of about 1 m, but so far no lunar bedrock has ever been sampled.

The Moon has no significant atmosphere and no liquid water. Because of this lack (and perhaps also because of the exceedingly slow rotation rate), the surface is exposed to sharp temperature extremes. At the *Apollo 11* landing site, the maximum midday temperature was a broiling 70°C. However, the temperature drops as low as −50°C during the long lunar night. At the lunar equator it can get as hot as 105°C, above the boiling point of water.

CRATERS

Perhaps the most conspicuous features on the lunar surface are the craters. They are nearly circular in shape and are generally surrounded by rings of mountains. On the inside there is a level floor that is depressed below the outside ground level. There is often a mountain or group of mountains at the crater center. Often there is a blanket of rubble and loose debris spreading out from the crater for a distance of several crater diameters. Groups of smaller secondary craters can sometimes be found near a major crater. Some of the larger craters on the Moon have extensive ray systems extending radially outward for hundreds of kilometers (Figures 5.22 and 5.23).

Craters range in size from 120 km in radius down to sizes less than a few meters across, and there are many hundreds of thousands of them on the Moon. The more prominent craters on the Moon bear the names of individuals who have made important contributions to science and astronomy (Kepler, Copernicus, Tycho, Plato, and so on).

FIGURE 5.22 *Apollo 16* photograph of the crater King on the lunar far side. The crater is 90 km in diameter. (Photograph courtesy of NASA)

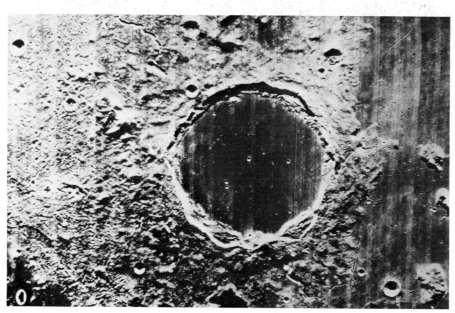

FIGURE 5.23 *Orbiter 4* photograph of crater Plato, a prominent dark-floored crater. The dark plain to the right of Plato is Mare Imbrium, the Sea of Rains. Molten lava from Imbrium has covered over the floor of Plato, leaving only the rim visible. (Photograph courtesy of NASA)

The origin of the craters was a subject of hot controversy for a long time, but it is now almost universally accepted that they were originally formed by the impacts of large meteoroids. When a high-velocity meteoroid strikes the surface of the Moon (or any other sizable body), a large amount of surface material is pulverized by the force of the impact. Most of the fractured material is violently ejected from the impact region, leaving a deep crater behind. Following the ejection of material, the floor of the crater rebounds, often forming a distinct mountain peak at the center. Some of the material blown out by the crater impact can be seen as a continuous blanket of loose rubble, extending outward for one or two crater diameters. In some of the more energetic impacts, the chunks of ejected material are of sufficient size to dig significant craters on their own when they fall back to the surface. These secondary craters can be found as much as 10 to 30 crater diameters distant from the primary.

THE HIGHLANDS

The density of large craters varies considerably over the surface. Some regions are relatively smooth and have few craters, whereas others are nearly saturated with large numbers of extensively overlapping craters. The most densely cratered regions on the entire Moon are known as *highlands*. The highlands tend to be the very brightest regions on the Moon and they contain chains of mountains that average a couple of kilometers higher than the surrounding terrain. A few lunar mountains tower nearly 5 km high, rivaling the highest mountain ranges on Earth. Many of the lunar highlands are given the names of some of the Earth's more prominent mountain ranges (Alps, Apennines, Caucasus, and so on).

The lunar mountains were not thrust up by geological processes even remotely similar to those responsible for terrestrial mountain building. There is no evidence that crustal shifting or continental drift phenomena ever took place on the Moon. Instead, the mountains of the Moon were formed by the extensive overlapping of impact craters.

The Apollo landings returned several highland rocks to Earth. The age of the lunar highlands can be determined by performing radioisotope analyses of the returned rock samples. The results indicate that the highlands are typically 4.2 billion years old. The oldest material found on the Moon is a fine rock dust (breccia), which probably covers much of the surface. It is 4.55 billion years old and is probably left over from the time just after the Moon first formed a solid crust.

Lunar highland rocks are generally of a type classified by geologists as *anorthositic*. Anorthositic rocks have a high abundance of plagioclase feldspar minerals. They are relatively poor in iron and magnesium but are rich in aluminum. Such minerals are also found in abundance on the Earth. The anorthositic rocks of the highlands have the highest melting point of any rocks yet found on the Moon, which probably means that they were the first to solidify at the time the Moon's permanent crust formed. They must have cooled rather slowly.

Another type of mineral is also present in the lunar highlands, one which is almost nonexistent on Earth. It has been named *KREEP norite*. "KREEP" is an acronym standing for "Potassium(K), Rare Earth Elements, and Phosphorus," so chosen for the unusually high concentrations of these elements found in these rocks. KREEP norite minerals are somewhat younger than anorthositic rocks and may have been formed by the partial melting of anorthositic crust that took place at the time of the meteor impacts that formed the highland mountain ranges. The lunar soil also contains large numbers of tiny glass beads. These were probably formed by the rapid cooling of crustal material that was melted during meteor impacts.

The lunar highlands are the scars left over from a violent period of intense meteoroid bombardment that took place more than 4 billion years ago. Shortly after the Moon and the planets condensed out of the primeval solar nebula, the early Solar System was filled with numerous chunks of material that had not combined to form large planets. During the first few hundred million years of the history of the Solar System, there was a heavy bombardment of planetary surfaces by these loose bodies. This rain of meteoroids continued up until about 4 billion years ago, by which time most of the larger chunks had been swept up by a collision with one or the other of the planets. The rate of meteoric bombardment declined to a relatively low value very shortly thereafter, and has remained low ever since. The Earth shows few traces of this early bombardment, because the craters that were excavated at this time were erased many years ago by the combined effects of water erosion and geological activity. Since the surface of the Moon has been so much less active, most of the craters have survived.

Before the Apollo mission, it was hoped that some pristine material left over from the formation of the Solar System might be found on the Moon. Such material would be invaluable to science, since it would contain a record of the conditions that existed in the primitive nebula at the time that the Moon and the rest of the planets condensed. However, all of the lunar material returned to Earth to date has undergone considerable thermal and chemical evolution subsequent to the Moon's formation. The geological history of the Moon is just as complex as that of the Earth. It may be that no primitive, unprocessed material left over from the initial solar nebula remains anywhere in the Solar System.

THE MARIA

To an observer from Earth, the Moon's surface appears to be covered with irregularly shaped dark regions that are hundreds of kilometers in width. Many of them are roughly circular in shape. They are known as *maria,* the Latin word for seas. They were given their name by Galileo, who at first thought that they might be bodies of water. The maria are actually plains of solidified dark-colored lava (Figure 5.24).

The floors of the lunar maria are typically 2 to 3 km below the surrounding terrain. Maria are often surrounded by rings of mountains, and craters with darkly colored floors can sometimes be found at their edges. Even though the lunar maria are perfectly dry, they have given exotic aquatic-type Latin names: Oceanus Procellarum (the Ocean of Storms), Mare Tranquilitatis (the Sea of Tranquility), Sinus Iridum (the Bay of Rainbows), Lacus Somniorum (Lake of the

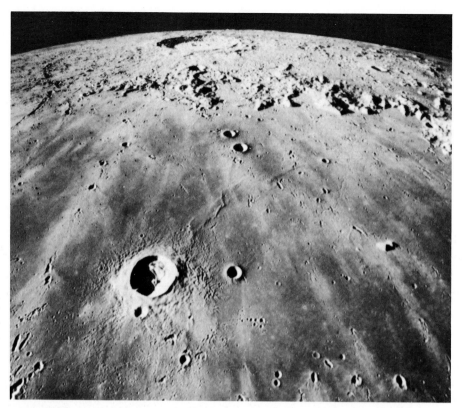

FIGURE 5.24 *Apollo 17* photograph of Mare Imbrium, the Sea of Rains. This view looks toward the south. The giant crater Copernicus is on the far horizon. The large crater in the foreground is Lambert. The chain of hills located between Lambert and Copernicus is the Carpathian Mountains. (Photograph courtesy of NASA)

Dreamers), and so forth. The largest mare on the Moon is Oceanus Procellarum, an irregular-shaped region over 2000 km wide.

Perhaps the most interesting mare on the Moon is Mare Orientale (the Eastern Sea), located near the edge of the Moon's visible face and visible only at extremely oblique angles from Earth (Figure 5.25). It consists of a very dark, nearly circular, central depression, surrounded by a set of concentric rings of high mountains. The central depression is approximately 300 km in diameter and its floor is nearly 8 km below the surrounding terrain. The mountainous ring that encircles the dark central mare consists of two main chains: the Rook Mountains at a distance of 250 to 300 km from the center and the Cordillera Mountains at a distance of 500 km. The Cordillera Mountains tower approximately 2 km above the outside plain. The total diameter of the Orientale mare and its associated ring formation is 1000 km. Several other maria have a hint of a multiringed structure but none so pronounced as that of Orientale. Because of its relatively well preserved state, Mare Orientale is probably the youngest of the major lunar maria.

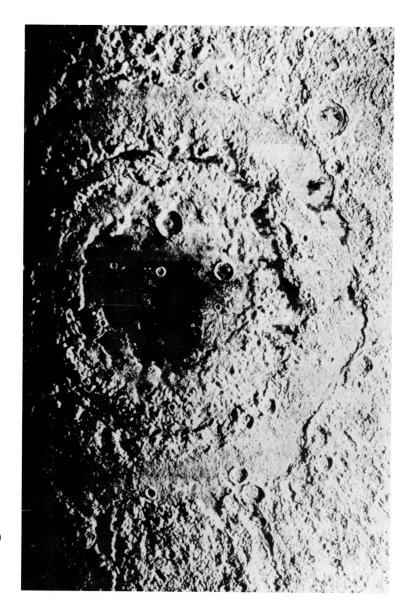

FIGURE 5.25 *Orbiter 4* photograph of Mare Orientale. The dark interior is 300 km across. The inner ring is the Rook Mountains, the outer ring is the Cordillera Mountains. (Photograph courtesy of NASA)

The lunar maria are much less densely cratered than the highlands, which must mean that they formed after the intense period of meteoric bombardment that formed the highlands had come to an end. The few craters that are present in the maria seem to be relatively young, with sharp rims and extensive ray systems. Copernicus is the best known example. It may be less than 1 billion years old.

Several rocks from the lunar maria were returned to Earth for study. These rocks are dark in color and basaltic in origin. They must have been deposited by lava flows. The lava in the lunar maria differs from terrestrial basalt in having a higher iron content and a lower sodium and potassium content. There is 10 times as much titanium in lunar basalts as there is in terrestrial basalts. This accounts for their dark color. Lunar basalts contain an especially large amount of oxidized iron, which gives them a rather high density of 3.3 g/cm^3. The high iron content of lunar basalt gives it an extremely low viscosity when molten, enabling volcanic eruptions to spread lava over considerable areas. The depth of the lava in most maria is only a few hundred meters and the outlines of previously existing features such as mountains and craters can sometimes be seen jutting out of the dark lava plains. An analysis of the relative amounts of radioactive isotopes present in these samples indicates that the lava in the maria solidified approximately 3.5 billion years ago, considerably later than the formation of the highlands.

The maria on the Moon are enormous depressions in the lunar surface whose floors were covered over by flows of volcanic lava approximately 3.5 billion years ago. These large, quasi-circular depressions have been termed *basins* by planetary scientists. Some basins have dark, lava-covered floors that are relatively free of impact craters. These basins are relatively young, and hence easy to identify. However, other basins are heavily scarred with impact craters and have little or no dark lava in their interiors. These basins are often so badly degraded as to be all but unrecognizable. These degraded basins must be considerably older. At least 30 basins with diameters of 300 km or more are known to exist on the Moon. Fourteen others are probably there but are so badly degraded that their identity is uncertain.

The quasi-circular geometry of the lunar basins is extremely interesting. It indicates an impact origin. Analysis of the rocks brought back to Earth by the Apollo astronauts suggests that the giant lunar basins were created by the impacts of particularly large bodies early in the history of the Solar System, perhaps 4.2 to 4.4 billion years ago. Some of these giant meteoroids must have been 100 km or more across, typical of the larger asteroids still present in orbits beyond Mars. The impacts of these large asteroid-sized bodies were severe enough to excavate large depressions in the lunar surface and penetrate a significant distance into the crust. Often, the impacts were sufficiently violent to throw up concentric rings of high mountains stretching out as far as 1000 km from the central impact point. Mare Orientale is the best preserved example of such a ringed impact basin, although many others also have at least a hint of a multiring structure. The distinction between a crater and a basin is somewhat artificial, being simply a matter of the size of the impacting asteroid that excavated the depression. A basin is nothing more than a particularly large crater. However, the term "basin" is now understood to refer specifically to a large impact-excavated depression that is surrounded by a concentric multiring structure.

Hundreds of millions of years after most of the lunar basins had been excavated, the Moon went through a period of intense volcanism. At that time these giant impact basins were filled up with molten lava forced upward from the mantle. Many preexisting craters and other features in the interiors of these basins were covered over with lava and were obliterated. Shortly thereafter the lava solidified, leaving the darkly colored "seas" seen today on the Moon. The onset of lava flows could not have been contemporaneous with the excavations of the basins themselves, since none of the maria show any evidence for the presence of large amounts of basin ejecta strewn over their surfaces. In addition, many maria have numerous large impact craters buried under the lava, indicating that a considerable amount of time (hundreds of millions of years) must have elapsed between the last of the large basin-forming impacts and the onset of volcanic activity. The energy released by the meteor impacts that excavated the giant ringed basins cannot be responsible for the flood of lava that filled up their interiors. A schematic overview of the primary events in the formation and evolution of the lunar maria is shown in Figure 5.26.

Many of the maria have prominent series of long, sinuous *scarps*. These are probably lava flow fronts associated with the volcanic activity that formed the maria. A number of smooth, low domes are scattered throughout the maria.

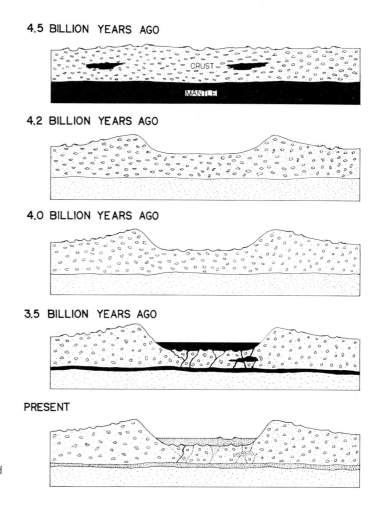

FIGURE 5.26 The formation of the lunar maria. The black areas are regions of molten lava; the shaded areas are regions of solidified lava.

These domes often have summit craters and are probably volcanic in origin. There are a few structures that look like volcanic vents or fissures. Although Earth-based telescopic observers have reported seeing some surface glows and perhaps some vapor clouds, there is no credible evidence that any region on the Moon is currently volcanically active.

There are some long, sinuous channels known as *rilles* that are sometimes 3 or 4 km wide and 1 km deep. Some are up to 300 km long. Rilles can cut through mountain ranges as well as through lowlands but they are usually found near the edges of lunar maria. The best known of these features is Hadley Rille, located in the Apennine Mountains near the edge of Mare Serenitatis (Figure 5.27). It was visited by the team of *Apollo 15* in 1971 (Figure 5.28). Rilles cannot be the dried-up remains of ancient riverbeds, as it is almost certain that there was never any liquid water on the Moon. It is much more likely that these channels were cut by the flow of low-viscosity molten lava that was set in motion at the time of the formation of the lunar maria.

The Lunar Orbiter series of satellites discovered that there are large concentrations of matter just below the surfaces of the maria. These concentrations, called *mascons,* are large enough to deflect the path of any satellite that flies over them. A mascon is probably made up of the remains of the asteroid-sized body that excavated the maria depression in the first place, as well as the lava that subsequently flowed upward from the interior to fill it up. Mascons are found only under the large circular maria such as Crisium and Serenitatis but not under the irregular maria such as Oceanus Procellarum.

FIGURE 5.27   Photograph of Hadley Rille taken by *Apollo 15* command module. North is at the top. The lunar Appenine Mountains are to the east. (Photograph courtesy of NASA)

FIGURE 5.28 *Apollo 15* landing site at edge of Hadley Rille. The lunar landing module is on the left, the four-wheeled Lunar Rover is on the right. (Photograph courtesy of NASA)

THE LUNAR FAR SIDE

Almost everyone had assumed that the far side of the Moon would look much the same as the side visible from Earth. However, spacecraft observation found this not to be the case (Figure 5.29). The far side is almost totally devoid of large maria, consisting mostly of densely cratered highlands. Almost all of the maria (as well as the mascons) are on the near side of the Moon. The Lunar Orbiter series of satellites also found that the lunar crust is significantly thicker on the hidden side (by 40–50 km) than it is on the near side. This difference is large enough to offset the lunar center of mass 2 km closer to the Earth than the geometrical center.

Why this asymmetry in the two faces of the Moon? One theory proposes that the mascons were formed at a time before the Moon's rotation rate had become locked into synchronization with its revolution. By pure chance, more mascons happened to form on one side of the Moon than the other. The "heavier" side of the Moon was attracted more strongly to the Earth than was the "lighter" one, so it eventually ended up facing the Earth. Another theory assumes that the Moon was already tidally coupled to the Earth at the time of the most intense meteoric bombardment and that the presence of the Earth provided some sort of "focusing" effect that directed more large meteoroids toward the Earth-facing side than to the opposite side. A third possibility is that the Moon became tidally coupled to the Earth while it was still largely molten. At this time, the Moon was much closer to the Earth, and strong tidal forces distorted the lunar interior to such an extent that the crust became significantly

FIGURE 5.29 *Apollo 16* photograph of lunar far side. The north pole is on the upper left. Some of the near-side maria are visible at the western limb. Mare Crisium (the Sea of Crises) is the dark circular spot near the limb on the upper left. The other two dark regions below Crisium are Mare Smythii (Smyth's Sea) and Mare Marginis (The Marginal Sea). These two maria are visible only at extremely oblique angles from Earth. (Photograph courtesy of NASA)

thinner on the near side as the Moon began to solidify. The thicker crust on the far side was much more difficult for a large meteoroid to penetrate, and very few of these objects were able to break through into the molten mantle and release the lava flows that resulted in the formation of the maria.

THE MOON'S GEOLOGICAL HISTORY

The Apollo landings produced enough data so that a crude early history of the Moon could be established. The Moon appears to have condensed out of the primitive solar nebula about 4.7 billion years ago, at approximately the same time that the Earth itself did. The Earth and Moon (and probably all of the other planets as well) are thus about the same age. Within the first few hundred million years of the Moon's existence, the decay of radioactive elements within the interior caused virtually its entire mass to melt, the heavier metallic material sinking inward to the interior. Large parts of the surface of the Moon were probably molten at this time, although the Moon was probably never completely covered with a global ocean of molten lava at any time in its history. All throughout this early period, the Moon's surface must have been continually bombarded with large meteors. Because of the molten, fluid nature of the surface, the impact scars from these early meteors must have been erased almost as soon as they were formed.

About 4.5 billion years ago the Moon began to cool and the first permanent solid crust began to form. The solidification of the lunar crust must have been an exceedingly slow process; it was probably not fully formed until 4.2 billion years ago. At this time most of the Moon's highland craters were formed and the large impact basins that later filled up with lava were excavated. The rate of meteoric bombardment during this period must have been much higher than it is now, perhaps 1000 to 10,000 times the present rate. It was so great that the regions of the lunar surface that have survived from this period are completely saturated with large basins and craters. Many of the basins and craters formed at this time were quickly obliterated by fresh impacts almost immediately after they were excavated. Consequently, only the record of the very latter stages of the meteorite bombardment epoch survives, even on the most ancient parts of the lunar surface.

The oldest impact basins on the Moon that still survive in any readily recognizable form are the depressions that form the Nectaris and Serenitatis maria. The impacts that formed these basins probably took place between 3.9 and 4.2 billion years ago. The formation of the Imbrium and Orientale basins came somewhat later — approximately 3.8 billion years ago. Shortly thereafter, the rate of arrival of large meteoroids began rapidly to decline, and the great epoch of lunar basin formation and crater excavation came to an end. The supply of meteoroids had been steadily depleted all through this epoch as these objects one by one were swept up by a collision with one or the other of the planets or moons in the young Solar System. By approximately 3.5 billion years ago most of them were gone. The cratering rate fell to a relatively low level at that time and has remained low ever since.

About 3.5 billion years ago the Moon suddenly became geologically active, with intense volcanism breaking out and forcing lava from the still-molten mantle upward to fill in the low-lying impact basins. Previously existing craters and other landforms in these low elevations were covered up and obliterated when this lava solidified. A fresh, uncratered surface was created, forming the darkly colored lava plains that cover the bottoms of deep impact basins. The era of lunar geological activity was short-lived, although there is evidence that some lunar volcanism persisted as recently as 2.1 billion years ago. Since then the Moon has been geologically dead.

By the time that most of the dark maria on the Moon were in place, the rate of meteoric bombardment had declined to a low value, perhaps nearly as low as the present-day meteoroid flux. Consequently, most of these lava plains have survived relatively unscarred by impact craters to the present day. Even though the rate of meteoric bombardment has remained quite low for the last 3.5 billion years, it has not been precisely zero. Even today, an occasional large meteoroid will strike the Moon. These recent impacts formed the spectacular ray craters that are found in the maria. The large crater Copernicus is an example; it may have formed only 900 million years ago. The crater Tycho (which lies in a southern-hemisphere highland region) is probably the youngest major crater on the Moon. The Tycho impact may have taken place as recently as 100 million years ago. At that time dinosaurs were walking on the Earth. The event must have been quite spectacular to observe.

THE MOON'S ORIGIN

Although the Apollo landings provided much detail about the early history of the Moon, they could not satisfactorily resolve the question of the Moon's origin. Four distinctly different theories of lunar origin have been proposed: *fission, capture, condensation,* and *impact.*

The fission theory of lunar formation was first proposed in 1879 by George Darwin, the second son of the famous naturalist. It proposes that the Earth and Moon were originally part of the same body. The proto-Moon split off from the proto-Earth early in the history of the Solar System, most likely at the time when both were largely molten (Figure 5.30). This theory has some attractive features. For one, the average density of the Moon is the same as that of the Earth's mantle, indicating that the lunar material could have come from there. It was even suggested that the Pacific Ocean is the "hole" left behind on Earth by the

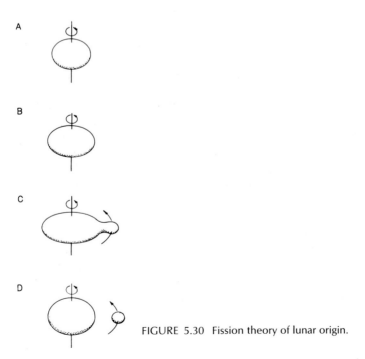

FIGURE 5.30 Fission theory of lunar origin.

removal of the Moon, since there are no continental landmasses located there at present. However, continental drift has reworked the surface of the Earth so much over the past 4 billion years that any record of such an early event would have been lost many years ago.

However, in order for the Moon to split off from the Earth in this fashion, the proto-Earth would have to have had an improbably high rate of spin. Furthermore, the Moon presently has 80 percent of the angular momentum in the Earth–Moon system, and some as-yet-unknown mechanism would have to be invoked to explain the transfer of the bulk of the angular momentum away from the Earth to the newly formed Moon. Another problem is the Moon's present orbit, which makes an angle with the terrestrial equator, ranging between 18° and 28°. If the Moon originally split off from a rapidly spinning Earth, it should have entered an orbit lying much closer to the plane of Earth's equator. The composition of the lunar and terrestrial crusts are sufficiently different to throw serious doubt on the possibility of the Moon ever having been part of the Earth; the differing proportions of volatile elements seem to suggest that the two worlds were probably gravitationally separate bodies from the very beginning.

The capture hypothesis involves the initial formation of the Moon in a different part of the Solar System and its subsequent "capture" by the Earth during an accidental close approach (Figure 5.31). The Moon is significantly less dense than Earth, which could mean that it initially formed much farther away from the Sun, perhaps as far out as the Asteroid Belt. In addition, the lunar orbit currently lies much closer to the ecliptic plane than it does to the plane of Earth's equator. Since an errant asteroid is most likely to approach Earth along a path that lies close to the ecliptic plane, the Moon's odd orbit could conceivably be a primordial link to an early capture event. The capture event, if it actually occurred, must have been catastrophic for the Moon. It would have completely reworked the lunar surface and perhaps even melted the interior. Consequently, capture could not have happened more recently than 4.2 billion years ago, as the lunar mountains and basins had already formed by that time.

Although capture theories do have some attractive features, there are serious problems. For one, the capture of a lunar-sized body requires that a whole series of individually improbable events take place in exactly the right sequence. Furthermore, there is as yet no demonstrated mechanism by which the Earth can capture in a stable orbit a smaller body that makes a single close approach. Such a body would simply fly right past the Earth, leaving at the same speed with which it approached.

The simultaneous (but separate) formation of the Earth and Moon in approximately their present locations is currently favored by most investigators. That is, the Earth and Moon condensed from a single cloud or ring of gas, dust, and debris gathered around the proto-Sun nearly 5 billion years ago. The original proto-Earth must have been a giant ball of gas and dust that was many times its eventual size. As the proto-Earth contracted, the heavier material sank inward to the center, leaving the lighter material at the periphery to form the Moon (Figure 5.32).

The high-temperature refractory materials were the first to condense from

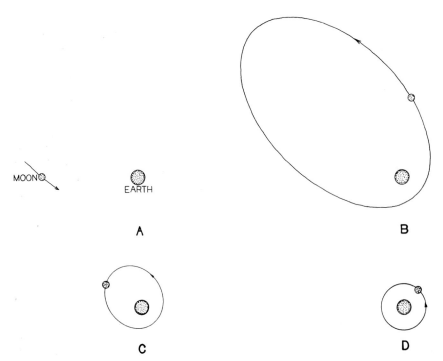

MOON

EARTH

A

B

C

D

FIGURE 5.31 Capture theory of lunar origin.

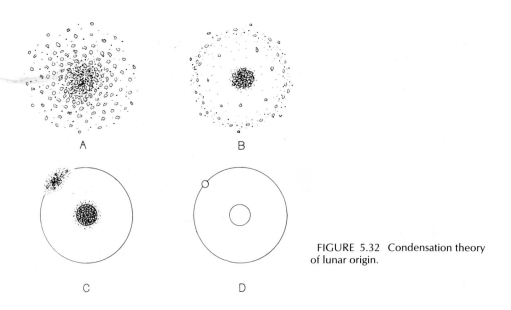

A

B

C

D

FIGURE 5.32 Condensation theory of lunar origin.

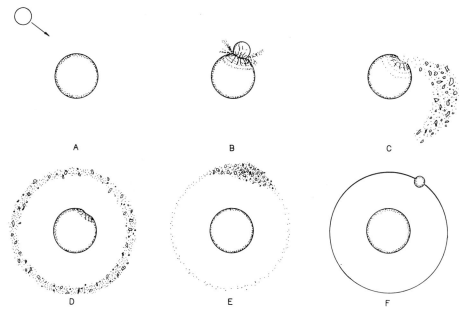

FIGURE 5.33 Impact theory of lunar origin.

the solar nebula, and the newly formed Earth/Moon dual planet system must have initially been rich in these materials. The more volatile materials condensed somewhat later, after the temperature of the solar nebula had cooled still further. The proto-Earth happened to grow much faster than the proto-Moon, and the Moon may have been unable to sweep up very much of this later-forming volatile material due to competition from the rapidly growing Earth.

The most recent contender in the competition among lunar origin theories is the impact hypothesis, which combines some features of the capture and the fission hypotheses. This theory proposes that the Moon is a product of a gigantic collision between the Earth and a Mars-sized body (Figure 5.33) early in the history of the Solar System. The force of this collision released an enormous cloud of vapor, dust, and debris, most of which subsequently escaped into space. However, a small fraction of the debris remained in geocentric orbit, forming a prelunar disk from which the Moon later accreted. Some of the more vexing questions about the compositional differences and similarities between the Earth and Moon are thus neatly explained, since the Moon is a composite body, made up of material that initially came from the Earth as well as of matter that originally accumulated elsewhere. The hypothesis also neatly resolves the angular-momentum controversy, since a glancing collision between Earth and a Mars-sized object is more than adequate to produce the angular momentum presently found in the Earth – Moon system. The impact hypothesis is, however, currently in the very earliest stage of study, and much more work is needed before it can be confirmed or refuted.

The Moon — Earth's Nearest Neighbor **115**

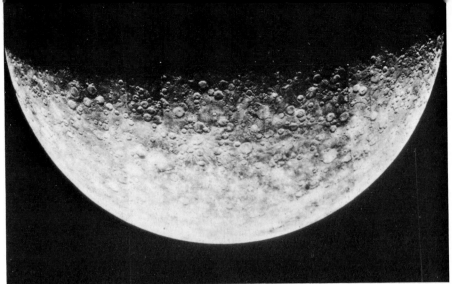

6

Mercury — the Land of Fire

Mercury is the planet closest to the Sun. It is an extremely difficult object to observe and can only be seen at twilight or early dawn under extremely favorable conditions. It is said (perhaps apocryphally) that the great astronomer Copernicus died without ever having seen the planet (Figure 6.1).

MERCURY'S ELONGATED ORBIT

Mercury's orbit has an average distance of 0.39 AU from the Sun, so it can never be seen more than 22° distant from the Sun in the sky (Figure 6.2). Because of its proximity to the Sun, Mercury is moving so fast that the length of its year is only 88 Earth days. Because of its great speed, the ancient Greeks named the planet after Hermes (and the Romans, subsequently, after Mercury), the winged messenger of the gods.

With the exception of Pluto, Mercury has the most eccentric orbit (e=0.206) of any of the planets, as well as the orbit with the highest inclination with respect to the ecliptic (7°). Because of Mercury's high orbital inclination, it can be seen crossing the solar disk only rarely. Such events, when they do occur, are called *transits*. Curiously, the plane of the Mercurian orbit just about

View of Mercury taken by the *Mariner 10* probe during approach to the planet. This photograph is a mosaic of 18 separate photographs, all taken by the probe when it was at a distance of 200,000 km from the planet. In this view, about two-thirds of the planet seen here lies in the southern hemisphere. (Photograph courtesy of NASA)

FIGURE 6.1 *Mariner 10* photograph of Mercury. This is a composite view of the planet, assembled from several different photographs taken during the spacecraft's initial approach to Mercury. At the time the photographs were taken, the spacecraft was 200,000 km from the planet. (Photograph courtesy of NASA)

matches that of the solar equator. This may be a primordial link to the processes of planetary formation nearly 5 billion years ago.

In the nineteenth century, precise measurements of the motion of Mercury around the Sun turned up an odd anomaly. It was discovered that the planet's perihelion was slowly rotating around the Sun in the prograde direction, advancing at a rate of about 574 arc seconds per century. At that rate, the perihelion of Mercury makes a full circuit of the Sun once every 225,000 years. This effect came to be known as the precession of the perihelion of Mercury.

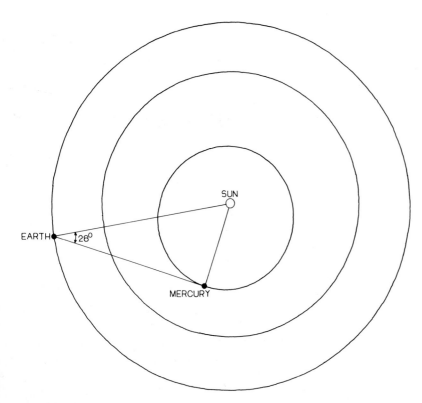

FIGURE 6.2 Mercury is never more than 28° from the Sun.

It was initially assumed that gravitational perturbations of Mercury's elliptical orbit by nearby Venus and Earth were responsible for the effect. However, even after all possible planetary interactions were taken into account, there was still a residual precession (about 42 arc seconds per century) which could not be explained. Astronomers then proposed that an unseen planet (named Vulcan) located inside the orbit of Mercury was responsible for these effects. Repeated searches for Vulcan turned up nothing, although there were a few tantalizing false alarms.

The answer to the dilemma finally came in 1916, with the announcement of Albert Einstein's general theory of relativity. This was a new theory of gravity, which proposed that the gravitational force is a result of a warping of the geometry of space-time in the vicinity of massive bodies. The predictions of the general theory of relativity agree quite well with those of Newton's gravitational theory for relatively small gravitational fields, such as those produced by the Sun and the planets. However, Einstein's theory also predicts the precession of the perihelion of Mercury to amazing precision. Because of the striking success of Einstein's theory, no modern astronomer has felt any need for the existence of the planet Vulcan, and serious searches for this fiery world have ceased.

THE ENIGMATIC ROTATION

Since Mercury's orbit lies inside Earth's, the planet appears to go through phases, just as the Moon does. Repeated telescope observations from Earth could detect no periodically varying surface markings, so it was concluded that Mercury keeps the same hemisphere perpetually turned toward the Sun, just as the Moon always keeps its same face toward the Earth. One hemisphere of Mercury should perpetually be hot, whereas the other should always be at a temperature very near absolute zero ($0°K, -273°C$). The synchronous rotation of Mercury was accepted virtually without question for so long that it entered the folklore of astronomy and became the basis for many excellent science fiction stories.

Radio astronomy measurements made during the early 1960s began to cast serious doubt on the synchronous rotation of Mercury. In 1962, W. E. Howard and his colleagues from the University of Michigan detected thermally produced radio waves emanating from the dark side of Mercury, indicating that the temperature there had to be at least $-100°C$. This was much warmer than would be expected if this side were in perpetual darkness. In 1965 Rolf Dyce and Gordon Pettengill used the radio telescope facility at Arecibo, Puerto Rico to beam a powerful radar signal at Mercury. By observing the slight difference in frequency of returning radar pulses reflected from various parts of the planetary disk, they were able to measure the rotational period of Mercury. Mercury is not in synchronous rotation; its day is 58.7 Earth days long. The rotation is in the direct sense, with the spin axis being very nearly perpendicular to the orbital plane.

Shortly after the reports of the radar measurement of Mercury's spin rate, an Italian dynamicist by the name of Giuseppe Colombo noticed that the rotational period is suspiciously close to exactly two-thirds of the length of the Mercurian year. As Mercury goes around the Sun twice it spins on its axis three times (Figure 6.3). It is extremely unlikely that such a day–year relationship is accidental. Tidal interactions with the Sun must have locked the planet into its present resonant rotational state very early in the history of the Solar System.

MISSION TO MERCURY

In 1974–75, the spaceprobe *Mariner 10* visited Mercury. This mission was one of the most successful exercises in interplanetary billiards ever performed. The craft was launched from Cape Canaveral on November 3, 1973, initially aimed toward Venus. Upon encounter with Venus, the craft was deflected into an elliptical orbit that took it close to Mercury. At the point of closest encounter, Mariner was at perihelion (closest approach to the Sun), whereas Mercury was nearly at aphelion (the point farthest from the Sun). The path followed by *Mariner 10* is shown in Figure 6.4.

Although it was not initially planned, the period of the new orbit of *Mariner 10* was exactly twice the length of Mercury's year. The probe returns to the

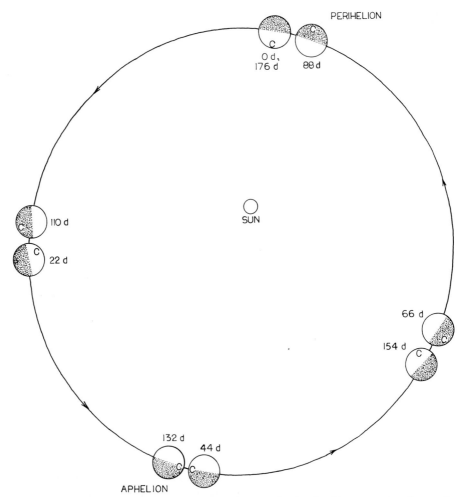

FIGURE 6.3 Resonant rotation of Mercury. In the time that it takes Mercury to travel around the Sun twice, it rotates on its axis three times. The letter C designates the location of the Caloris Basin, the most prominent feature yet seen on Mercury.

vicinity of Mercury once every 2 Mercurian years. Since Mercury itself spins on its axis exactly three times during this period, the same face of the planet is illuminated during each successive pass of the probe. As a result the craft's cameras could only photograph one hemisphere of the planet during the three encounters that the probe was able to make before its attitude control jets ran out of fuel.

THE INTERIOR OF MERCURY

The diameter of Mercury is 4880 km, about 38 percent of Earth's, but 1.4 times larger than the Moon's (Figure 6.5). By observing the deflection of the trajectory of *Mariner 10* during encounter with Mercury, the mass of the planet could be

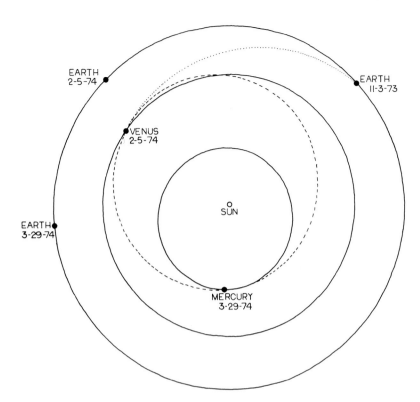

FIGURE 6.4 *Mariner 10* mission to Mercury. The spacecraft was launched from Earth on November 3, 1973. It encountered Venus on February 5, 1974 and was deflected into an orbit that encountered Mercury for the first time on March 29, 1974. The period of *Mariner 10's* orbit is exactly twice that of Mercury's.

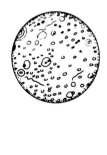

FIGURE 6.5 Relative sizes of Earth and Mercury.

measured. The mass turned out to be 5.6 percent of that of Earth, which makes Mercury nearly 5 times more massive than the Moon. The average density of Mercury is unexpectedly high (5.4 g/cm), about the same as Earth's. So Mercury and the Earth are probably made of the same material and their internal structures must be similar, with an outer crust, an inner mantle, and a central core (Figure 6.6).

However, the near equality of the densities of the two planets is misleading. The large mass of the Earth causes its interior to be appreciably compressed. Mercury is much less massive than Earth, so the effects of internal compression are not nearly so important. The uncompressed density of Mercury would be 5.3 g/cm³ (as compared with 4.04 g/cm³ for Earth). Therefore, Mercury is an intrinsically more dense object than the Earth. In order to account for the difference, the core of Mercury must be proportionately larger than the terrestrial core. Calculations indicate that the core has a radius of 1800 km and comprises 60 to 70 percent of the planet's mass. Mercury's dense metallic core is twice as massive (by percentage) as that of any other planet in the Solar System. The outer 650 km of the planet is probably made up of a mantle and crust rich in silicates.

To everyone's surprise, *Mariner 10* discovered that Mercury has a magnetic field, albeit only 1 percent as strong as Earth's. The magnetic axis makes an angle of about 11° with the spin axis, and the field has the same polarity as Earth's. Like the Earth, Mercury has a magnetosphere, although not nearly as large and intense. Much of the magnetosphere actually lies within the bulk of the planet, with the field lines extending out only to about 1.5 Mercurian radii. The field is not strong enough to trap any appreciable number of charged particles within radiation belts, and it is sufficiently weak so that some of the solar wind particles ejected from the Sun can actually reach the surface of the planet. During particularly intense solar outbursts, the pressure of the solar wind can actually compress the field lines on the sunward side of Mercury almost to the surface of the planet.

Mercury's magnetic field presents planetary scientists with somewhat of a problem, since a planet of such small size should long ago have cooled suffi-

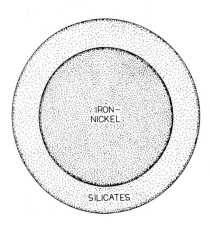

FIGURE 6.6 The interior of Mercury.

ciently so that its core has by now either partially or completely solidified. If so, there no longer could be any convective currents within the interior to support magnetism. One possibility is that Mercury's present magnetic field is only a remnant field "frozen" into the surface crust, rather than a true dipolar field coming from the interior. This surface field is hypothesized to be still present long after the original interior dipolar field that created it has vanished. Similar events may have transpired on the Moon.

THE SURFACE OF MERCURY

During its three flybys of Mercury, *Mariner 10* was able to photograph the planet. A composite of the photographs taken is shown in Figure 6.1. Unfortunately, the spacecraft saw exactly the same hemisphere every time it approached Mercury. Consequently, only half of the total surface area could be seen. The part of the planet that was photographed superficially looks a lot like the Moon, being dominated by craters, impact basins, and smooth plains. The planet is like the Earth on the inside but like the Moon on the outside.

The craters on Mercury look very much like those on the Moon. They range in diameter from hundreds of kilometers down to 100 m (the smallest resolvable by the Mariner cameras). The smaller craters on Mercury are bowl shaped, but the larger ones have prominent central peaks and exhibit extensive terracing on their inner walls. Often there are concentric rings within the interiors of the large craters. In naming the major craters on Mercury, it was decided to honor prominent figures in literature, art, and music (Dürer, Mozart, Tolstoy, Goethe, and so on) rather than scientists (Figure 6.7).

The craters on Mercury are in varying states of preservation. Some of them are relatively fresh, with sharp rims and well-defined ray systems that extend for hundreds of kilometers. Others are extensively degraded, with very low or discontinuous rims. Many large craters have extensive ejecta deposits and secondary craters surrounding them, formed by debris that was thrown out from the crater at the time of the violent impact. However, the ejecta deposits surrounding Mercurian craters are generally less extensive than those surrounding similar sized craters on the Moon. This is undoubtedly a result of the larger gravitational field of Mercury, which limited the range of the debris thrown out by the impact.

Unlike the Moon, no sample of Mercury has yet been returned to Earth for study. Consequently, we have no way of directly determining the age of any part of its surface. However, the most heavily cratered terrain on Mercury has about the same density of large craters as the lunar highlands. The same population of objects that bombarded the Moon must have also scarred the surface of Mercury. Consequently, the densely cratered regions of Mercury are probably of the same age as the lunar highlands — 4.2 billion years.

There are also a number of impact basins on Mercury, although not as many as are present on the Moon. The basins are roughly circular in shape and can often be recognized by concentric rings of mountains at their periphery.

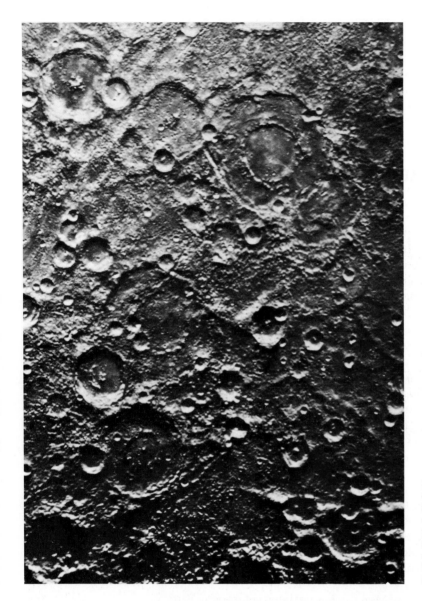

FIGURE 6.7 *Mariner 10* photograph of heavily cratered terrain on Mercury. North is at the top. The double-ringed crater at the top of the photograph is Bach. It is 200 km across. To the west of Bach is the crater Wagner and to the west of Wagner is the crater Chopin. (Photograph courtesy of NASA)

The interiors and surrounding areas of many of these basins are typically much less densely cratered than the surrounding terrain.

Perhaps the most distinctive feature on Mercury is the Caloris Basin (Figure 6.8), a multiringed impact basin 1300 km wide that resembles Mare Orientale on the Moon. It was given its unusual name because the basin happens to lie almost exactly at the subsolar point during alternate perihelion passages. It must be particularly hot at that spot on Mercury at that time. Unfortunately, half of Caloris was in darkness during each of the three *Mariner 10* encounters, so it has never been seen in its entirety.

Caloris is bounded by a ring of high mountains (Caloris Montes) that rise 2 to 3 km above the floor of the basin. The interior of Caloris is extensively ridged and fractured. Most of the ridges predate the fractures. Generally, the density of fracturing increases toward the middle of the central plain. The ridges may be compressional features formed by the subsidence of the basin floor; the fractures may represent tensional stresses associated with the subsequent uplift of the central basin. The landforms within and surrounding the Caloris Basin seem to have a wide range of crater densities, indicating that the geological resurfacing of this terrain occurred long after the event that formed the basin itself.

FIGURE 6.8 *Mariner 10* composite photograph of Caloris Basin. (Photograph courtesy of NASA)

The Caloris Basin must have been excavated by a particularly violent meteor impact that took place near the end of the period of the most intense bombardment. There is some particularly hilly terrain antipodal to Caloris that appears to have violently disrupted preexisting landforms. It may have been formed by focused seismic waves resulting from the Caloris impact.

Other basins on Mercury are more highly degraded than Caloris and have higher densities of craters. This must mean that they are considerably older. Some of the more ancient basins are so badly degraded that they can be distinguished from surrounding cratered terrain only by their roughly circular shape. There are extensive crater-free plains lying near the north pole of Mercury. These could be evidence for the existence of an ancient basin centered in the dark hemisphere of Mercury in the Mariner photographs. It may be as much as 1000 km across.

The sparsely cratered terrain found in and around the impact basins on Mercury is extremely puzzling. These landforms have been termed *smooth plains* by planetary scientists. The smooth plains of Mercury are superficially similar to the lunar maria, but are not nearly as dark. Unlike the lunar maria, the reflectivity (or albedo) of the smooth plains of Mercury is quite similar to that of the heavily cratered highlands. Smooth plains are usually found in and around large basins, but smaller tracts of smooth terrain are found almost everywhere on the portion of Mercury viewed by *Mariner 10*. The smooth plains bear the names given to Mercury in various languages (Odin, Sobkou, Suisei, and so on).

The low density of craters in the smooth plains must mean that these landforms are considerably younger than the highlands, perhaps only 3.5 billion years old. It is uncertain exactly what produced them. The light color of these plains might mean that they are not volcanic in origin, as are the dark maria on the Moon. Since their albedo closely resembles that of the heavily cratered highlands, it is thought that the smooth plains could be the result of an accumulation of massive amounts of ejecta thrown out by crater and basin impacts. However, the present surface of Mercury does not seem to have nearly the number of large impact basins required to have produced the massive supply of ejecta needed to cover such large areas. However, it is possible that the shortage of readily identifiable sources of ejecta may be due to the destruction of old basins by the excavation of new ones.

On the other hand, Mercury is much denser than the Moon. This could mean that the mantle of the planet is quite different in structure and in chemical composition from that of the Moon. Consequently, volcanic lava on Mercury may be quite different in composition than lunar basalt and it could very well have a higher albedo. If the smooth plains on Mercury are indeed vast beds of solidified lava, they must have been formed during a particularly violent episode of volcanism in the distant past, one which was so intense that it involved almost every region on the planet. Molten lava flowing upward from the interior filled up the floors of impact basins and other low-lying areas, obliterating preexisting craters and creating a "fresh" crust. In some cases, the "ghosts" of previously existing craters can be found jutting out of the plains that cover the interiors of basins. This suggests that a considerable amount of time elapsed between the impact event that excavated the basin and the subsequent emplacement of smooth plains in its interior.

The question of the true origin of Mercury's smooth plains will probably not be resolved until a sample of its surface can be returned to Earth for study. Perhaps the answer to the dilemma lies on the other face of Mercury — the one that was in darkness during the Mariner encounters.

The most widespread terrain type on Mercury is the *intercrater plains*. These are gently rolling landforms with a high density of craters less than 15 km

FIGURE 6.9 Photograph of Discovery Scarp. North is at the top. The scarp cuts through the small crater Rameau near the center of the photograph. The double crater south of Rameau is Hesiod. (Photograph courtesy of NASA)

in diameter, but few or no larger craters. This type of terrain is rather scarce on the Moon. Intercrater plains are found scattered throughout the densely cratered highlands. Many of the craters in these plains are clustered together, suggesting a secondary impact origin. The intercrater plains are clearly much older than the smooth plains; their formation spans a range of times coincident with the era of heavy bombardment that formed the highlands. They must be among the oldest features on Mercury.

A unique feature on Mercury is a series of shallow, scalloped cliffs that run for hundreds of kilometers across the surface (Figure 6.9). These scarps are given the names of ships of exploration and discovery, both sea- *(Discovery, Endeavour, Victoria, Santa Maria)* and spacefaring *(Vostok)*. These scarps generally trend either northwest to southeast or else northeast to southwest, which suggests that the tidal despinning of Mercury may have had something to do with their formation. They could also have been a result of an early period of crustal shortening that took place on a global scale at the time Mercury's core began to cool and contract.

There is no significant atmosphere, although Mercury does have a thin shell of hydrogen and helium gas that probably originates from the Sun. It is more properly termed an "exosphere" rather than an atmosphere; it may have originally been produced by solar wind ions that were entrapped by Mercury's magnetic field and neutralized by collision with the planet's surface.

The surface environment on Mercury is probably the harshest of any planet in the Solar System. The intensity of sunlight at the surface is 7 times that at Earth, and during the day the temperature can rise to as high as 430°C. The dark side can cool to temperatures as low as −170°C. Given such a severe environment, it is almost certain that no living creatures exist on the surface of Mercury.

7

Venus — the Shrouded Inferno

Venus is the second planet from the Sun. With the obvious exceptions of the Sun and the Moon, Venus is the brightest object in the sky. It can actually be seen during daylight hours at particular times, provided one knows when and where to look. Venus has fascinated and intrigued humans for many years. Its appearance in the sky is so spectacular that the ancient Greeks and Romans gave the planet the name of the goddess of love. It is therefore tempting to view Venus as a serene planet, somehow friendly and comforting. However, modern astronautical science has found that this planet more closely resembles Hell than does any other spot in the Solar System (Figure 7.1).

THE ORBIT OF VENUS

Venus is in a nearly circular orbit (the least eccentric of any of the planets), with an average distance from the Sun of 0.72 AU. With the exception of the Moon (and a few small asteroids), Venus comes closer to the Earth than does any other heavenly body (4.1×10^7 km). The point of closest approach is called the *inferior conjunction*. The orbital plane of Venus makes an angle of 3.4° with the ecliptic plane, and transits of the planet across the solar disk as seen from Earth are relatively rare. When transits do occur, they occur in pairs, always 8 years apart.

Photograph of the surface of Venus taken by the Soviet lander *Venera 13*. (Photograph courtesy of USSR Academy of Sciences)

129

FIGURE 7.1 *Mariner 10* photograph of Venus in ultraviolet light. The wave patterns in the upper atmosphere can be clearly seen. In visible light Venus is a completely featureless disk. (Photograph courtesy of NASA)

Transit pairs are separated from each other by long intervals. The last transit of Venus occurred in 1882; the next one is not due until 2004.

Since Venus's orbit lies between the Earth and the Sun, the planet always appears close to the Sun in the sky, never more than 46° away (Figure 7.2). Venus always appears either in the western sky just after sunset or in the eastern sky just before sunrise (Figure 7.3). At one time it was thought that the morning and evening apparitions of Venus were two different planets. Venus was known as Vesperus or Hesperus when seen as an ''evening star,'' and as Phosphorus or Lucifer when seen as a ''morning star.'' Pythagoras may have been the first to recognize that the two apparitions of Venus were one and the same object.

Because Venus's orbit lies within Earth's, the planet is seen by ground-based telescopic observers to go through phases, just as the Moon does. Whenever Venus and the Earth are on opposite sides of the Sun, Venus is always seen nearly in full phase. When Venus and Earth are on the same side of the Sun, Venus appears in crescent phase. This effect was first noted by Galileo in 1610

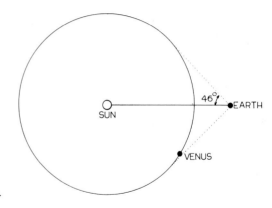

FIGURE 7.2 Venus is never more than 46° from the Sun.

and was striking evidence in favor of the Copernican model of the Solar System. If the Ptolemaic model were right, it would never be possible to see Venus in full phase (Figure 2.6).

THE VEXING VENUSIAN ROTATION— AN EARTH RESONANCE?

Venus appears in the telescope to have a yellowish-white color, but no markings of any kind can be seen. It has been known for a long time that the reason for this featureless appearance is the presence of a dense atmosphere, which completely shields the surface from view. This atmosphere was first spotted by the Russian astronomer Mikhail Lomonosov (1711–1765) during Venus's transit of the Sun in 1761. He noticed that the Venusian disk appeared to have a fuzzy outline when seen up against the Sun, and correctly concluded that a dense atmosphere was present. The existence of this atmosphere has made the study of Venus so frustrating that up until the advent of interplanetary space travel we

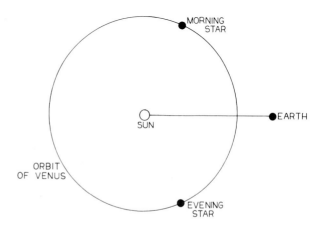

FIGURE 7.3 The morning and evening "stars." Venus appears in the sky either as a bright star in the western sky after sunset or as a bright star in the eastern sky just before sunrise.

knew far more about distant galaxies than we did about Earth's nearest planetary neighbor.

The lack of visible surface detail makes the measurement of Venus's rotation rate extremely difficult. Some astronomers have spotted a few transient dark features in the clouds (especially when using ultraviolet light) that could be watched as the planet rotated. Generally, these features do not last long enough for any meaningful measurements to be made. However, on rare occasions the same dark features would reappear after 4 days, suggesting that Venus might have a rotational period 4 Earth days long. On other occasions, no changes in the dark features could be noted at all, suggesting that Venus's rotation rate could be exceedingly slow.

Frustrations with attempts to make direct observation of the rotation rate led astronomers to try indirect measurement techniques. These involved the use of measurements of the light reflected from different parts of the Venusian disk. The technique is quite similar to that used in the measurement of the rotatation of the Sun (Figure 3.6). As Venus rotates, one edge (or limb) will be moving toward Earth at any given time, whereas the opposite limb will be moving away. Consequently, the light reflected from the limb of Venus that is moving toward us should have a slightly shorter wavelength than that reflected from the limb moving away. It should be possible to use the slight differences in the wavelengths of prominent spectral lines reflected from different parts of the Venusian disk to calculate the rate at which the planet is spinning. Again, inconsistent and contradictory results were obtained. Some measurements gave rotational periods as small as 4 Earth days. Others found the rotational period too long to be estimated.

The controversy over Venus's rotational rate raged for nearly a century. In retrospect, the problem was due to the fact that all of these measurement techniques were of necessity looking at the top of the atmosphere rather than the surface. It was not until 1962 that the correct answer was finally obtained. In that year radar waves were transmitted from Earth and were reflected from the surface of the planet. A detailed analysis of the Doppler shifts of the radar pulses reflected from various parts of the Venusian surface made it possible to measure the rotation rate. To everyone's surprise it was found that Venus rotates very slowly in the retrograde direction, completing one rotation every 243 Earth days. The equator makes an angle of 2.6° with the plane of the orbit, so seasonal effects should not be important on Venus.

Unlike in the case of Mercury, there does not appear to be any simple numerical relationship between the lengths of the Venusian day and year. However, there might be a relationship between the Venusian day and the motion of the Earth. Whenever Earth and Venus are in inferior conjunction, Venus turns the same hemisphere toward our planet (Figure 7.4)!

How could such a Venus – Earth resonance ever arise? Tidal forces exerted on Venus by the Earth should be far too weak at such extreme distances to produce any significant amount of coupling. Nevertheless, several theoretical papers appeared in the astronomical literature attempting to explain the effect. However, later, more precise measurements seem to indicate that the Venus –

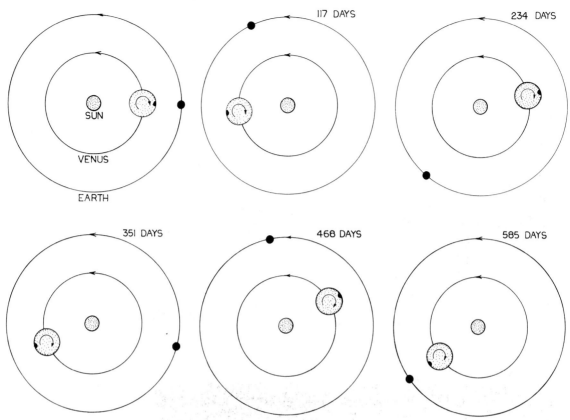

FIGURE 7.4 Resonant rotation of Venus. Every 585 Earth days Venus returns to inferior conjunction. Venus rotates once every 243 Earth days in a retrograde direction. This rotation rate is such that the same hemisphere of Venus is presented toward Earth at every inferior conjunction.

Earth rotational relationship is not exact. Enough of a difference exists between the true Venus rotational rate and that required for a Venus–Earth resonance that the apparent effect can probably be dismissed as a coincidence. Perhaps at one time Venus's retrograde spin rate was much faster than it is now. Over the years the spin rate has been gradually slowed by tidal braking forces exerted on the planet by the Sun. If so, then Venus must have passed through several "resonant" spin states, during which either the same or alternating hemispheres of the planet faced Earth at successive inferior conjunctions. There is then nothing all that significant about this particular "resonance."

THE INTERIOR OF VENUS

Venus is sometimes called the Earth's twin, since it is very nearly the same size (Figure 7.5). The diameter is 12,110 km, which makes Venus only 600 km smaller than the Earth. There is no noticeable oblateness. Since Venus has no moons, its mass is difficult to determine. A rough estimate of the mass can be

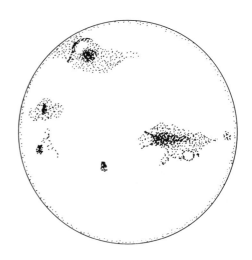

FIGURE 7.5 Relative sizes of Earth and Venus.

made by studying the small perturbing effects that the presence of Venus has on the orbits of Earth and Mercury, but truly reliable estimates can only be provided by observing the trajectories of spacecraft as they fly near the planet.

Venus turns out to have a mass 82 percent that of Earth's. This gives the planet a density of 5.27 g/cm³, only slightly less than that of Earth. The interior of Venus must then be quite similar to Earth's, with a dense metallic core, a mantle, and a silicate crust (Figure 7.6). However, the slightly smaller density of Venus may mean that the planet has relatively smaller amounts of iron and/or sulfur than does Earth, and hence a smaller core.

Spacecraft have detected only an exceedingly weak magnetic field, 0.001

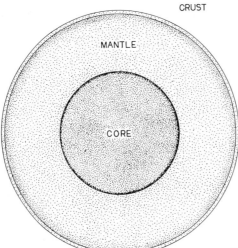

FIGURE 7.6 Diagram of the interior of Venus.

the intensity of Earth's. The weak field may be a result of Venus's slow rotation rate or it may be a reflection of an inner core that is much less fluid than Earth's. The true reason is not yet known.

THE VENUSIAN ATMOSPHERE—A RUNAWAY INFERNO

Up until the 1950s the popular view of Venus was that its surface environment was similar to Earth's, although appreciably warmer and probably a lot more moist. Some even imagined Venus to be covered with vast swampy forests, as the Earth was more than 300 million years ago. However, no water vapor could be detected in the atmosphere, so this picture of the Venusian atmosphere was more a product of wishful thinking than a result derived from hard scientific data.

Radio astronomy studies of Venus begun during the 1950s suggested that a "wet" model for Venus could not possibly be correct. In 1956, U. S. Naval Research Laboratory workers Cornell H. Mayer, Timothy P. McCullough, and Russell M. Sloanaker detected radio emissions coming from Venus that seemed to indicate that its surface temperature could be as high as 330°C. This temperature was at first thought to be so inconceivably hot that it must obviously be an error. Maybe the unusual radio waves did not come from the surface at all, but from some point in the upper atmosphere. It was not until 1962 that this question was to be answered. In December of that year, the American probe *Mariner 2* passed within 34,000 km of the surface of Venus and verified that the radio emission was indeed coming from the surface, and that the temperature there had to be as high as 400°C.

The Soviet Union subsequently made the exploration of Venus a prime goal of its space program. Several probes were launched that carried landing capsules designed to explore the surface environment of Venus. After a series of frustrating failures, the Soviets finally succeeded with *Venera 7* in 1970. The lander parachuted to the surface and survived long enough to radio back readings of an incredibly high temperature of 460°C. This high temperature has been repeatedly confirmed by subsequent Soviet as well as American atmospheric-entry and surface-exploration probes. It is hot enough at the surface to melt zinc, and the temperature is so high that the surface should actually glow a dull red at nighttime. The temperature is fairly constant over the entire surface; it is just as warm at the poles as it is at the equator, and it does not cool off at night.

In 1932 Walter Adams and Theodore Dunham, both of the Mount Wilson Observatory, obtained good infrared reflectance spectra of Venus and found that carbon dioxide (CO_2) gas was a major atmospheric constituent. Further details about the Venusian atmosphere had to await the advent of space travel. The Venus atmospheric entry and landing probes confirmed that CO_2 was indeed present in great abundance. The atmosphere is (by volume) 96.4 percent carbon dioxide, 3.4 percent nitrogen, 0.02 percent sulfur dioxide (SO_2), and 0.14 percent water vapor. There are trace amounts of oxygen, argon, neon, hydrochloric acid (HCl), hydrogen fluoride (HF), and hydrogen sulfide (H$_2$S). In addition, there is some carbon monoxide, helium, and atomic oxygen in the upper atmosphere. There does not appear to be any methane or hydrogen. The surface pressure is very high, about 90 atm. The high pressure coupled with the high temperature makes for a surface environment so hostile that no spacecraft has been able to remain operating there for more than a couple of hours.

The vertical profile of the Venusian atmosphere is shown in Figure 7.7. The Venusian troposphere extends from the surface to an altitude of 90 km. The surface temperature is 460°C, but the temperature declines 10°C for every kilometer of altitude. Venus has neither a stratosphere nor a mesosphere, but it does have a thermosphere, which starts at an altitude of 100 km. It is heated by the absorption of ultraviolet light from the Sun and cools down at sunset to temperatures far below that of the troposphere.

The upper atmosphere has a layer of clouds 15 km thick starting at a height of 45 km above the surface. The clouds are so thick that only about 1 percent of the sunlight incident on Venus actually reaches the surface, so it is never brighter there than it is on Earth during a dark and rainy day. Because of the thick clouds, most of the incident sunlight is absorbed in the upper parts of the atmosphere rather than at the surface. The composition of the clouds is uncertain, but they probably consist of water droplets rich in sulfuric acid. The sulfuric acid droplets may be responsible for giving Venus's clouds their characteristic yellow tinge. This sulfuric acid is probably formed by the sulfur dioxide in the upper air, becoming dissolved in the water droplets in the upper cloud layers, a process similar to the mechanism that produces "acid rain" in the terrestrial atmosphere.

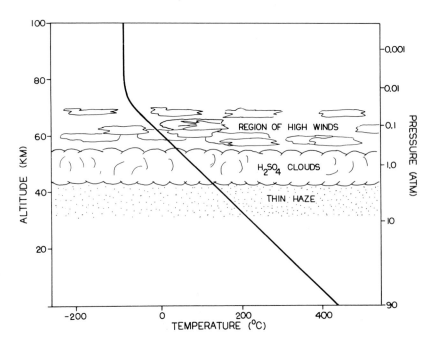

FIGURE 7.7 Vertical profile of Venusian atmosphere.

There is probably a steady "rain" of sulfuric acid droplets in the upper Venusian atmosphere, which may account for a layer of haze that landing spacecraft have spotted beneath the cloud deck extending to within 30 km of the surface. None of this rainfall could ever reach the surface, as it is far too hot there for liquid water to exist. The Soviet landers *Venera 11* and *12* noted a continuous pattern of lightning bolts below the clouds that may cast an eerie glow over the surface. These may be a result of the acid rain in the upper atmosphere, perhaps stimulated by local regions of intense volcanism or other types of geological activity.

Surface winds on Venus appear to be rather gentle, but there are rather violent winds in the upper atmosphere that are planetwide in scope. These winds all blow from east to west (in the same direction as the planet's rotation) and begin at a height of 10 km above the surface. The wind speed increases steadily as one rises, reaching a maximum velocity of 100 m/sec at the top of the clouds, 60 km above the surface. These high winds were first noted in ultraviolet-light photographs taken of Venus by *Mariner 10* as it flew past the planet on its way to Mercury (Figure 7.1). These upper-atmosphere winds cover the entire planet and result in the entire weather pattern being carried around the planet once every 4 days. The upper atmosphere rotates 60 times faster than the planet itself and may indeed have been the source of the earlier difficulties in measuring the Venusian rotational rate.

Superimposed upon the east–west wind pattern is a far more gentle equatorial–polar circulation that is driven by ultraviolet absorption in the upper atmosphere. Rising warm air over the equator spreads out laterally and flows toward the poles, where it cools and sinks back down. The *Pioneer-Venus* orbiter of 1978 noted a "hole" in the cloud layer 1100 km wide over the north pole. This may be caused by a downward flow of air over these polar regions. The lower atmosphere does not appear to participate to any significant extent in this large-scale circulation, remaining fairly stagnant and maintaining a constant temperature over the entire surface of the planet.

Why is it so hot on the surface of Venus? It is true that sunlight is 1.9 times brighter at Venus than it is at the Earth, but this fact by itself cannot account for such high temperatures. Unlike on Earth, most of the solar energy is absorbed high in the dense Venusian cloud layer rather than at the surface, so it could actually be rather cool at ground level. In fact, the *effective temperature* of Venus (as measured from a spectral analysis of the infrared light emitted by the planet) is actually cooler than that of Earth because of Venus's higher reflectivity.

Although there is still a good deal of controversy about the reason for the heat, most planetary scientists now accept Carl Sagan's theory of a runaway greenhouse effect as the cause. It is possible that Venus, during its early history, was much like the Earth is today, being a lot cooler and having a much thinner atmosphere. It may have even had a large amount of liquid water flowing on its surface. The closer proximity of the Sun, however, caused the surface temperature to become so warm that a large percentage of this water was forced into the atmosphere, resulting in an enhanced opacity to infrared radiation and a sharp temperature rise. Eventually the surface became so hot that an appreciable amount of carbon dioxide vapor was baked out of the carbonate materials in the surface rocks. This outgassing, in turn, produced still more trapped heat, forcing still more carbon dioxide out of the rocks, creating yet a further temperature rise. The process did not stop until virtually all of the carbon dioxide originally stored in surface rocks was forced out into the air, producing the exceedingly hot and dense atmosphere present on Venus today. As the Sun continues to increase its luminosity over the next few billion years, similar events may take place on the Earth. It is interesting to note that the amount of carbon dioxide gas currently found in the Venusian atmosphere is approximately equal to the total inventory of terrestrial carbon dioxide. The difference between the two worlds is that on Earth most of the carbon dioxide is stored within sedimentary rocks and on Venus it is all present in the atmosphere.

VENUS'S DISAPPEARING WATER

Compared to the Earth, Venus is almost completely lacking in water. If the Earth were suddenly to become as hot as Venus, all of its oceans would boil away, forcing all the water into the atmosphere as superheated steam. The pressure of the water vapor in the atmosphere would rise as high as 150 atm. However, there is currently only enough water in the Venusian atmosphere to produce a pressure of 0.14 atm, which means that the planet has less than 0.001 the amount of water that the Earth does. Why is Venus so dry? There are two schools of thought on this question. The first hypothesis is that the region of space in which Venus initially formed was too warm for any appreciable amount of water to condense. Venus has so little water today because it never had very much to begin with. The other line of reasoning involves the assumption that Venus did indeed originally have about the same amount of water as Earth does now, but has lost most of it over the years.

Some evidence for an originally "wet" Venus was provided by the American *Pioneer-Venus* atmospheric-entry probe of 1978. Instruments aboard the probe were able to measure the relative abundance of deuterium (or "heavy hydrogen") in the water droplets in the upper Venusian cloud deck. Venusian water was found to have 100 times more deuterium than terrestrial water does. This difference has been interpreted as evidence that Venus was at one time nearly as wet as Earth is now but that most of its water was photochemically destroyed. Ultraviolet light splits water molecules into their constituent hydrogen and oxygen atoms. Most of the ordinary hydrogen produced by the decomposition of water rapidly escaped into outer space, but the heavier deuterium isotope was preferentially retained by Venus and recombined with oxygen radicals to produce the "heavy water" currently found in unusual abundance in the upper atmosphere.

At one time Venus may have been covered by oceans of liquid water as deep as Earth's present oceans. Perhaps some primitive life forms existed in the primeval Venusian seas. However, the runaway greenhouse disaster drove all the water on Venus into the atmosphere as superheated steam. In the upper atmosphere of Venus the water vapor was exposed to strong ultraviolet light from the Sun. As the years passed, most of it was photochemically decomposed into its constituent hydrogen and oxygen atoms. The hydrogen atoms quickly escaped into outer space, whereas the oxygen atoms reacted with the chemicals in the Venusian soil. At present, almost no water is left.

Such a catastrophe did not take place on our planet because of two factors. First, Earth was sufficiently distant from the Sun so that most of its water remained in the liquid rather than the vapor state. Second, the Earth developed an efficient ozone screen against harmful ultraviolet radiation very early in its history. The appearance of an effective ultraviolet ozone screen in the terrestrial atmosphere may have been an event much more critical for our survival on this planet than we have even dared to imagine. Not only did the ozone blanket make it possible for life on Earth to survive outside the water, it now seems likely that it also may have played a crucial role in permitting Earth to retain any appreciable amount of its water in the first place.

THE ISOTOPE ARGUMENT

The Venus landers were also able to measure the isotopic abundances of some of the gases in the Venusian atmosphere. Such measurements are especially important, as they can shed much light on the processes by which a planet originally formed and subsequently evolved.

The relative abundances of the isotopes of carbon and the isotopes of oxygen are the same as they are in the terrestrial atmosphere, suggesting that conditions in the primitive solar nebula must have been quite similar for both planets at the time of their formation. However, the isotopes of argon tell a much more interesting story.

Argon 36 and argon 38 are primordial isotopes. They must have originally been present within the primitive solar nebula out of which the planets condensed. To everyone's surprise, the atmosphere of Venus was found to have almost 100 times as much primordial argon as that of the Earth. The reason for the difference is not known, but most experts in planetary formation assume that it must somehow be a reflection of sharp differences in chemical composition and in temperature that must have existed in the regions of the primitive solar nebula in which Venus and Earth initially formed.

The most abundant isotope of argon in both the Venusian and terrestrial atmospheres is ^{40}Ar. It is created by the beta decay of ^{40}K atoms within surface rocks. When the surface rocks are heated or fractured, some of the ^{40}Ar thus created is driven out of the rocks into the atmosphere. Both the Venusian and the terrestrial atmospheres must have been produced by secondary outgassing from the interior. However, it was found that ^{40}Ar in the Venusian atmosphere is only 30 percent as abundant as it is in the terrestrial atmosphere. This may mean that Venus has been somewhat less geologically active than Earth has been over most of its history.

THE CLOUD-SHROUDED SURFACE

Because of the thick clouds, there is not much known about the surface morphology of Venus. Earth-based radar installations have been able to map some of the surface features down to a horizontal resolution of 10 to 20 km. In some isolated cases, a 2-km resolution has been achieved. However, such mapping techniques are practical only when Venus is in inferior conjunction with the Earth. Unfortunately, the same face of Venus is always turned toward the Earth at that time. The radar coverage is therefore effectively restricted to only one hemisphere. In addition, the polar areas of Venus are at highly oblique angles with respect to Earth and are hence quite difficult to map.

In 1978 *Pioneer Venus 1* went into orbit around Venus. Among other instruments, the satellite carried a radar altimeter capable of probing the surface of the planet as it flew overhead. Although the best horizontal resolution that could be achieved (30 km) was not as good as that which could be obtained by Earth-based radar, the entire planet between 74°N and 63°S could be covered. The radar map that was obtained is shown in Figure 7.8.

In 1984 the Soviet Union placed a pair of satellites (*Veneras 15* and *16*) in orbit around Venus. They each carried sophisticated side-looking, synthetic-ap-

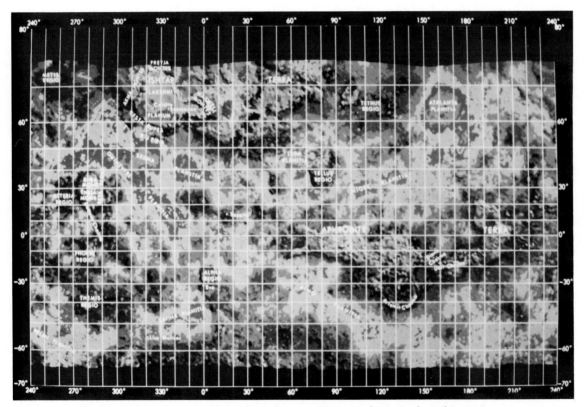

FIGURE 7.8 *Pioneer Venus 1* radar map of the surface of Venus. (Courtesy of U.S. Geological Survey)

erture radar equipment that was capable of mapping the surface of Venus down to a horizontal resolution of a few kilometers. Because of the nature of their orbits, these two craft were only able to provide detailed coverage of the northern one-third of the planet's surface.

Since Venus is named after a goddess, most of the prominent terrain features identified so far have been given the names of female deities. Other landforms have been named after women who have achieved distinction in science, politics, literature, or the arts. A few features are given the names of fictional female characters who have appeared in novels, in legends, or in myths.

The radar maps show that Venus has a surface that is on the average somewhat flatter than that of the Earth. At least 60 percent of the surface of Venus falls within a height interval of only 1 km. Three distinctly different types of terrain have been identified on Venus: *lowlands*, *rolling plains*, and *highlands*.

The first of these are the lowlands, comprising approximately 27 percent of the surface. A prominent lowland feature is Atalanta Planitia, at 65°N, 165°E. It is a roughly circular depression approximately the size of the Gulf of Mexico. It may be the remains of an ancient impact basin, but the resolution of the radar maps is far too poor for us to be certain. There is a pair of lowland depressions, named Sedna Planitia and Guinevere Planitia, arranged in a Y-shaped pattern, ranging from 0 to 50°N and 270 to 300°E.

The second type of terrain on Venus is the rolling plains, which occupy 65 percent of the surface. Several large plateaus have been identified, for example, Alpha Regio (25°S, 0–10°E) and Phoebe Regio (10°S, 280°E). These plains contain several large troughs that are of possible tectonic origin. The Soviet radar mappers *Venera 15* and *16* found that the entire north polar region is covered with a large, flat plain known as Aurora Planitia.

The third type of Venusian landform is the highlands. These may be analogous to the large continental landmasses that appear on Earth. There are several Venusian "continents" that rise a few kilometers above the mean surface level, but these comprise only about 5 percent of the total surface area. In contrast, 35 percent of the Earth's surface lies within the boundaries of continental plates.

A large highland feature, named Ishtar Terra, lies at a latitude of 60 to 75°N and is centered at about 0°E. It is approximately the size of the United States. A prominent feature of Ishtar is a huge mountain named Maxwell Montes that towers 11 km above the surrounding plain. Maxwell is 2 km higher than Mt. Everest. There is a dark, circular feature that is 100 km across located near the summit of Maxwell. This feature, named Cleopatra Patera, may be volcanic caldera. To the west of Maxwell is a rimmed, pear-shaped, smooth feature called the Lakshmi Plateau. It is fringed on the north and west by high mountains.

Another large highland feature, named Aphrodite Terra, lies just south of the equator at 70 to 140°E. It is nearly as large as Africa. Just to the east of Aphrodite is a giant rift valley, known as Diana Chasma, that is 900 km long and 200 km wide. It is 4.8 km deep, much deeper than any canyon on Earth. The floor of Diana Chasma is probably the lowest point on Venus. Nearby is a large quasi-circular trough, known as Artemis Chasma, which may be the remains of a giant impact crater.

Beta Regio (at 38°N, 284°E) is an oval-shaped highland feature that extends for about 2500 km in a north–south direction and is approximately 2000 km wide. Beta is dominated by two topographic features, which rise 4.5 km above the mean planetary radius. These have been named Theia Mons and Rhea Mons. They may be volcanoes.

Some of the features seen on Venus seem to suggest the presence of plate tectonic activity on the planet. Many areas of the rolling plains do contain large numbers of complex linear features, which may be geological faults caused by stresses exerted by crustal motions. Other rolling plains have extensive networks of intersecting ridges and grooves, each 10 to 20 km long. These odd looking ridges and grooves may be evidence for the presence of tectonic activity, but since there does not seem to be anything quite like them anywhere else in the Solar System, it is difficult to be certain. The region in and around Maxwell is characterized by a series of longitudinal ridges and valleys. These may have been created by the actions of intense compressive strains that were present during the massive crustal uplifting that formed this gigantic mountainous feature. High-resolution radar images of Beta Regio made by the Arecibo radio telescope show a major rift system cutting through the center of this highland feature, suggesting a major crustal shifting episode in the relatively recent past.

Even though there are regions on Venus suggestive of plate tectonic activity, the steady motions of large crustal plates seem not to have played the dominant role in altering the surface that they have on Earth. There is nothing on Venus that resembles the mid-oceanic ridges that are so prominent on Earth. Venus is on the average much flatter than the Earth. The sharp distinction in elevation that exists between the high continental landmasses and the low oceanic basins of Earth is entirely absent on Venus. There are far fewer continents on Venus than there are on Earth, and their distribution over the surface is quite different.

Careful studies of the orbit of *Pioneer-Venus 1* as it flew over Aphrodite and Ishtar indicate that these "continental" landmasses are of about the same density as the surrounding terrain. This distinguishes them from terrestrial continents, which are significantly less dense than the surrounding oceanic basins. The phenomenon of isostasy, in which low-lying dense crust hydraulically supports a lighter, more elevated crust, may be impossible on Venus.

Most planetary geologists suspect that the crust of Venus is significantly thicker than Earth's, so thick in fact that it never did break up into numerous mobile plates. The crust under Aphrodite may be as much as 160 km thick. Large-scale plate tectonics and crustal subduction may be impossible on Venus, although there was certainly some localized uplifting of a few isolated regions caused by the circulation of hot magma deep in the interior of the planet. Such uplifting may have been responsible for creating the large continental landmasses such as Aphrodite and Ishtar.

The dominant force shaping the surface of Venus may have been volcanic activity. The entire Maxwell highland feature may be an enormous *shield vol-*

cano, created by the successive buildup of massive lava flows. (The term comes from the resemblance of such structures to warriors' shields.) It is possible that Cleopatra Patera is the caldera associated with the Maxwell volcano. However, Cleopatra is not precisely at the summit of Maxwell, and it is probably more likely that this feature is actually a gigantic impact crater. Beta Regio, with its two high mountains Theia and Rhea, may be an enormous volcanic structure, formed by the steady accumulation of massive amounts of lava over many years. The Lakshmi Plateau west of Maxwell appears to have been formed by a gigantic flood of lava. Perhaps this lava flood took place at the same time as the volcanic episode that created the Maxwell feature itself.

Many areas of the rolling plains in the northern hemisphere, as well as many regions of Ishtar itself, are covered with some rather strange-looking, oval-shaped features that range in size from 300 to 500 km across. Soviet scientists examining the first radar maps from *Venera 15* and *16* have termed these landforms *ovoids*. Many of the ovoids are sharp and distinct, but others are heavily eroded and seem to be filled with volcanic lava flows. It is tempting to identify the ovoids as the remains of meteorite impact craters, but the uniformity in their sizes suggests a different origin. It is possible that they were originally formed by rising bubbles of molten mantle that forced up large areas of crust into gigantic volcanic domes, which then collapsed into jumbles of concentric and radial features. Much further study is needed before a positive identification of their origin can be made.

There are a couple of places (Beta Regio and the Scorpion Tail region of Aphrodite Terra) where volcanism may currently be present. This may account for the flashes of lightning that were sighted over these particular regions by Soviet landing probes. The *Pioneer-Venus* orbiter found that there was an overall 5-year decline (by more than a factor of 10) in the sulfur dioxide abundance at the cloud tops. This may be evidence for a massive injection of SO_2 into the atmosphere by a major volcanic eruption that took place only a few years ago!

It is hard to determine the age of the Venusian surface, given the lack of detail. If some of the quasi-circular images that have been noted on the radar maps actually are the remains of ancient impact craters, much of the Venusian surface may be several billion years old. On the other hand, if the intense heat produced by the runaway greenhouse effect has been present over most of Venus's history, it is likely that viscous relaxation would have erased any impact craters or basins older than 3.8 billion years. Many of the surface features that are seen could then have a volcanic origin and might actually be only a few hundred million years old. The *Venera 15* and *16* radar maps show only 138 easily identifiable impact craters in the entire northern third of the planet, suggesting that this region of Venus is not as old as the lunar maria or highlands. Perhaps it is less than 1 billion years old. We can only speculate until more detailed surface maps become available.

Although there is still a great deal of uncertainty, the consensus reached from a study of the crude surface maps seems to be that Venus has been, on the whole, much less geologically active than Earth throughout most if its history. The large-scale continental-drift activity so characteristic of Earth seems to be absent on Venus. The Venusian atmosphere has only 30 percent as much ^{40}Ar as does Earth's, which must mean that there was considerably less recycling of molten rock between the mantle and outer crust on Venus than there was on Earth. There appear to be significant numbers of large impact craters present on Venus, which suggests that there are many regions on the planet that have remained essentially unaltered for 2 billion years or even longer. However, it also appears likely that the surface of Venus has been significantly more active than that of the Moon. Probably no landform on Venus is nearly as old as the lunar highlands. Venus appears to be an intermediate world, one not nearly as lively and active as the Earth, but one not quite as cold and dead as the Moon.

THE VENERA LANDERS

In 1975 the Soviet Union landed a pair of spacecraft (*Venera 9* and *10*) on the surface of Venus. They survived long enough to send back photographs, the first ever taken on the surface of another planet (Figure 7.9). The *Venera 9* photographs depict a rather stony landscape, possibly located on the slope of a hill. Many stones 60 to 100 cm across with sharp edges could be seen in the field of view. Many of these stones are pitted, and a few are fractured. The whole area seems to be geologically rather young. The *Venera 10* site is smoother, with many stony elevations being noticeable. Some of these elevations have smooth edges and are partially covered with a darker, fine-grained soil. This suggests some sort of atmospheric chemical interaction with the rocks that has gradually broken them down. This plain is probably a good deal older than the *Venera 9* site.

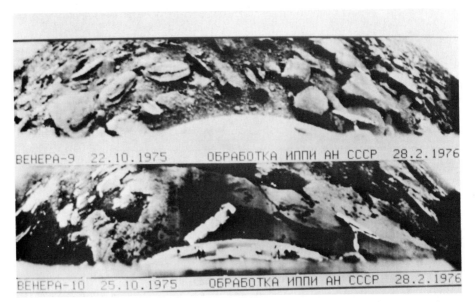

FIGURE 7.9 *Venera 9* and *10* photographs from surface of Venus. (Photograph courtesy of USSR Academy of Sciences)

The Venera landers carried gamma-ray spectrometers capable of measuring the radioactivity produced by uranium, thorium, and potassium in the Venusian soil. Both landers found that the radioactivity was similar to that found for terrestrial basalts produced by relatively incomplete differentiation of the original planetary material.

Better photographs were returned by the *Venera 13* and *14* landers in 1982 (Figure 7.10). Some of these were in color. The landing sites were chosen after a detailed study of the first radar maps of Venus; *Venera 13* was directed toward a rolling plain, whereas *Venera 14* was aimed toward a low-lying basin area, both sites near the Phoebe Regio mountainous region. Before the landings, it had been proposed that *Venera 14's* basin had been formed by relatively recent volcanic activity, whereas *Venera 13's* rolling plain was part of Venus's original crust. Before they were destroyed by the intense heat, the spacecraft performed some quick analyses of the elemental composition of the surface rocks. *Venera 13* found a chemical composition similar to that of high-potassium alkali basalt rocks on Earth, whereas *Venera 14* found a rock similar in composition to tholeiite, the most common volcanic rock found on Earth. Both sites seem therefore to be volcanic in origin. Perhaps *Venera 13* overshot its target and actually set down in a volcanic region.

The color photographs show that the surface has a distinctly reddish cast, perhaps caused by the presence of lots of Fe_2O_3 in the soil. Mechanical penetrometers aboard the Venera landers prodded the soil near the landing sites and

FIGURE 7.10 *Venera 13* and *14* photographs from surface of Venus. (Photograph courtesy of USSR Academy of Sciences)

established that the average surface density on Venus is of the order of 2.7 to 2.9 g/cm³, typical of surface rocks on Earth. This is considerably less than the overall Venusian density, indicating that Venus also underwent a period of internal differentiation, during which many of the heavier metallic elements sank to the interior of the planet.

FUTURE PLANS

Many details about the Venusian surface remain to be investigated. NASA had hoped to launch a spacecraft later in the 1980s that would be capable of mapping the entire surface of Venus to a much higher resolution. Termed Venus Orbiting Imaging Radar (VOIR), the spacecraft was to be launched by the space shuttle and placed in a nearly polar circular orbit around Venus. It was to carry an advanced synthetic-aperture radar system that would be capable of making a radar map of Venus with a resolution comparable to the performance of some of the television-camera-carrying spacecraft that have studied the surfaces of the Moon and Mars. Unfortunately, budgetary problems caused this project to be canceled in January 1982. NASA then proposed a scaled-down mission, named Venus Radar Mapper (VRM), which would use radar imaging to map 92 percent of the surface at 0.2-km resolution. The VRM project has been funded and assigned the name *Magellan*. The *Magellan* mission had originally been scheduled for launch in April of 1988 aboard one of the space shuttles, but the *Challenger* tragedy will almost certainly force a postponement. Alternatively, the spacecraft could be shifted to an expendable rocket booster.

The Soviet Union undoubtedly will continue its extensive program of Venus study. In 1985 the Soviets announced that they will collaborate with the French government in a joint asteroid exploration mission. The project has been given the name Vesta and is scheduled for launch in the year 1992. Two separate spacecraft are to be constructed. According to the initial plans, both Vesta craft were to be directed to fly past Venus at a close distance, using the gravity of that planet to propel them outward to the Asteroid Belt. One of the pair of craft was to be dedicated solely to an exploration of the asteroid Vesta, but the other was also to carry an advanced lander that would be dropped off during the Venus flyby phase of the mission. This lander was to be capable of imaging the surface of Venus from an altitude of several thousand meters as it slowly descended through the dense atmosphere. Once on the surface, the lander would make a detailed chemical analysis of the soil. There were also plans for the lander to deploy an enormous kite that would suspend a package of instruments several kilometers above the surface.

In March of 1986, the Soviets announced a shift of emphasis in their planetary exploration program, refocusing their attention on Mars rather than Venus. The Vesta mission concept was modified to give it a Mars flyby trajectory rather than the Venus flyby originally planned. The future of the Venus lander that was originally being considered is uncertain. Perhaps it will be part of a separate dedicated Venus mission, as yet unannounced.

8

Mars — a Frozen Miniature Earth

Of all the other worlds in the Solar System, Mars has for long been thought to be the one that had a surface environment most like Earth's and the one other planet most likely to harbor some sort of life. For this reason the planet has been the object of intense interest and speculation ever since the origin of modern astronomy. Mars has a distinct reddish appearance in the sky, which led the ancients to name the planet after the god of war (Figure 8.1).

The first astronomical studies of Mars seemed to indicate that the planet might be a lot like Earth. The planet appears in the telescope to have a dense Earth-like atmosphere, perhaps one capable of sustaining some form of life. The length of the Martian day is 24 hours, 37 minutes, about the same as Earth's day. The planet rotates in the prograde direction, just as the Earth does. The equator of Mars is inclined at an angle of 25° with the plane of its orbit, so the planet goes through seasons, just like the Earth. Earthbound astronomers can watch seasonally varying polar caps, and they can even see a seasonally changing surface texture and color. This was taken as evidence for the presence of dense vegetation, which advanced in summer and retreated in winter. Some observers even imagined that they saw canals on the surface, evidence of technological civilization on Mars. There were even speculations that the two tiny moons of Mars were actually giant artificial satellites!

Viking 2 photograph of the rock-strewn surface of Utopia Planitia on Mars. The rocks may be ejecta from the crater Mie, located 200 km to the east of the landing site. (Photograph courtesy of NASA)

FIGURE 8.1 *Viking 1* photograph of Mars. The giant basin Argyre can be seen at the terminator. Part of the Valles Marineris complex can be seen near the terminator at the top. The bright south polar cap can be seen at the bottom. (Photograph courtesy of NASA)

MAR'S EGG-SHAPED ORBIT

Mars is 1½ times farther from the Sun than is the Earth, having an orbit with a semimajor axis of 2.28×10^8 km (1.52 AU). The length of the Martian year is 687 Earth days. The orbit of Mars is significantly more elliptical than Earth's (with an eccentricity of 0.09), and the planet swings 4.0×10^7 km closer to the Sun at perihelion than at aphelion (Figure 8.2). This large orbital eccentricity may have long-term effects on the Martian climate.

The closest approach that Mars makes to Earth (inferior conjunction) is 5.55×10^7 km. At its most distant point (superior conjunction), Mars is 3.78×10^8 km from Earth. Since the Martian orbit lies entirely outside Earth's orbit, the apparent disk of Mars as seen in the telescope can never depart from full phase by more than $47°$. It is impossible to see a crescent Mars from Earth.

THE INTERIOR

Since Mars has two small moons, the mass of the planet can readily be determined. Mars is a good deal smaller than the Earth, with a mass only 0.1 as large (Figure 8.3). Mar's diameter is 6760 km, so its average density is 3.86 g/cm³. This is quite a bit smaller than the densities of the inner three planets and is more nearly equal to the density of Earth's Moon.

The internal structure of Mars (Figure 8.4) is probably quite similar to that of Earth. There is an inner metal-rich core, surrounded by a rocky mantle 1000 to 1500 km thick. The outermost 200 km comprises the crust. Because of the planet's lower density, the Martian core must be far less extensive than the terrestrial core. The core is estimated to be 1300 to 2000 km in radius. It takes up 19 percent of the total planetary mass, as compared to 35 percent for Earth's core. Mars's smaller core is undoubtedly a reflection of the conditions of the

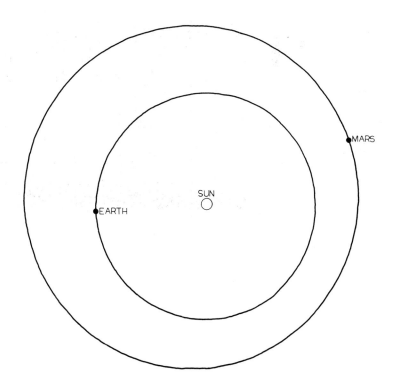

FIGURE 8.2 The orbit of Mars. Mars has an orbit that is appreciably more eccentric than that of Earth.

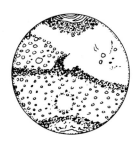

FIGURE 8.3 Relative sizes of Earth and Mars.

primitive solar nebula at the time of its formation. Since Mars must have accumulated in a cooler region of the primitive solar nebula than did the Earth, it is expected that Mars should be poorer in metals and refractory elements and richer in volatile elements than is the Earth.

Mars does not appear to have any significant magnetic field. This is initially surprising, since Mars does have a metallic core and spins on its axis almost as rapidly as the Earth does. The Martian core is, however, significantly smaller than Earth's, and some models of the Martian interior predict a core that long ago cooled off and became either partially or fully solidified. If so, then there is no internal dynamo currently operating to sustain any permanent magnetic field. The absence of a magnetic field exposes the upper atmosphere to the full

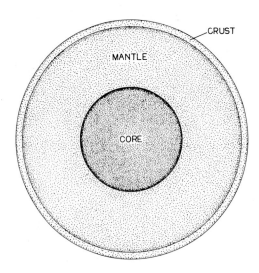

FIGURE 8.4 The interior of Mars.

force of the solar wind on the sunward side of the planet, which may have played a role in the long-term evolution of the Martian atmosphere.

THE PHANTOM CANALS

Surface features on Mars are difficult to observe in any detail from Earth because of the turbulence of the terrestrial atmosphere. Most observers were able to distinguish the variable polar caps, the reddish areas, and the dark regions on the surface of the planet. In 1877, the Italian astronomer Giovanni Schiaparelli reported that he saw a number of "canali" or "channels" criss-crossing the surface. The Italian *canali* was unfortunately mistranslated as *canals* by the press, and an orgy of argument and speculation about intelligent life on Mars followed.

Many respectable scientists lent their voices to the controversy, in particular the American astronomer Percival Lowell. Lowell built a plausible-sounding scenario in which an ancient Martian civilization, desperately in need of water, melted some of the ice in the polar caps and then dug a series of canals to transport this water from the poles to the arid equatorial regions. Vegetation grew up around the shores of the canals, making them visible from space. The planetwide scope of the canal system implied that the Martians were unimpeded by nationalistic competition and were capable of engineering feats far in advance of humanity.

Many astronomers observed these canals in detail and even drew elaborate maps of Mars with the canals prominently displayed. However, many other astronomers were unable to see any canals at all, and the mysterious lines stubbornly refused to show up on telescopic photographs. The debate about the Martian canals became quite acrimonious, and the heat of the controversy was enough to drive many scientists completely out of the field of planetary astronomy. Fruitless debates about the Martian canals were still taking place as late as the 1950s.

The answers to the canal controversy and to the question of intelligent life on Mars came only with the advent of space travel and the launching of probes that were able to fly close enough to the planet to take photographs and make physical measurements. The first successful Mars probe was *Mariner 4,* which flew within 9830 km of the surface in July 1965 and returned 21 television pictures to Earth. It was followed by *Mariner 6* and 7 in 1969.

The photographs show no canals, nor any sort of evidence for any large-scale engineering works whatsoever. The origin of the canals was purely psychological, the human mind subconsciously arranging random surface features into ordered patterns of lines. When our biases are strong, we are likely to fool ourselves.

THE CRATERS OF MARS

Although no canals were found, the first spacecraft photographs did find something rather interesting: craters. They cover much of the surface but are far more numerous in the southern hemisphere than in the northern (Figure 8.5). Almost

FIGURE 8.5 Viking orbiter photograph of heavily cratered region of Martian southern hemisphere. (Photograph courtesy of NASA)

all of the larger craters are shallow and have flat floors. The bright ray craters that are so prominent on the Moon are almost completely absent on Mars.

In general, Martian craters are softer in appearance than those on the Moon, perhaps due to erosion produced by wind and blowing dust. Many of the craters have associated bright and dark streaks that extend outward for many kilometers. These streaks were probably formed by erosion and deposition of airborne dust in the downwind direction. Some craters appear to be relatively fresh, but others are badly degraded. The sharply varying degrees of crater preservation suggests a complex history of surface erosion and geological activity having taken place on Mars in the distant past.

In the southern hemisphere, the density of craters larger than 30 km across is about the same as it is in the lunar highlands, suggesting that this part of Mars is approximately 4 billion years old. Compared to the lunar highlands, though, there is a relative shortage of smaller craters. Many smaller craters must have been covered up and obliterated by lava flows, which formed extensive intercrater plains. The northern hemisphere is much less densely cratered and, therefore, must be much younger. The few craters that are present in the northern hemisphere are relatively well preserved.

The southern hemisphere has several large basinlike areas. These consist of flat, lightly cratered lowlands surrounded by one or more rings of mountains. The three most prominent of these features are Hellas, Argyre, and Isidis. Hellas is an oval-shaped lowland feature nearly 2000 km wide. It is surrounded by an irregular rim of mountains, ranging in width from 50 to 400 km. Argyre (Figure 8.6) is approximately 900 km wide, with a broad pattern of surrounding mountains extending out as far as 1400 km from the center. As many as five concentric rings of mountains may be present. Isidis is 1100 km wide and has a 300-km-wide ring of mountains at its periphery. About 20 smaller two-ring basins have been identified. The lightly cratered interior regions of Isidis, Hellas, and Argyre have numerous wrinkle ridges and flow fronts, which suggests that they probably have a volcanic origin. Like the basins on the Moon, the Martian basins were most likely excavated by the impacts of giant asteroids more than 4 billion years ago. The interiors of these basins were subsequently filled by massive flows of molten volcanic lava.

Martian basins are generally much more badly degraded than similar-sized basins on the Moon. Over the years, many of their prominent features have been gradually softened or completely erased by the erosive action of wind and blowing dust. Furthermore, the lava plains in the interiors of Martian basins are not nearly as dark as the lunar maria, possibly a consequence of billions of years of steady deposition of wind-borne dust.

THE MOUNT OF OLYMPUS—THE VOLCANOES OF MARS

By chance, the three Mariner flyby craft had managed to photograph only the heavily cratered regions of Mars, leading scientists to conclude that Mars was a dead and uninteresting world, much more like the Moon than the Earth. This conclusion was premature.

In 1971 the spaceprobe *Mariner 9* arrived in Mars orbit. It set out on an extensive photographic survey of the entire surface of the planet. To the surprise of almost everyone, the photographs showed that the northern hemisphere is covered with large numbers of extinct volcanoes, most much larger than any on Earth. These giant Martian shield volcanoes all have a broad dome-like structure, with a large caldera at the top.

The most spectacular volcanoes are located in the Tharsis region, where

FIGURE 8.6 Viking oribter photograph of Argyre basin, taken at a distance of 19,000 km. The basin is approximately 1000 km across. The large crater on the rim of Argyre is Gale (300 km across). At the horizon a thin layer of atmospheric haze can be seen. (Photograph courtesy of NASA)

there are three large ones (Arsia, Pavonis, and Ascraeus) lying in a row near the equator. One thousand kilometers away lies the huge Olympus Mons, 500 to 600 km across and towering 23 km (~73,000 feet) above the surrounding plain. Olympus has rough-textured flanks that have many radially elongated lobes and remnant lava channels. The edge of the mountain is marked by an escarpment 2 km high. Olympus Mons is one of the most striking features in our entire Solar System (Figures 8.7 and 8.8).

The entire Tharsis region is actually a huge bulge of solidified lava 5000 km wide that sticks out about 7 km from the rest of the Martian disk. The low-viscosity lava flowing out of these volcanoes covered up and obliterated any craters that previously existed in these northern plains. The Martian lava plains look

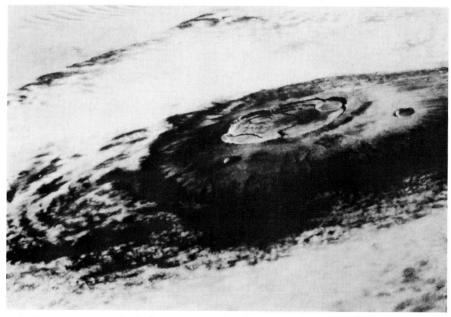

FIGURE 8.7 Photograph of Olympus Mons taken by Viking orbiter. It is 26 km high and more than 500 km wide at the base. Here the base is hidden by thick clouds. (Photograph courtesy of NASA)

superficially like the maria on the Moon. Many of them have flow marks characterized by long, lobed escarpments that formed when the lava solidified. Unlike the lunar maria, the Martian lava plains are brighter than the surrounding terrain. Perhaps weathering and depositional processes have produced a brighter sediment overlying a much darker basalt.

The era of Martian volcanism was so intense that virtually all of the northern hemisphere became covered by solidified lava flows. There are only a few craters scattered throughout the lava plains in the northern hemisphere, indicating that the episode of Martian volcanism must have occurred at a time much later than the era of intense bombardment that scarred the southern hemisphere. Judging from the relative sparsity of craters on their slopes, the three Tharsis volcanoes probably range in age from 400 to 800 million years. Olympus Mons may be only 200 million years old. Other volcanic features are a good deal older. A large 1600-km fracture ring north of Tharsis, known as Alba Patera, is the severely degraded remains of an extinct volcano perhaps 1 to 2 billion years old. There are also some volcanoes in the southern hemisphere (particularly near the Hellas region), but these are much older, perhaps as much as 3.5 to 4.0 billion years in age. The era of Martian volcanism began almost immediately after the period of the most intense meteoric bombardment came to an end and continued until quite recent times. However, there is no evidence that any of the Martian volcanoes are currently active.

FIGURE 8.8 Viking orbiter photograph of olympus Mons caldera. It is approximately 90 km wide. Features as small as 100 m across can be seen. Steep cliffs surround the flat central lava plain. Prominent lava flow fronts can be seen on the inside of the caldera, and an extensive system of cracks can be seen radiating outward from the rim. (Photograph courtesy of NASA)

THE VALLEY OF THE MARINERS— CONTINENTAL DRIFT INTERRUPTED?

In spite of the clear evidence for intense volcanism, Mars seems on the whole to have been much less geologically active than the Earth. There is no evidence for the presence of any type of sustained plate tectonics or continental-drift activity anywhere on the planet. Much of the surface is quite ancient; craters and basins that formed 4 billion years ago are still present on Mars today. On Earth such features were totally erased many years ago by the combined effects of weathering and geological activity.

There may be one place on Mars where continental-drift activity started many millions of years ago and then abruptly stopped. *Mariner 9* discovered that there is an enormous complex of canyons east of Tharsis. This feature was given the name Valles Marineris, in honor of the spacecraft responsible for its discovery. It is 2700 km long, and individual canyons in the complex are up to 200 km wide and 6 kilometers deep. Valles Marineris easily dwarfs the Grand Canyon of Arizona and would stretch across the entire width of the United States (Figures 8.9 and 8.10).

The canyon system is subdivided into a series of linear, steep-walled troughs that are typically 300 to 1000 km long and 50 to 150 km wide. The troughs are aligned roughly parallel to each other, two or four wide, separated by ridges or plateaus. Some troughs are joined laterally to each other to form composite features up to 300 km wide. Others are joined lengthwise to form features as long as 2500 km. Large regions of the valley floor are covered by extensive areas of chaotic, hummocky terrain. These may be the result of massive landslides, formed by the collapse of the steep canyon wall into the broad interior. Some canyon walls are gullied, whereas others are landslide scarred or

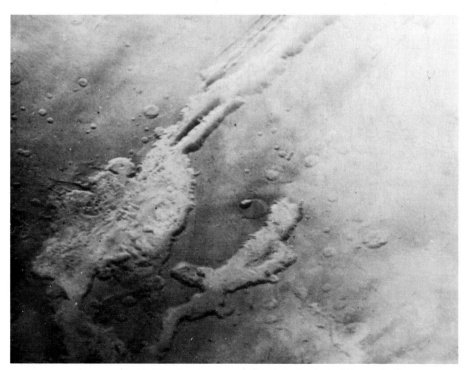

FIGURE 8.9 Photomosaic of Valles Marineris made by U. S. Geological Survey facility at Flagstaff, Arizona from 15 individual photographs. The total width of the mosaic is approximately 2000 km. The circular forms outside the canyon complex are impact craters that have been degraded by wind erosion and by the deposition of sediment. (Courtesy of USGS)

FIGURE 8.10 Viking orbiter photograph of Ganges Chasma region of Valles Marineris. Several landslides at the edges of the walls of this canyon can be seen. The crater at the edge of the far wall is 16 km across. Note that this crater is cut nearly in half by the canyon edge, indicating that the formation of the crater preceded the opening of the canyon. (Photograph courtesy of NASA)

are dissected by tributary canyons. Other parts of the floor are smooth and have numerous ridges or plateaus rising at least 2 or 3 km high.

Valles Marineris cuts through upland surfaces of heavily cratered terrain, so it must have formed considerably later than the episode of intense meteoric bombardment. The floor of the valley itself is very sparsely cratered, suggesting that its formation is a relatively recent event. Unlike the Grand Canyon of Arizona, Valles Marineris could not have been cut by running water. More likely, it was created by geological fracturing of the Martian surface. Nothing like it appears anywhere else on Mars. It may have been formed by the onset of a forced separation of continental plates induced by the volcanic bulge that formed the Tharsis region. This plate motion did not last for long. Shortly after the opening of Valles Marineris, all volcanic activity and crustal motion ceased, leaving Mars a cold, dead world forever after.

THE VIEW FROM THE SURFACE

In 1976 two Viking spacecraft entered Mars orbit. After a careful survey of the surface, they dispatched landing craft, which successfully touched down in the lightly cratered northern lava plains of Mars. The television images that were sent back to Earth (Figure 8.11) showed a terrain that did not look at all like the lava plains on the Moon. They looked more like some terrestrial desert areas. The surface has a lot of loose siltlike material that seems to have been deposited by strong winds. The landscape is strewn with large boulders, many of which exhibit heavy pitting. The pitting may be caused by the erosion produced by blowing dust and sand. The rocks appear to be basaltic in nature, as would be expected in a region known to have been deposited by lava flows.

The Viking landers carried sophisticated analytical instruments that performed elemental analyses of the surface soil. Martian soil is not very different in composition from that of soils found on the Earth or on the Moon. The most abundant heavy element is silicon (15–30 percent), followed by iron (12–16 percent), magnesium (5 percent), calcium (3–8 percent), sulfur (3–4 percent), aluminum (2–7 percent), chlorine (0.5–1.0 percent), and titanium (0.5–2.0 percent). There is undoubtedly a lot of oxygen chemically combined with these elements, but the instruments were not able to measure its relative abundance directly. Magnetic materials are present in the Martian soils in significant quantities, as would be expected when large amounts of iron are present. The elemental analyses are consistent with a regolith composed of iron-rich clays, magnesium sulfate, iron oxides, and carbonates. The analytical instruments of Viking established that the reddish dust responsible for Mars's characteristic color is hydrated ferric oxide, which is also quite common on Earth. Oddly enough, Mars appears to have 100 times as much sulfur as does Earth, but only one-fifth as much potassium.

The Viking instruments showed that Martian soil does differ from terrestrial soil in one very important respect. There is no measurable amount of

FIGURE 8.11 *Viking 1* lander photograph taken from the surface of Chryse Planitia. (Photograph courtesy of NASA)

organic material on the surface of Mars. Organic molecules should actually form quite readily whenever an atmosphere of carbon dioxide, water vapor, and nitrogen is steadily exposed to strong ultraviolet light, and it is initially rather surprising to find them completely absent on Mars.

The cause of this scarcity is still uncertain but it may be due to Mars's relatively thin atmosphere. There is no effective screening against short-wavelength ultraviolet light emitted by the Sun. As a result, any organic molecules that may have initially been synthesized on Mars were immediately disrupted by strong ultraviolet light. Martian organic molecules, therefore, were probably never synthesized in any more than trace amounts. Life may never have gotten a chance to start.

THE AIR OF MARS

The early telescopic observations of Mars seemed to indicate that it had a more-or-less Earth-like atmosphere, with a surface pressure estimated to be as high as 0.1 atm. Carbon dioxide and water vapor had been identified spectroscopically, but neither gas appeared to be abundant enough to account for more than a small fraction of the presumed total atmosphere. The most likely candidate for the unseen component was nitrogen, in view of its abundance in Earth's atmosphere. Surface conditions were believed to be relatively mild, although Mars was thought to be appreciably cooler and dryer than Earth.

Reliable information about Mars, its surface, and its atmosphere did not come until the advent of the space age. The first hint that we were fundamentally wrong about Mars's atmosphere came with the flight of *Mariner 4* in 1965. As the craft flew past Mars, it momentarily passed behind the Martian disk. As the spacecraft's radio signal grazed the rim of Mars, it underwent refraction and provided a measure of the density of the planet's atmosphere near the surface. The surface pressure was surprisingly low, less than 0.01 atm. Carbon dioxide was clearly the main component of the atmosphere, with nitrogen being present only in small amounts if indeed it was present at all. The overestimates of the surface pressure appear to have been caused by the presence of a lot of dust particles in the atmosphere, making the Martian air appear to terrestrial astronomers to be a lot denser and thicker than it really is.

Measurements performed by the Viking landers established that the average surface pressure is only 0.008 atm and that 95.6 percent (by volume) of the atmosphere is carbon dioxide, with 2.7 percent nitrogen and 1.6 percent argon. There is only about 0.1 percent oxygen. The atmosphere has trace amounts of water vapor, carbon monoxide, neon, and xenon. Some ozone has been found in the air, particularly over the polar regions during the winter.

The trace amounts of water vapor in the Martian atmosphere vary with the seasons and with the time of day. The atmospheric water vapor in each hemisphere reaches a maximum concentration during the local summer. Oddly enough, the northern hemisphere appears to have a much higher concentration of atmospheric water vapor than does the southern. Particularly large amounts of water vapor seem to congregate around the edges of the polar caps, especially during the springtime. During the night, some water vapor freezes onto the surface, covering the ground with a thin layer of frost. With the coming of daylight, the frost evaporates. During the early morning hours, dense layers of ground fog can be seen accumulating in certain low-lying areas. During the Martian winter, a permanent layer of frost gathers on the surface, evaporating only with the coming of summer. Clouds of ice crystals are sometimes seen in the upper atmosphere, and in the cold polar regions clouds of solid carbon dioxide (dry ice) crystals sometimes appear. Clouds can sometimes be seen forming complex wave patterns on the downwind side of high surface obstructions.

The Viking landers found that the maximum summer daytime surface temperature in the northern hemisphere is about $-10°C$. At night, the temperature cools off to $-85°C$. During the winter, the highest daytime temperature that was reached was $-85°C$. It got down as low as $-125°C$ during the winter night.

The vertical profile of the Martian atmosphere is shown in Figure 8.12. As on Earth, the temperature of the lowest component of the Martian atmosphere declines with increasing altitude. It is therefore a troposphere. The temperature of the Martian troposphere declines at a rate of $1.6°C$ per kilometer of altitude. Like the terrestrial troposphere, it is heated by the absorption of sunlight at the ground. However, the atmosphere of Mars is so thin that there is no measurable greenhouse effect. The top of the troposphere is at an altitude of 40 km. The temperature there is $-130°C$. Unlike Earth, Mars lacks a stratosphere. The temperature of the atmosphere remains fairly constant above 40 km, until it eventually merges into outer space. However, Mars does have an ionosphere, created by the by-products given off by the ultraviolet photochemistry of carbon dioxide, water vapor, and nitrogen molecules in the thin upper air.

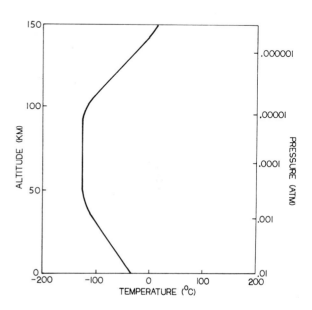

FIGURE 8.12 Vertical profile of the Martian atmosphere.

The Martian sky had a distinctly pinkish cast, caused by the large amounts of dust that perpetually hover in the atmosphere. This dust is picked up from the surface and carried away by gusts of wind. Some winds can reach speeds as high as 120 km/hour, although the average speed is much slower. Once every Martian year, an enormous planetwide dust storm takes place, which conceals the entire surface of the planet from view. These storms originate in the southern hemisphere during the local summer. At that time, Mars is near perihelion. In a typical storm the first dust clouds appear near the Hellas region. These clouds quickly spread out in a westerly direction until they envelop the entire planet. Martian dust storms typically last about 100 days. The first Mars orbiter, *Mariner 9,* happened by chance to arrive during the height of one of these storms. The spacecraft had to wait several weeks before the air was clear enough for any of the surface to be photographed.

THE DISAPPEARING MARTIAN ATMOSPHERE

The thin Martian atmosphere was somewhat of a disappointment to scientists, since they had secretly hoped that the planet might turn out to be much more like Earth. However, there is persuasive evidence that at one time in the distant past the Martian atmosphere was far more dense than it is today.

Like the atmospheres of most planets, the Martian air shows a rich pattern of photochemistry. The traces of oxygen, ozone, and carbon monoxide that are presently found in the Martian atmosphere were all created by the disruption of water vapor and carbon dioxide molecules in the upper Martian atmosphere. Ultraviolet light from the Sun breaks up water molecules into their hydrogen and oxygen components. The hydrogen atoms quickly escape into space, but the oxygen atoms combine with each other to form molecular oxygen. A few of these oxygen atoms manage to react with molecular oxygen, forming ozone. Carbon dioxide is also destroyed by ultraviolet light, producing carbon monoxide molecules and oxygen atoms. Finally, nitrogen molecules are broken up into their nitrogen atomic components. Similar photochemical reactions take place in the terrestrial atmosphere, creating and sustaining Earth's protective ozone shield.

However, the Martian gravitational field is only one-third as strong as Earth's. Because of the weaker Martian gravity many of the energetic atoms and molecules created by the ultraviolet photochemistry in the upper atmosphere can escape into space, never to return. Acting over many millions of years, ultraviolet photochemistry must have destroyed a significant fraction of the Martian atmosphere.

The best evidence for the existence of a denser primitive Martian atmosphere may be found in the relative abundance of the ^{14}N and ^{15}N isotopes found in the current Martian atmosphere. In the Martian atmosphere, the $^{15}N/^{14}N$ abundance ratio is 1.7 times larger than it is in the terrestrial atmosphere, which might mean that some sort of process existed on Mars in the distant past that stripped the planet of most of its nitrogen. One possible loss mechanism is the breakup of nitrogen by the absorption of ultraviolet light from the early Sun. This light must have photoionized numerous nitrogen molecules in the upper Martian atmosphere. Since the Martian gravitational field is much weaker than the terrestrial field, many of these energetic nitrogen ions were able to escape into outer space, never to return. Because of their smaller mass, larger numbers of ^{14}N isotopes than ^{15}N isotopes escaped from Mars. Detailed calculations lead to estimates that Mars must have lost as much as 150 times its current amount of nitrogen over the past 4 billion years.

Similar losses of carbon dioxide and water must also have taken place. However, the relative abundances of the various isotopes of carbon and of oxygen in the Martian atmosphere are the same as they are in the terrestrial atmosphere. This suggests that the atmospheric carbon dioxide and water vapor must periodically exchange with a larger reservoir located on the surface. Substantial reserves of carbon dioxide and water must still be left there.

NOBLE GASES—A RECORD OF EARLY MARTIAN HISTORY

From the standpoint of studies of planetary origin, the abundances of the noble gases (neon, argon, xenon) in the Martian atmosphere are especially interesting. Noble gases do not chemically react with any other elements. Consequently, their concentrations in a planetary atmosphere can shed much light on the origin of the planet and its early evolution, independent of subsequent geochemical processes.

It was found that the Martian atmosphere has nearly two orders of magnitude less noble gas (normalized to the planetary mass) than the Earth does. In particular, the primordial argon isotopes ^{36}Ar and ^{38}Ar are far less abundant in the Martian atmosphere than they are in the terrestrial atmosphere. Mars's noble gas inventory continues a trend that was first noted in the initial measurements of the Venusian atmosphere. There is a general decrease in noble gas abundance as one moves outward in the Solar System from Venus to Earth to Mars. The Venusian atmosphere contains a good deal of noble gases, Earth an intermediate amount, and Mars very little. At the time of the initial formation of the Solar System, there must have been a steep gradient in the concentration of noble gases within the solar nebula, with larger concentrations existing closer to the Sun than farther away. The reasons for the existence of this gradient are uncertain.

The relative abundance of ^{40}Ar in the Martian atmosphere is especially interesting. The isotope ^{40}Ar is created by the radioactive decay of ^{40}K in the crust, and its abundance in a planetary atmosphere can be used as a rough measure of the intensity of the geological activity that has taken place on that planet in the past. Compared to the Earth, Mars has less than 0.1 the amount of ^{40}Ar. Since Mars has a relative shortage of radiogenic argon in its atmosphere, it is reasonable to conclude that the red planet has been far less geologically active than has Earth over most of its life.

THE POLAR CAPS

Mars has two prominent polar caps, one at the north pole and the other at the south pole. The reflectivity of the caps is so high (43 percent) that they can be seen from Earth even in relatively small and inexpensive telescopes. They alternatively advance and recede with the seasons, much like the polar ice caps on Earth. As spring comes to the northern hemisphere, the north polar cap shrinks, the process reversing itself in the fall.

Even though *Mariner 9* and the later Viking orbiters returned many photographs of these caps (an example is shown in Figure 8.13), there is still a considerable amount of controversy about their composition. The first spacecraft observations seemed to indicate that they were largely frozen carbon dioxide (dry ice). The current view, however, is that they are actually made up largely of water ice intermixed with windblown dust. Some parts of the polar caps may be as much as several kilometers thick (thick enough to obscure any underlying craters). There is probably enough ice in the polar caps to flood the entire Martian surface with an ocean of water a few meters deep were it suddenly to melt.

The dust in the caps was probably transported from the equatorial regions to the poles by strong prevailing winds. The dust intermixed with water when it froze to the surface at the poles. The Martian polar caps show an extensive structure with several distinct layers. In the oldest deposits the ice and dust are nonlayered. The younger deposits show alternating layers, each layer being a

FIGURE 8.13 *Viking 2* orbiter photograph of north polar region of Mars. (Photograph courtesy of NASA)

few tens of meters thick. Interspersed within the frost-covered layered deposits is a pinwheel-shaped pattern of largely frost-free areas.

The age of the polar caps is difficult to estimate. A heavily cratered plain lies buried under the ice caps and layered deposits. There are no fresh craters in the layered deposits, nor are there any in the ice-covered or ice-free areas of the polar caps. The ice and dust in the polar caps must have been deposited somewhat after the era of intense meteoric bombardment had come to an end. However, the large amount of dust present in the polar caps suggests that the bulk of the polar deposits must have formed at a time when the atmosphere was much denser than it is today. It does not seem likely that the present thin atmosphere is capable of transporting such large quantities of material.

At present, both polar caps are being eroded by the wind and blowing dust. The material that has been removed has preferentially settled at middle latitudes surrounding the polar caps, where it forms a thin veneer on the surface and partly fills up the interiors of small craters.

Spacecraft observations indicate that the temperature at the poles is cold enough to keep the water in the caps completely frozen all year long. The seasonally variable component of the Martian polar caps is believed to be an outer layer of solid carbon dioxide that freezes onto the surface during the exceedingly cold winter and evaporates during the summer. The temporary layer of frozen CO_2 may reach a thickness of a few meters during the winter. It is estimated that approximately 20 percent of the carbon dioxide in the Martian atmosphere is cycled back and forth between the atmosphere and the polar caps every year.

There are some rather peculiar differences between the northern and southern polar caps, the causes for which are still poorly understood. The north permanent cap is significantly larger than its southern counterpart. The temperature during local summer at the Martian north pole is $-68°C$, but the summer temperature at the south pole is a much colder $-110°C$. Virtually all of the dry ice seems to evaporate from the northern polar cap during the summertime, but some dry ice may remain frozen to the south polar cap even during the warmest periods. In addition, the average amount of water vapor in the air over the northern polar cap is far greater than that found over the southern cap. This asymmetry may be a result of the high eccentricity of the Martian orbit, which causes the planet to be significantly closer to the Sun during the northern-hemisphere winter than it is during the southern-hemisphere winter. Consequently, the southern hemisphere has a longer and colder winter than does the northern.

THE WATER OF MARS— RUNAWAY GLACIATION

Even though the polar caps are the most visible evidence for Martian water, they are believed to represent only a relatively small fraction of the total amount of water currently present on Mars. Earth-based spectroscopic studies, as well as measurements by the Viking landers, indicate that there is at least 10 times as much water ice stored in the outer crust of the entire planet as there is in the polar caps. There is probably enough water ice currently lying frozen in the soil to cover the entire surface of Mars with an ocean of water 30 to 300 m deep were it all to melt. The rise and fall of the atmospheric water vapor content with the seasons seems to be caused by an exchange of water with the ice frozen into the soil rather than with ice stored in the polar caps.

All Martian water is found either in the vapor phase in the thin atmosphere or in the solid phase frozen to the surface. Because of the low surface pressure, water cannot possibly exist in the liquid state anywhere on Mars. Although there is presently no liquid water on Mars, there is ample evidence for its presence in the past. The *Mariner 9* photographs showed numerous sinuous channels that appear to have been cut by running water. The largest of these is 1000 km long and as much as 200 km wide (Figure 8.14). These long channels tend to be found near the equator, presumably because the temperature is warmest there. There are areas with complex braided patterns that were perhaps produced by the deposition of silt and debris from running water. There are even a few teardrop-shaped "islands" that were probably produced by water flowing around obstacles such as craters (Figure 8.15).

Where did this rush of water come from, and what happened to it? The primeval Martian atmosphere of 4 billion years ago was much denser than it is today. It might have been a lot like Earth's primitive atmosphere, being rich in ammonia, methane, water vapor, nitrogen, and carbon dioxide. This early atmosphere may have trapped a considerable amount of heat via the greenhouse effect, and the pressure and temperature at the surface of Mars may have been

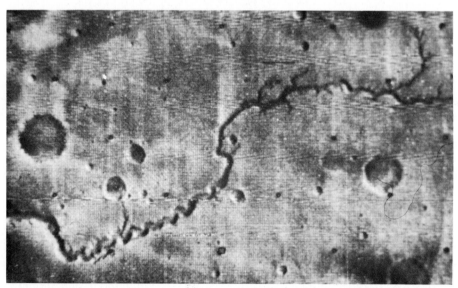

FIGURE 8.14 Photograph taken by *Mariner 9* of Nirgal Vallis, a long riverlike channel located in the southern hemisphere between Valles Marineris and the Argyre basin. The total length of the channel is approximately 1000 km. (Photograph courtesy of NASA)

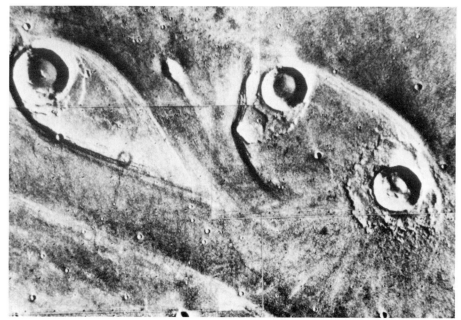

FIGURE 8.15 Photograph of channel "islands" near mouth of Ares Vallis taken by the Viking orbiter. Each island is 40 km long. This terrain is a part of the northern hemisphere Chryse plain. North is at the bottom right in the photograph. (Photograph courtesy of NASA)

high enough for extensive rainfall to have occurred. Lakes, oceans, and rivers of liquid water may have covered much of the Martian surface.

However, ultraviolet photochemistry slowly but steadily destroyed the gases in the upper atmosphere. Over the past 4 billion years, virtually all of Mars's nitrogen and a large fraction of its carbon dioxide and water vapor have disappeared. The vanishing Martian atmosphere stripped away the protective blanket of the greenhouse effect, causing the surface temperature to decline. As the surface became steadily cooler, the water that remained on the planet began to freeze. Since the poles were the coldest places on the planet, the water vapor in the atmosphere first began to freeze to the surface in these regions. The water remains there to this day, perpetually frozen into solid ice. This process by which a planet's entire water supply freezes is known as runaway glaciation; it is essentially the inverse of the runaway greenhouse effect believed responsible for the high surface temperature on Venus.

However, an early dense Martian atmosphere cannot explain all of the water erosion features seen on Mars. Some of the erosive features actually seem to be quite young, showing signs of being only millions of years old rather than billions. The Martian water channels seem to have formed at irregular intervals over a rather extended time period, ranging from 3.5 billion years ago to as recent as 500 million years ago or even less. These features probably have a catastrophic origin (Figures 8.16 and 8.17).

FIGURE 8.16 Viking orbiter photograph of Capri Chasma, a channel emerging from chaotic terrain. The channel is 20 km wide. It connects to Simud Valles, a part of the Valles Marineris complex of canyons. (Photograph courtesy of NASA)

The terrain on Mars that shows the clearest sign of having been produced by water erosion appears to have been created by a sudden and catastrophic release of massive amounts of water (for example, by a flood) rather than by rivers, oceans, or lakes of standing water being present for a long period of time. Sand bars and eroded cliffs, features that are caused by lots of liquid water being present for relatively long periods, are completely absent on Mars. Many of the Martian water channels appear to emerge from particularly chaotic regions where extensive fracturing of surface rocks has taken place. In addition, many water floods seem to have originated either from volcanoes or from impact craters. Many Martian craters have surrounding ejecta blankets that seem to have been deposited by a flood of muddy water that was explosively blown outward by the force and heat of the meteoroid impact (Figure 8.18).

The vast majority of Mars's water is known to be present in the form of frozen ice intermixed with the surface soil and rocks. It is thought that the sudden melting of this subsurface ice was the cause of these intense flooding episodes. Violent events in the past, such as volcanic eruptions, meteorite impacts, or even landslides, generated intense localized heating that melted some of the subsurface ice. The melting ice released a sudden flood of water that swept over the surrounding terrain, cutting deep channels and depositing massive amounts of silt. Within a matter of perhaps only a few weeks, the liquid water released by the flooding evaporated into the thin Martian atmosphere, ultimately to end frozen solid in the polar caps.

THE CYCLES OF CLIMATE

There are some theories of the Martian climate that predict the periodic reappearance of a denser, thicker atmosphere, one that would permit water to exist in the liquid state. Perhaps the best indication of a cyclic Martian climate is the multilayered aspect of the polar ice/dust deposits. There must have been several distinct episodes of ice and dust deposition at the poles, separated by relatively inactive periods.

The oscillating climate is thought to be driven by a periodic change in the

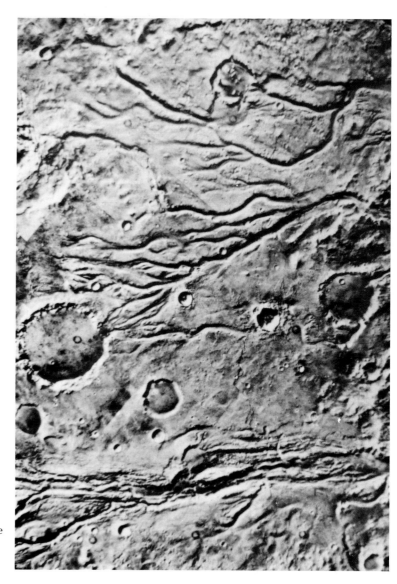

FIGURE 8.17 Viking orbiter photograph of channels in Lunae Planum region of Mars. This particular terrian is located in the northern hemisphere, west of Chryse and north of Valles Marineris. The total width of this view is 200 km. The channels at the top are named Vedra Valles. Those at the bottom are Maja Valles. (Photograph courtesy of NASA)

FIGURE 8.18 Two craters on Mars. On the left is Arandas (28 km across) with a pancake-shaped flow pattern surrounding it. This material appears to have flowed across the ground rather than having been forcibly ejected from the crater by the force of the impact. Yuty (18 km in diameter) is on the right. It has lobate-shaped ejecta.

angle of the Martian axial tilt. This angle is currently 22.5°, but William Ward of Harvard University has proposed that it can vary from a minimum of 15° to a maximum of 35° in response to tidal forces exerted by the Sun. The time period of the variation is estimated to be somewhere between 100,000 and 1 million years. There may also be a periodic variation in the Martian orbital eccentricity, ranging from a minimum of 0.004 to a maximum of 0.141 in response to the motion of the planet Jupiter. Similar variations take place in the Earth's motion, although they are not nearly as extreme. Such effects seem to be the cause of the waxing and waning of the terrestrial ice ages.

An increased orbital eccentricity will bring more solar heat to the Martian cap that is tilted toward the Sun at perihelion than to the one that is tilted away. This increased heat will drive some frozen carbon dioxide and water ice out of the polar caps and into the atmosphere, increasing the air pressure. Perhaps the pressure rises to 0.03 atm or even higher. A denser $CO_2 - H_2O$ atmosphere will produce an enhanced greenhouse effect, which will in turn warm the surface even more. Perhaps pressure and temperature both rise high enough to permit water vapor to condense as a liquid, allowing rainfall to take place. We may simply be observing Mars during the middle of one of its periodic cold and dry phases.

All of these cyclic theories of the Martian climate assume a polar cap made largely of frozen carbon dioxide rather than water ice. The temperature increase required to drive all of the frozen carbon dioxide in the polar caps back into the atmosphere is rather moderate, but the temperature at the poles would have to get at least 70°C warmer in order to force any appreciable amount of water into the air. Such a dramatic temperature rise is rather unlikely. If the Martian polar caps are indeed largely water ice, the planet has probably been as cold and dry as it is now ever since the end of the era of meteoric bombardment over 4 billion years ago.

MARS'S MISSING VOLATILES—AN EXERCISE IN COMPARATIVE PLANETOLOGY

One of the most important things to know about a planet is the amount of volatile material that is present on its surface and in the atmosphere above. The most abundant volatile materials in the Solar System are water, carbon dioxide, nitrogen, methane, and ammonia. These compounds generally exist in either the gaseous or the liquid phase at temperatures typically found on planetary surfaces, whereas refractory materials such as iron or silicon dioxide are solid at these temperatures. It is generally thought that the planets that condensed in warm regions close to the Sun have proportionately larger amounts of refractory materials and smaller amounts of volatiles than do those planets that condensed in cooler regions farther from the Sun.

Mercury is so close to the Sun that its surface lacks any appreciable amount of volatile material. The planet has no significant atmosphere, and any nitrogen, water, carbon dioxide, methane, or ammonia initially present must have been driven into space many years ago by the intense heat.

The planet Venus is twice as far from the Sun as is Mercury. Venus now appears to have almost as much nitrogen and carbon dioxide as the Earth does, which must mean that the two worlds condensed with comparable amounts of volatile material. However, all of the carbon dioxide and nitrogen presently on Venus are in the atmosphere, with none present on the surface. These gases were forced out of the soil and rocks many years ago at the time of the runaway greenhouse disaster that produced rapid and uncontrolled temperature rises. However, the temperature never got hot enough to drive all of these vapors into space. In contrast to the moist Earth, the Venusian atmosphere is bone dry. However, there is persuasive evidence that Venus did possess a considerable amount of water in the distant past. Most of Venus's water was destroyed many years ago by the action of ultraviolet photochemistry in the upper atmosphere. Almost none is left.

As planets go, Earth was very fortunate. It was far enough from the Sun so that appreciable amounts of volatiles could condense onto the surface, but not so far away from the Sun that its water all froze solid. Over the long history of the Earth, most of our original supply of carbon dioxide has been incorporated into living creatures and into the minerals that form sedimentary rocks. Relatively little still remains in the atmosphere. On Earth, most of the nitrogen is in the atmosphere, but a substantial fraction has been incorporated into crustal rocks. On the surface of our planet, most of the water has remained in the liquid state, permitting life to survive.

What about Mars? In many respects Mars is much more like the Earth than it is like Venus. Like Venus, Mars has an atmosphere rich in carbon dioxide. However, the Martian atmosphere is quite thin, and it appears likely that vastly larger amounts of carbon dioxide are permanently bound into the rocks and soil at the surface. Mars has trace amounts of water vapor in its atmosphere, but there are much larger amounts of water ice permanently frozen in the polar caps and in the surface soil. Like the Earth, most of the Martian volatiles are on the surface rather than in the atmosphere.

How much volatile material does Mars currently possess? Since we have looked in detail only at two spots on the surface, this is difficult to determine. However, crude estimates can be made by studying the remote-sensing measurements made by the Viking orbiters. These estimates lead to a surprising result. By summing up the best estimates of the amounts of volatile material stored in the atmosphere, the polar caps, and the surface soil, it turns out that Mars has significantly lesser amounts of volatile material than the Earth does. Compared to the Earth, Mars has proportionally less than 3 percent the total amount of nitrogen, only 1 to 2 percent the total amount of water, and only 10 to 40 percent the total amount of carbon dioxide.

Mars's relative shortage of water, carbon dioxide, and nitrogen is initially surprising, since the planet condensed in a cooler region of space than Earth and would therefore be expected to have formed with proportionally larger amounts of volatile material than did our planet. However, it is probable that Mars's atmosphere was at one time much thicker than it is today. The atmosphere has slowly evaporated into space, under the combined effects of ultraviolet photochemistry and weak gravity. Over the years, Mars must have lost almost all of its original supply of nitrogen, and a sizeable fraction of its original inventory of water and carbon dioxide.

However, the atmospheric-loss hypothesis cannot be the entire explanation for the volatile shortage on Mars. Assuming that Mars originally had at least as much water, carbon dioxide, and nitrogen as the Earth did, evaporation into space can account for only a small fraction of the overall volatile deficit. Perhaps substantial amounts of water, nitrogen, and carbon dioxide are still hidden away deep under the Martian surface, where they are difficult to detect from space. The relative scarcity of radiogenic argon in the Martian atmosphere seems to mean that the planet has on the whole been far less geologically active than the Earth has. If so, then there must have been much less outgassing of material from the Martian interior, and Mars probably has substantial reserves of volatiles still locked away deep under the surface. The answer requires further exploration.

THE MOONS OF MARS

Mars has two tiny moons, named Phobos and Deimos. They are named after the mythological horses Phobos (Fear) and Deimos (Flight) that pulled the Greek war god's chariot. Both were discovered in 1877 by the American astronomer Asaph Hall of the U. S. Naval Observatory after a long search.

SWIFT'S REMARKABLE PRECOGNITION

This discovery was one of the most puzzling coincidences in modern astronomical history. In 1726 the English author Jonathan Swift gave a fictional account of the discovery of two small Martian moons in his novel *Gulliver's Travels*. His description of these objects was quite detailed and very closely parallels that of the actual moons discovered nearly a century and a half later. How did he know? Some more imaginative investigators have speculated that Swift must have had access to ancient knowledge handed down from long-vanished "supercivilizations" such as Atlantis or Mu. Perhaps he had even been given information originally supplied by extraterrestrial space travelers who had recently visited the Earth.

The truth of what actually happened is probably much more mundane. It now appears that the origin of the mystery lies in the work of the famous astronomer Johannes Kepler. Kepler, like most of his contemporaries, incorporated a good deal of mysticism into his scientific work. He believed that there had to be some sort of orderly progression in the number of moons that a planet possessed. In the seventeenth century, it was known that the Earth had one moon, Jupiter had four moons, and Saturn had five. For some reason, the as-yet-undiscovered planet presumed to be located between Mars and Jupiter was imagined to have three moons. It then stood to reason that Mars should have two moons. Numerological arguments carried much more weight in the seventeenth century than they do now, and Kepler's reputation was such that his views were generally accepted. It turned out that he was right.

RENDEZVOUS WITH A PAIR OF ROCKS

Phobos and Deimos are situated in circular orbits located approximately in the plane of Mars' equator. Phobos has a period of 7.6 hours and is 9300 km from Mars whereas Deimos has a period of 30.3 hours and is 23,500 km distant (Figure 8.19). Both moons orbit Mars in the direct sense, that is, counterclockwise as seen from above the Martian north pole. They thus travel around Mars in the same direction that the planet rotates on its axis. Because the Martian day is about 24 hours long, the two moons appear to travel across the Martian sky in opposite directions.

The *Mariner 9* and Viking orbiters passed close enough to these moons to photograph their surfaces. These photographs show that Phobos is rather irregular in shape; it is shaped more like a potato than a sphere. It is 27 km long and 20 km thick (Figure 8.20). Deimos is roughly egg shaped and is 15 km long and 11 km wide. Martian tidal forces have long ago locked both moons into synchronous rotation, and they perpetually point their long axes toward the planet.

Phobos is heavily scarred with impact craters (Figure 8.21). The freshest craters are bowl shaped, and, because of Phobos's small size and mass, there are no craters with central peaks or rings, and there are no secondary craters. There is a particularly large crater named Stickney (in honor of Asaph Hall's wife, who was a major driving force behind her husband's discovery of the Martian moons), which is about 10 km across. Near Stickney is an odd pattern of parallel striations that appears to be associated in some way with the crater itself. These

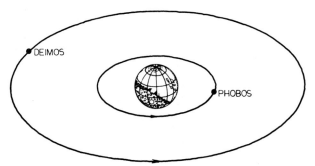

FIGURE 8.19 The orbits of the moons of Mars.

FIGURE 8.20 Comparison view of sizes of the moons of Mars. They are each approximately the size of New York City.

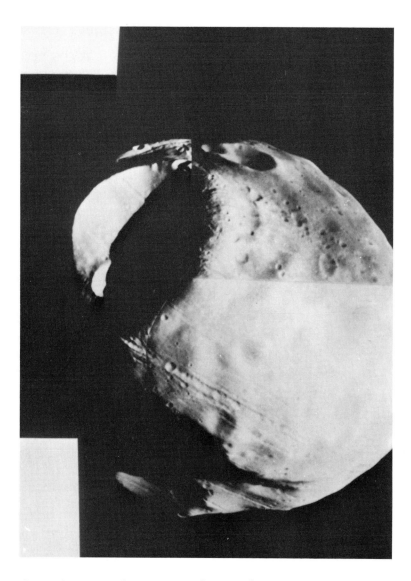

FIGURE 8.21 Viking orbiter photograph of Phobos. The range is 600 km. The large crater on the facing side is Stickney. Long parallel grooves can be seen running outward from Stickney. (Photograph courtesy of NASA)

may be indications of deeper cracks in the internal structure — fissures that might actually have come very close to disrupting the tiny moon at the time of the Stickney impact.

Deimos (Figure 8.22) appears to be much smoother than Phobos and has no craters any larger than 2 kilometers across. The largest crater on Deimos is Voltaire, a severely eroded structure 2 km across. Nearby is a 1-km crater, named Swift, which has a somewhat sharper outline. Deimos has conspicuous bright patches that cover about 10 percent of its surface, whereas Phobos exhibits very little variation in reflectivity.

The Viking spacecraft were directed to pass close enough to these moons so that their masses could be determined. Their densities are of the order of 2

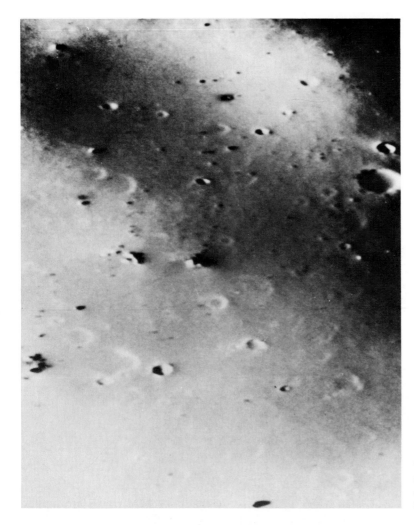

FIGURE 8.22 Viking orbiter photograph of the surface of Deimos. The area shown is 2 km across, and the boulders on the surface are approximately as large as houses. (Photograph courtesy of NASA)

g/cm³, appreciably less than that of Mars itself. This may mean that they are made of an entirely different sort of material than is Mars. The density is much too small for the material to be any sort of basalt. The moons also differ from Mars in that they have colors that are more gray than red. The reflectivity spectra of Phobos and Deimos very closely match those of large asteroids such as Ceres or Pallas as well as certain rare meteorites known as carbonaceous chondrites. Carbonaceous chondrites are rich in hydrated silicates, which are interspersed with much organic material. They are believed to be among the oldest objects still left in the Solar System.

CAPTURE OR ACCUMULATION?

If Phobos and Deimos are so different in structure from Mars, where then did they originally come from? The sharp differences that appear to exist between their compositions and that of Mars itself suggest to some theorists that Phobos and Deimos did not originally accumulate in their present orbits, but rather first formed much farther out in the Solar System. Perhaps they were originally asteroids that were captured by Mars during accidental close approaches.

A significant flaw in this theory is that both moons are currently located in nearly circular direct orbits lying close to the plane of Mars's equator, an unlikely configuration for objects captured at random from nearby space. Phobos is sufficiently close to Mars so that tidal forces acting over 4 billion years of time could have gradually converted a highly elliptical orbit into a circular one, but Deimos is much too distant.

Donald Hunten of the University of Arizona has proposed that gas drag was responsible for the initial capture and subsequent circularization of the orbits. At the time of the initial formation of the Solar System, there must have been a dense ball of gas and dust gathered around the newly accumulated Mars, a sort of protoatmosphere. This protoatmosphere may have been able to produce sufficient viscous drag to capture a couple of errant asteroids unfortunate enough to pass too close to the newly formed Mars. Over the years, the viscous drag gradually converted the orbits of these moons into more nearly circular ones lying close to the plane of the Martian equator. Eventually, this dusty Martian protoatmosphere was stripped from the planet by the intense solar wind given off by the Sun when it first began to burn, leaving the two moons orbiting in the now gas-free space near the planet.

On the other hand, Phobos's and Deimos's circular orbits may actually mean that these objects originally formed right where they are currently found. The heavier material originally present in the proto-Martian cloud of gaseous and dusty debris may have fallen inward to form the planet itself, leaving only the lighter material in the periphery, where it could form the moons. It is possible that the carbonaceous material presumed to be present is only an outer coating, with material much more like the Martian crust on the inside. The abnormally low densities of Phobos and Deimos may only mean that they are made of loosely packed material that happened to accumulate in a low-gravity environment, rather than their being made of intrinsically low-density material. The answer must await further exploration.

ARE PHOBOS AND DEIMOS ARTIFICIAL?

In 1945 the astronomer B. P. Sharpless of the U. S. Naval Observatory reported that the orbital velocity of Phobos is slowly increasing with time, indicating that this moon is slowly but inexorably coming closer and closer to Mars. If it keeps falling at the same rate, it will crash onto the surface within 100 million years. What is causing this decay? Friction with the thin Martian atmosphere at the extreme height of Phobos's orbit should be far too weak a force ever to produce any significant orbital decay for so massive an object.

In the late 1950s the Soviet astronomer Iosif Shklovsky (working at the Space Research Institute of the Academy of Sciences of the USSR) suggested that atmospheric friction could explain the decaying orbit if Phobos is actually very much less massive than expected for an object of its size. Such might be the case if Phobos were hollow! It is difficult to imagine how a natural astronomical object could possibly be hollow, but an artificial object certainly could be. Shklovsky speculated that Phobos and Deimos are both giant artificial satellites that had been launched by a Martian civilization in the distant past. They continue to orbit the planet long after the society that built them has vanished. Other workers speculated that the Martian moons had been placed in orbit much more recently, their sudden discovery after many years of fruitless search indicating that they had actually been launched only a few years before Asaph Hall first saw them!

Arguments and speculations about the Martian moons continued in the popular press until the first Earth-launched Martian orbiters returned detailed photographs of these objects. These pictures clearly show the Martian moons to be entirely natural bodies. Although there is still controversy about the origin of the decay of Phobos's orbit, the cause is now believed to be related to tidal friction generated within the Martian core. The gravitational forces exerted by Phobos cause the Martian core to bulge outward ever so slightly towards the tiny moon. Because Mars happens to rotate on its axis considerably more slowly than Phobos revolves around the planet, internal friction within the Martian interior causes the tidal bulge to lag slightly behind Phobos's position in the sky. This lag produces a small but steady force on Phobos in a direction opposite to its orbital motion (Figure 8.23). Over many years of time, this dragging force beings Phobos's orbit closer and closer to Mars. However, long before Phobos reaches the top of the Martian atmosphere, tidal stresses will become so strong

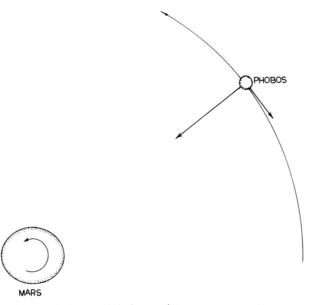

FIGURE 8.23 Decay of Phobos's orbit.

that they will probably tear the unfortunate moon completely apart, leaving only a narrow ring of particles circling the planet.

FUTURE EXPLORATION

There are many questions left to be answered about Mars. NASA is currently making plans for a Mars Observer mission, scheduled for launch in the early 1990s. The craft would orbit the planet for a year, gathering information on surface composition, magnetic field (if any), and seasonal cycles involving carbon dioxide, water vapor, and windblown dust. An important goal of the mission is an accurate estimate of the total amount of water and carbon dioxide on Mars. Little emphasis will be placed on surface imaging.

In 1986 the Soviet Union announced that it was changing the focus of its planetary exploration program from Venus to Mars. An extremely advanced Mars mission is currently being planned, scheduled for launch in 1988. The mission involves two separate craft, each of which will be placed in orbit around Mars. Both craft will take remote-sensing measurements of the Martian atmosphere and surface. When the first of these craft is finished with its Mars observations, it will be directed to approach within 100 m of the surface of Phobos. Upon arrival, it will direct laser and ion beams upon the moon in order to vaporize some of the surface soil, providing a quick chemical analysis. The craft is also scheduled to dispatch a landing vehicle that will actually touch down on the surface of Phobos. This lander will return photographs and other data, perhaps even performing an on-site chemical analysis of the soil. If the Phobos encounter is successful, the second craft may be diverted to Deimos.

The European Space Agency (ESA) has also expressed an interest in Mars. They have begun to explore the possibility of placing a spacecraft in Martian orbit to study the planet's magnetosphere and atmosphere. The project has been given the name *Kepler,* in honor of the famous seventeenth-century astronomer. The Kepler mission has not yet been officially approved. Even assuming that funding is made available, it probably will not be launched until the mid-1990s.

The primary flaw with the Viking landers was that they were immobile. They could see no farther than the local horizon. The Soviet Union is in the early stages of planning for a Mars rover vehicle that can wander from one location on the Martian surface to another. The Mars rover would be similar in concept to the remotely controlled Lunokhod vehicles, which the Soviets placed on the lunar surface in the early 1970s. A Martian rover could sample the soil and air in many different place on Mars and send back pictures of many different types of surface terrain. Perhaps Martian life lurks in some as-yet-unexplored ancient river valley. The mission is tentatively scheduled for the mid-1990s.

The Soviets have also begun the initial definition of an end-of-century mission to return a sample of Martian soil to Earth for study. This would involve an automated spacecraft that would travel to Mars, land in a favorable spot, scoop up a soil sample, place it in a special canister, and fire it back to Earth. Similar missions were performed during the early 1970s by the Soviet Luna landers, in which capsules containing lunar soil samples were returned to Earth. However, the presence of the Martian atmosphere, the appreciably stronger Martian gravity, and the much greater distances involved make the task at least an order of magnitude more difficult and expensive.

An automated Martian airplane has also been discussed. It would have a huge wingspan, which would permit it to fly in the thin Martian atmosphere. Perhaps it could fly to the summit of Olympus Mons to see if this volcano is still active. It might even be able to soar above the enigmatic polar caps, uncovering important data that could reveal the secret of their origin.

When will man land on Mars? A journey by humans to Mars and a return to Earth would take at least 3 years. The technology required for such a mission is probably already at hand. Some of NASA's long-range planners have proposed that flights by Americans to Mars and back should be the next primary national goal in space exploration, following the completion of the Earth-orbiting space station. However, there is at present no perceived national need for such a mission, at least none sufficiently pressing to induce the government to spend the billions of dollars that would be required.

If America does not send people to Mars, perhaps some other nation will. There are persistent rumors that the ultimate goal of the Soviet space program is to send astronauts to Mars. Cosmonauts have remained aboard the Soviet *Salyut* space station in near-Earth orbit for several months at a time. This might mean that much longer-duration missions to more distant points are being contemplated. However, Soviet cosmonauts have never even flown to the Moon, which suggests that flights to the far more distant Mars are an exceedingly remote prospect for them.

Flights by astronauts to Mars are at present far too expensive for any single nation to perform. Perhaps the time is ripe for the organization of an international Mars effort. Such a program would involve astronauts, scientists, engineers, manufacturers, and planners from many different nations, all acting in collaboration. The ultimate goal would be the exploitation and colonization of Mars for the benefit of all humanity. Such a program would involve virtually the entire human race and would be far more inspiring than the present sterile nationalistic competition that has thus far been characteristic of space exploration. In any event, it is unlikely that human beings will walk the sands of Mars until the beginning of the next century.

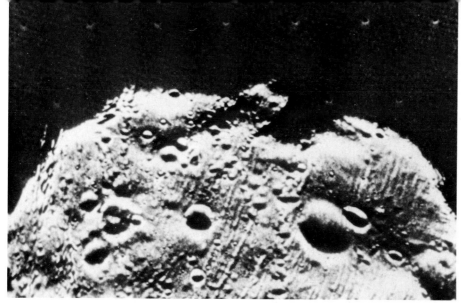

9

The Asteroids — the Debris of the Solar System

In 1766 the German astronomer Johann Titius (1729 – 1796) stumbled upon a mathematical sequence that gave a rather accurate description of the distances from the Sun of the six planets that were known at that time. He wrote down the sequence of numbers 0, 3, 6, 12, 24, 48, 192, and so on. He then added 4 to each number and divided the sum by 10. The resulting numbers gave a very good approximation of the actual locations of the planets as measured in astronomical units (AU). This is shown in Table 9.1.

At first the new law of planetary spacings attracted little attention. However, it caught the eye of Johann Bode (1747 – 1826), the director of the Berlin Observatory. Bode publicized the law and brought it to the attention of the general astronomical community. Consequently, Bode is often given credit for the law, although he had nothing to do with its creation.

When the Titius law was first proposed, the planet Uranus had not yet been discovered. When it was, the orbit fit Titius's formula almost exactly. However, in order to make the law fit the actual positions of the known planets, it was necessary to skip the fifth entry, which predicted a planet at 2.8 AU, between Mars and Jupiter. The success of the Titius law in predicting the loca-

Viking 1 photograph of the Martan moon Phobos. Phobos may have originally been an asteroid that happened to wander too close to Mars and was captured. (Photograph courtesy of NASA/JPL)

TABLE 9.1 TITIUS'S PLANETARY SPACING LAW

TITIUS'S SEQUENCE	PLANET	ACTUAL POSITION (AU)
$\dfrac{0+4}{10} = 0.40$	Mercury	0.39
$\dfrac{3+4}{10} = 0.70$	Venus	0.72
$\dfrac{6+4}{10} = 1.00$	Earth	1.00
$\dfrac{12+4}{10} = 1.60$	Mars	1.52
$\dfrac{24+4}{10} = 2.80$	?	?
$\dfrac{48+4}{10} = 5.20$	Jupiter	5.2
$\dfrac{96+4}{10} = 10.00$	Saturn	9.54
$\dfrac{192+4}{10} = 19.6$	Uranus	19.2
$\dfrac{384+4}{10} = 38.8$	Neptune	30.1
$\dfrac{768+4}{10} = 77.2$	Pluto	39.4

tion of Uranus gave impetus to a search for a planet lying between Mars and Jupiter.

In 1801 the Sicilian monk Giuseppe Piazzi (1746–1826) accidentally found an object at a distance of 2.77 AU from the Sun during a routine star search. He named it Ceres. At first, Ceres was hailed as the long-sought planet between Mars and Jupiter. However, it was only 1000 km in diameter, more nearly the size of a moon than a planet. In 1802 the German astronomer Heinrich Olbers (1758–1840) found another small world (later named Pallas) in this same region. In the next few years two more planetoids (Juno and Vesta) were found. With the advent of astronomical photography, many more were discovered. By the year 1890 the orbits of about 300 were known. Recent sky surveys indicate that there are probably 500,000 chunks of matter in this region of space that are large enough to show up on telescopic photographs. They are collectively known as *asteroids*.

The honor of naming an asteroid falls to the person or persons who discover it. At first, newly discovered asteroids were named after Greek and Roman deities. When these names were exhausted, the discoverers turned to heroines in Wagnerian operas, and then finally to friends, wives, flowers, cities, colleges, pets, and even favorite desserts. When the orbit of a newly discovered asteroid is sufficiently well determined so that it cannot be lost, it is assigned a number. The numbers are assigned sequentially, in the order of discovery. A milestone of sorts was reached in 1984, when the 3000th asteroid was officially numbered.

THE ORBITS OF ASTEROIDS

Most asteroids lie in orbits between Mars and Jupiter, at an average distance of 2.8 AU from the Sun. Nearly half of all asteroids have orbits within 0.25 AU of the average. This region of space between the orbits of Mars and Jupiter is known as the *Asteroid Belt* (Figure 9.1). Most asteroid orbits lie within a few degrees of the ecliptic plane, and all asteroids without exception are in prograde orbits. They travel around the Sun in the same direction that the planets do.

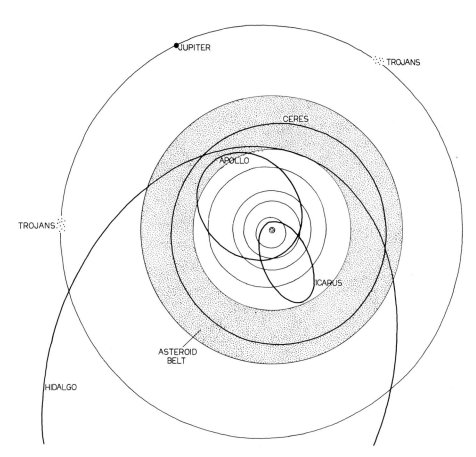

FIGURE 9.1 The orbits of the asteroids.

There are a few asteroids whose orbits depart greatly from the norm. For example, the tiny world 1566 Icarus (only about 1 km in size) is in a very eccentric orbit that brings it as close as 0.18 AU to the Sun, well inside the orbit of Mercury. The 30-km-wide asteroid 944 Hidalgo is in a highly elliptical orbit that swings out as far as 9.6 AU from the Sun, nearly out to the orbit of Saturn. The orbit of Hidalgo makes an angle of 43° with the ecliptic. However, the most interesting of these "abnormal" asteroids are those whose orbits cross that of Earth.

There are 28 known asteroids whose orbits cross that of the Earth. They are known as *Apollo* asteroids, named after the first example to be discovered (in 1932). Apollo asteroids are quite small, ranging in size from 0.4 to 8 km. Some have been seen only once, and their orbits are therefore somewhat poorly determined. A few are known only by numbers, pending the assignment of names. The total number of Earth-crossing asteroids may be as large as 1300.

Apollo asteroids can on occasion come quite close to the Earth. In 1937 the asteroid Hermes missed Earth by only 780,000 km, the closest known approach of any asteroid. It has not been seen since, so its orbit is somewhat uncertain. Apollo asteroids are of somewhat more than strictly academic concern. Once every quarter-million years, on the average, one of them collides with the Earth. Unknown Apollo asteroids constitute an ever-present danger, as appreciable damage and loss of life would occur should any one of them ever strike the Earth.

In the eighteenth century, the French mathematician Joseph Lagrange (1736–1813) demonstrated that points in a planet's orbit 60° ahead and behind the planet are positions of high stability where loose objects and debris will tend to accumulate. Since Jupiter is so massive, it was suspected that some asteroids might have accumulated at its *Lagrange points*, orbiting the Sun with the same period as the planet itself. In 1906 the asteroid Achilles was discovered at the Lagrange point 60° ahead of Jupiter. Subsequent discoveries were made of asteroids at both Lagrange points of Jupiter. These are collectively referred to as *Trojan* asteroids. They are all named for heroes from Homer's *Iliad*, the asteroids at the leading Lagrange point after Greeks and those at the trailing point after Trojans. For some reason, there are at least twice as many Greek Trojans (at the leading point) than there are Trojan Trojans (at the trailing point). The total number of Trojan asteroids may be as high as several hundred.

In 1867 The American astronomer Daniel Kirkwood (1814–1895) reported that there were a few places in the Asteroid Belt in which relatively few asteroids are found. These regions have come to be known as the *Kirkwood gaps*. He noted that the positions in the Asteroid Belt at which these gaps occurred happened to coincide with orbits whose periods had an exact numerical relationship with the period of the orbit of Jupiter. For example, there is a scarcity of asteroids at the point where a body would travel around the Sun exactly twice while Jupiter goes around the Sun once. This is called the 2:1 resonance point. The gravitational tug of Jupiter gradually sweeps the resonance points free of asteroids. Similar gaps occur at the 3:1, 4:1, 5:2, 5:3, and 4:3 resonance positions.

The details involved in the complex gravitational couplings that form the Kirkwood gaps are quite poorly understood. To compound the problem, it has been found that there are actually *concentrations* of asteroids at the 3:2 resonant position [and of course, at the 1:1 position (the Trojans)]. The 3:2 asteroids are known as the *Hilda* group. Much more theoretical and experimental work remains to be done before the orbital positions of these asteroids can be fully explained.

THE ASTEROID CLASSIFICATION SCHEME

As yet no spacecraft has flown close enough to an asteroid to photograph its surface or measure its density. Consequently, the only way we can determine their composition is by an analysis of their optical reflectance spectra. The only samples of extraterrestrial matter that we have (aside from Moon rocks) come from meteorites. Consequently, asteroid composition is usually determined by comparing their spectra with those of meteorites whose composition is known. These studies show that asteroids can be divided into several distinct classes, which are given the leters S, C, M, D, and U.

Three-quarters of all asteroids have reflectance spectra that are distinguished by relatively low reflectivities and grayish colors. These gray asteroids are designated by the letter C. Their spectra most closely resemble those of certain rare meteorites that are classified as carbonaceous chondrites. Carbonaceous-chondrite meteorites are rich in water-bearing silicate minerals but have very little nickel–iron metal. C-type asteroids dominate the outer parts of the main Asteroid Belt. The two large asteroids 1 Ceres and 2 Pallas fall under the C classification.

Ten percent of all asteroids have reflectance spectra that resemble stony-iron meteorites, with relatively high reflectivities and reddish colors. These are designated by the letter S. The largest S-type asteroid is 3 Juno, 244 km in diameter. Details of their reflectivity spectra are consistent with the presence of plentiful amounts of iron and magnesium silicates, intermixed with nickel–iron metal in uncertain proportions. S-type asteroids are generally 100 to 200 km in diameter and tend to congregate in that part of the Asteroid Belt nearest Mars. Most of the Apollo asteroids also fall into the S category.

A few asteroids are classified as M, for "metallic," because their reflectance spectra show evidence of large amounts of nickel–iron metal exposed on the surface but no sign of the presence of silicates. The asteroid 16 Psyche (diameter of 250 km) is the largest M-type asteroid. M-Type asteroids are found throughout the Asteroid Belt, but there is an unusually large concentration of them near the two Kirkwood gaps at the middle of the belt.

D-type asteroids (for "dark") have M-like reflectance spectra but very low albedoes. They are typically found in the very outermost parts of the main Asteroid Belt and in the Trojan family of asteroids coorbital with Jupiter.

Five to 10 percent of all asteroids have spectra that do not fall within any one of these S, C, M, or D classifications. These asteriods are usually assigned the letter U (for "unclassifiable"). The reflectance spectrum of 4 Vesta (600 km in diameter) indicates that its surface is covered with basaltic rocks that were at one time molten. The interior of Vesta may have undergone considerable melting and differentiation in the distant past. The 68-km-wide asteroid 44 Nysa is highly reflective and white in color like the enstatite achondritic meteorites. Others are difficult to reconcile with any meteorite type.

The composition gradient that seems to exist across the width of the Asteroid Belt is quite intriguing. Objects that are rich in high-temperature refractory minerals tend to lie closer to the Sun than do those that have lots of low-temperature volatile materials. This gradient must be a reflection of the temperature of the solar nebula at the time these objects condensed. The low-temperature minerals and the volatile carbon and water could only condense from the vapor at relatively cool temperatures, far from the Sun. The inner parts of the Asteroid Belt remained relatively deficient in these materials, and objects that formed there became much richer in high-temperature minerals.

ASTEROID ROTATION

Most asteroids have irregular shapes and thus vary in brightness as they rotate. The rotational periods of about 200 asteroids have been measured. They range from 3 to 18 hours, and most rotations appear to be in the direct sense. There is some evidence that, on the average, M asteroids rotate more rapidly than S asteroids, which in turn rotate faster than C asteroids. For asteroids larger than 120 km across, the mean rotational period decreases with increasing diameter. However, for asteroids smaller than 120 km across, the mean rotational period increases with increasing diameter. It is possible that the change in rotational behavior with size reflects two different populations of asteroids. The larger asteroids may be primordial bodies that have been relatively unaltered since their formation, whereas the smaller ones may simply be fragments of rock that were produced by collisions.

There is evidence that some asteroids have satellites revolving about them. A few asteroids may actually be double or multiple bodies. During the times of passages of larger asteroids in front of conveniently located stars, there have been occasional observations of "secondary occultations." This may mean that these asteroids have retinues of minor satellites that accompany them in their orbits. However, these observations are all poorly confirmed, so it is not yet certain if any asteroid actually has satellites. Speckle-interferometric observations of a few asteroids can also be interpreted as evidence of the presence of large companions. The large asteroid 2 Pallas may have a small satellite in orbit around it. The photometric light curve of the large Trojan asteroid Hektor can be interpreted as evidence that the asteroid may actually consist of two separate objects in such close proximity that they are virtually in contact.

PLANETARY FORMATION INTERRUPTED — THE ORIGIN OF ASTEROIDS

It was at one time assumed by almost everyone that the Asteroid Belt was formed by the catastrophic breakup of a larger planet that originally occupied an orbit between Mars and Jupiter, in the position predicted by the Titius law. However, it is estimated that the combined mass of all the matter in the Asteroid Belt is only 0.0004 times that of the Earth, a good deal less than any planet. It is therefore unlikely that asteroids are debris from any cataclysmic planetary breakup in the distant past, unless, of course, most of the material created by the disruption was swept up by a collision with a planet. The sharp differences in composition seen between objects located in different parts of the Asteroid Belt make it improbable that they could all have come from the same large planet-sized body. During their journey to Jupiter and beyond, the probes *Pioneer 10* and *11* found that the Asteroid Belt contains no more dust than any other part of the Solar System, which makes it unlikely that any sort of massive disruption ever took place in this particular region of space.

It is now believed that the asteroids are small objects that escaped incorporation into larger, planet-sized bodies at the time of the birth of the Solar System. According to the most widely accepted models of the origin of the Solar System, all of the planets were formed by the gradual accretion of particles of gas and dust that took place in the primitive solar nebula nearly 5 billion years ago. As the solar nebula contracted and cooled, the material in its outer parts gradually clumped together to form larger and larger objects. Some of the largest were perhaps over 100 km across.

The newborn Sun became surrounded by a vast, flat ring of asteroid-sized chunks. This sheet of loose rocks may have looked somewhat like Saturn's ring structure, but on a much larger scale. Collisions between such bodies must have been very frequent in the early history of the Solar System. In some collisions, the objects broke up into smaller fragments, whereas in others they coalesced to form larger bodies. Once formed, the larger bodies steadily grew in size as they swept up more and more objects. Eventually, they grew large enough to form planets.

Over the years, the newly formed large planets swept up most of the small objects still remaining free, so that the Solar System was almost entirely cleared of small chunks of matter. Most of the planets and moons in the Solar System still bear the scars of these early collisions.

The gradual accretion of smaller bodies into larger planet-sized objects seems to have taken place all over the Solar System, except in the region of the Asteroid Belt. Apparently, the strong tidal forces generated by the giant planet Jupiter prevented smaller asteroid-sized chunks of matter from combining to form a large planet-sized object in this region of space. The asteroids are therefore relatively old objects, left over from the era immediately prior to the complete formation of the planets. A close examination of an asteroid should provide many clues about the structure and composition of the primitive solar nebula during the time immediately before planetary formation was completed.

APPOINTMENT WITH AN ASTEROID

In 1984 it was found that the proposed Galileo mission to the planet Jupiter might just possibly be able to make a short detour and pass close to an asteroid. The object in question was 29 Amphitrite, a 200 km-wide S-type asteroid. After much deliberation, NASA decided to approve the detour. The encounter with Amphitrite was to take place on December 6, 1986, the spacecraft approaching to within 10,000 km of the asteroid. However, the destruction of the Space Shuttle *Challenger* has resulted in an indefinite delay in the American space program and has forced a postponement of the *Galileo* mission until at least 1989. Consequently, the appointment with Amphitrite was missed. It has not yet been decided if an encounter with a different asteroid is possible.

The Soviet Union and France have announced that they will collaborate on an advanced asteroid mission. The mission has been assigned the name Vesta. According to the plan, two craft will be launched in 1992, initially aimed toward Mars. The Vesta spacecraft will use close encounters with that planet to propel them outward toward the Asteroid Belt. The first of the two craft will be targeted toward the asteroid Vesta. This asteroid is of special interest because its surface has probably been volcanically active in the past. During the flyby of Vesta, the craft will release a lander that will touch down on the surface of the asteroid. The lander will carry instruments capable of making a chemical analysis of the topsoil of this intriguing world. The other spacecraft in the Vesta series will be targeted to a different asteroid, to be named later.

NASA is giving some serious thought to a combined comet-rendezvous/ asteroid-flyby mission, to be launched sometime during the early 1990s. The primary goal of the mission will be a rendezvous with a comet. However, while en route to the comet, the spacecraft will be directed to pass within a few thousand kilometers of an asteroid. The mission has not yet been provided with any new-start funding, so it is not yet certain which targets will be selected. A 1992 launch would make it possible to encounter asteroids Malautra and Hestia in 1993 and 1995, respectively, with a rendezvous with comet Tempel-2 in 1996.

The European Space Agency (ESA) has begun a serious study of an extremely sophisticated mission to the asteroid Vesta. Named *Agora* (which is an acronym standing for Asteroid Gravity, Optical, and Radar Analysis), the craft is to carry an advanced ion-propulsion system powered by a pair of enormous solar arrays. This propulsion system will make it possible for the craft to rendezvous with Vesta and enter orbit around it. The craft will be equipped with television cameras and other instruments for a detailed study of the surface of this world. After the Vesta study is complete, the Agora spacecraft could be directed to encounters with other asteroids. The mission is tentatively scheduled for launch in 1993, although no long-range funding has yet been approved.

PART THREE

PLANETS OF LIQUID AND GAS:
The Jovian giants

Voyager 2 close-up view of Saturn's A ring. (Photograph courtesy of NASA/JPL)

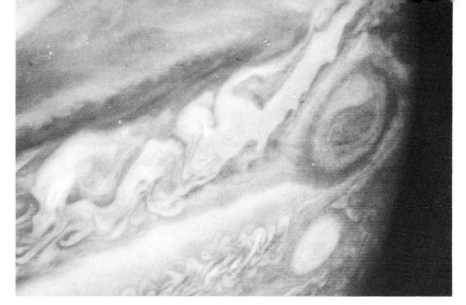

10
Jupiter — a Giant among Worlds

Jupiter is by far the most massive of the planets. Excluding the Sun, 35 percent of all the matter in the Solar System is contained within the body of Jupiter. Jupiter's mass is 318 times Earth's, but it is still only 0.001 that of the Sun. Jupiter lies in a nearly circular orbit that averages 5.2 AU from the Sun, over 3 times farther out than Mars. Jupiter is particularly interesting because its internal structure is completely different from that of the four inner planets. Unlike Mercury, Venus, Earth, or Mars, the giant planet Jupiter is not a solid body. It is actually made up mostly of liquids and gases (Figure 10.1).

In December of 1973 the American spaceprobe *Pioneer 10* flew past Jupiter, passing within 130,000 km of the cloud tops. After the Jovian encounter, *Pioneer 10* gained enough additional momentum to be expelled from the Solar System, never to return (Figure 10.2 *a* and *b*.) The virtually identical *Pioneer 11* passed within 42,800 km of the Jovian clouds a year later. The far more sophisticated Voyager probes followed in 1979. These probes increased our knowledge of Jupiter a millionfold.

THE STAR THAT FAILED

Because of its distance from the Sun, Jupiter is always seen very nearly in full phase from the Earth. It appears in the telescope as a slightly flattened disk, with

Voyager 1 photograph of the limb of Jupiter. The Great Red Spot can be seen near the limb. (Photograph courtesy of NASA/JPL)

FIGURE 10.1 *Voyager 1* photograph of Jupiter. This view shows the belts and zones as well as the Great Red Spot. (Photograph courtesy of NASA/JPL)

alternating light and dark bands running parallel to its equator. The equatorial diameter is 143,100 km, but the polar diameter is only 133,800 km (Figure 10.3). The cause of this oblateness appears to be the extraordinarily rapid rate at which Jupiter spins on its axis. A point on Jupiter's equator completes one turn every 9 hours, 50 minutes, making Jupiter the fastest spinning of any of the planets. Jupiter's equator makes an angle of only 3° with the plane of its orbit, so seasonal effects should not be important.

Different parts of the planet appear, however, to be spinning at slightly different rates, with polar regions turning slightly slower (by about 1 part in 120) than are equatorial areas. Because of the differential rotation, Jupiter cannot possibly be a solid body like the Earth. It must instead be made up primarily of more fluid substances such as liquids or gases.

The average density of Jupiter is only 1.4 g/cm³, much less than that of any of the inner planets. Jupiter must contain a high proportion of lighter elements rather than heavy metals or silicates. Theoretical calculations indicate that no elements other than hydrogen or helium could possibly give Jupiter such a low density at the enormous temperatures and pressures certain to be present in the interior of so massive a body. Jupiter is mostly hydrogen and helium, just like the Sun.

Jupiter receives less than 4 percent of the solar energy per unit area, as does the Earth. A body of Jupiter's albedo at that distance from the Sun should have an effective temperature of −165°C. However, it has been known since the late 1920s that Jupiter actually has an effective temperature that is significantly higher, the most recently measured value being −149°C, as determined during the Voyager mission. The high effective temperature indicates that Jupiter must have an appreciable internal energy source!

Detailed spacecraft measurements indicate that Jupiter radiates about 1.67 times as much energy into space as it receives from the Sun. Most of this excess energy is emitted in the infrared region of the spectrum between 10 and 100 μm in wavelength. The Earth has a solid mantle and crust that effectively traps its heat energy deep within the molten core, whereas Jupiter appears to have a largely liquid interior that permits convective heat currents to carry thermal energy from the interior out toward the surface. Theoretical calculations suggest that the energy source responsible for such intense internal heating is the slow contraction of the interior in response to the intense gravitational field. The planet is slowly cooling down from an initial high-temperature state.

Large as the internal heating is, Jupiter is still nine orders of magnitude less luminous than the Sun. Jupiter is sometimes called "the star that failed," as it seems to have approximately the same elemental composition as the Sun but is not nearly massive enough to produce interior temperatures sufficiently high to initiate thermonuclear fusion.

JUPITER'S INTERNAL STRUCTURE

Data acquired by the Pioneer and Voyager probes enabled a realistic model of the Jovian interior to be constructed (Figure 10.4). Both the pressure and the temperature in the interior steadily increase with depth and are high enough to maintain the hydrogen–helium mixture in all but the very outermost layers of the planet permanently in the liquid rather than the gaseous phase.

The outermost 30,000 km of Jupiter consists of a "mantle" made up of a mixture of liquid hydrogen and liquid helium. The liquid hydrogen in the mantle is in the molecular phase. It consists of a large number of hydrogen molecules, each composed of a pair of hydrogen atoms. Molecular liquid hydrogen is commonly manufactured in laboratories and factories on Earth.

However, at a depth of 30,000 km into Jupiter, something much more interesting seems to happen. According to theoretical calculations originally

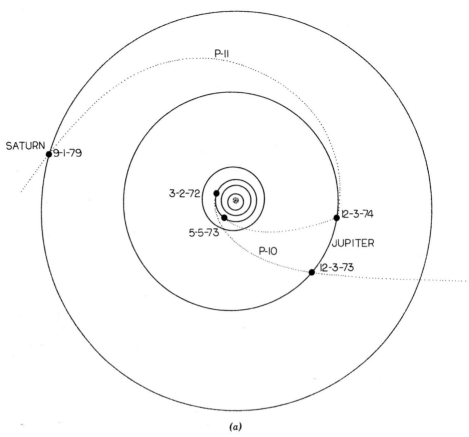

FIGURE 10.2 Spaceprobe missions to Jupiter and the outer planets. (a) Pioneer missions;
(b) Voyager probes.

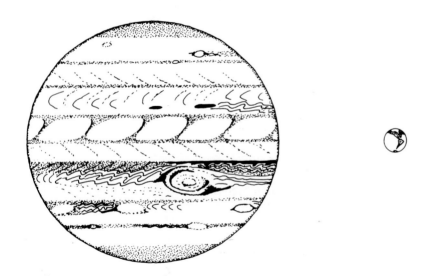

FIGURE 10.3 Relative sizes of
Jupiter and the Earth.

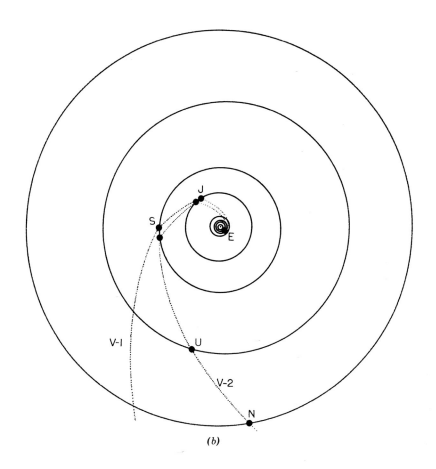

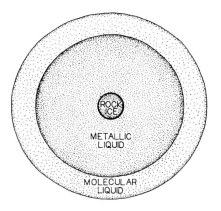

FIGURE 10.4   The interior of Jupiter.

performed by the physicists Edwin Salpeter and David Stevenson, an assembly of hydrogen molecules will break up into an electrically conductive mixture of protons and electrons when subjected to a pressure greater than 3×10^6 atm. A pressure of this magnitude is reached at a depth of 30,000 km inside Jupiter, where the temperature is 11,000°C. This electrically conductive, high-pressure phase of hydrogen is known as *liquid metallic hydrogen*. Liquid metallic hydrogen is the simplest of the alkali metals (also including sodium and potassium) but it can exist only at such elevated temperatures and pressures that it has not yet been created in laboratories on Earth. The liquid-metallic-hydrogen phase of Jupiter probably extends nearly all the way to the center of the planet. The temperature and pressure in the lower reaches of the liquid-metallic-hydrogen shell may be as high as 19,000°C and 3.6×10^7 atm, respectively.

Detailed mathematical models of the Jovian interior seem to indicate that the very innermost region of Jupiter is significantly denser than the rest of the planet. It is too dense, in fact, to be composed exclusively of hydrogen or helium, even at such elevated pressures. It must be rich in denser substances such as metals, silicates, or perhaps methane, ammonia, and water. This dense core comprises about 9 percent of the Jovian mass (which corresponds to about 28 Earth masses). If Jupiter had exactly the same composition as the Sun, the core should be only about 2.5 percent of the total mass.

Some theorists studying planetary origins have suggested that the dense core presently found at the center of Jupiter is the result of an initial accumulation of a large number of rocky and/or metallic planetesimals in the Jovian region at the time of the formation of the Solar System. This accumulated mass happened to be large enough to induce a catastrophic instability in the surrounding protosolar nebula, causing a massive amount of hydrogen and helium to collapse down to form the giant planet presently seen. Perhaps all Jovian-type planets are formed in this manner.

JOVIAN AIR

The part of Jupiter that can be seen from Earth consists of an enormously dense and thick atmosphere. This atmosphere is 81 percent hydrogen and 19 percent helium (by mass), approximately the same percentages as are found in the outer layers of the Sun itself. Jupiter and the Sun must have formed from the same material. Slight traces of methane (CH_4), ammonia (NH_3), water vapor (H_2O), phosphine (PH_3), germane (GeH_4), ethane (C_2H_6), and acetylene (C_2H_2) have been identified by the appearance of their characteristic spectra in the light reflected from the planet. The ethane and acetylene are probably not indigenous to Jupiter, but instead were created by the ultraviolet disruption of methane in the upper atmosphere.

Most pre-Voyager models of the Jovian atmosphere had assumed that the concentrations of ammonia, methane, hydrogen sulfide, and water vapor in the air would be in approximately solar proportions to the dominant hydrogen and helium. Remote-sensing measurements by the Voyager spacecraft indicate that this is not precisely the case. These measurements reveal that the Jovian atmosphere has at least twice the relative amount of carbon as the solar photosphere does. However, the Jovian O/H abundance ratio is less than 1/30 of the solar value. The N/H abundance ratio does seem to approximate the solar ratio in the upper regions of the atmosphere but is probably appreciably larger at lower depths, where some ammonia may be trapped in clouds in the form of NH_4SH or $NH_3 - H_2O$. And there seems to be absolutely no amount of H_2S present at all.

Why these differences? The enhancement of carbon in the Jovian atmosphere over the solar value can be used to argue for the initial formation of the planet from a gradual accretion of icy and/or rocky planetesimals rather than by a direct condensation from the substances within the solar nebula. Most of the ammonia, methane, and water presently inside Jupiter condensed onto the dense rocky core at a time before the final massive collapse of the giant hydrogen–helium "envelope." Over the years, much of the methane in the core has diffused outward into the liquid hydrogen mantle, but most of the water remains permanently trapped within the rocky core by its inability to pass through the layer of liquid metallic hydrogen. The elusive Jovian hydrogen sulfide may be present farther down in the atmosphere, hidden in the dense clouds.

When the Pioneer and Voyager spacecraft flew past Jupiter, they passed behind the planetary disk as seen from Earth, and their radio signals momentarily grazed the top of the atmosphere. By observing the fading of the signals, the temperature and density profile of the atmosphere could be determined. The atmosphere was so thick that the radio signal could only pass through the outermost 80 km.

The vertical profile of the upper Jovian atmosphere is shown in Figure 10.5. The upper atmosphere has two distinct components, a stratosphere and a

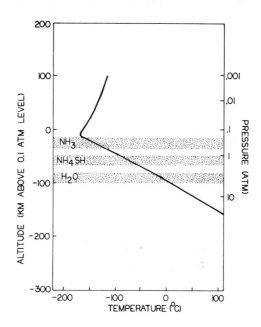

FIGURE 10.5 Vertical profile of the Jovian atmosphere.

troposphere separated by a tropopause. The Jovian tropopause lies at the level where the pressure is 0.1 atm and the temperature is $-160°C$. In the stratosphere, the temperature steadily increases with altitude, reaching a temperature of $-120°C$ at a pressure of 0.03 atm. At an altitude of 400 km above the tropopause, the temperature is $-70°C$ and the pressure is 10^{-8} atm. At 1400 km, the temperature is 830°C and the pressure is 10^{-10} atm. Below the tropopause, both the temperature and the pressure steadily increase with depth. The temperature increases by 2°C for every kilometer of descent, the limit of detection being reached at about 30 km below the tropopause, where the pressure was 0.6 atm and the temperature was $-120°C$. The difference between the polar and equatorial upper-atmosphere temperatures is less than 3°C (on Earth it is about 30°C), probably because of the heat energy flowing outward from the interior of the planet. There is no significant cooling of the atmosphere at night.

It is not at all certain whether or not Jupiter has any sort of well-defined surface beneath its massive atmosphere. The hydrogen-rich material in Jupiter has a temperature well above the critical point all the way down to the core. Some models of the Jovian interior predict that the gaseous atmosphere slowly and continuously blends into a more dense liquid phase as one descends, with no clearly defined interface existing between the two. However, some alternative models do predict a definite liquid-hydrogen surface beneath Jupiter's atmosphere, with a sharp density increase 1000 km below the tropopause at the point where the temperature and pressure are 2000°C and 500 atm, respectively. A liquid-hydrogen "ocean" may lie at this level, locked in perpetual darkness under crushing pressures and broiling temperatures.

It is highly unlikely that any spacecraft can be built in the near future that could actually reach such a liquid-hydrogen surface and survive long enough to send back any useful data. However, NASA plans to launch a Jovian atmospheric-entry probe (named *Galileo*), which will descend into the lower Jovian atmosphere suspended beneath a parachute. As the entry probe descends into the atmosphere, the data that are accumulated will be transmitted to a companion "bus" spacecraft for relay back to Earth. The probe is expected to have a useful life of only 1 hour in the Jovian atmosphere, probably failing at a depth of 130 km below the tropopause, where the pressure is 20 atm and the temperature is 170°C. After the mission of the entry probe is completed, the bus spacecraft will enter Jovian orbit to carry out a 20-month survey of the magnetosphere and the system of moons.

The *Galileo* mission was originally scheduled for launch from the Space Shuttle *Atlantis* in May of 1986. However, the *Challenger* disaster and the subsequent cancellation of the Centaur upper stage has forced a postponement until 1990 at the earliest.

JOVIAN WEATHER—
THE BELTS AND ZONES

There is a complex layer of thick clouds in the upper atmosphere. The Pioneer and Voyager flyby spacecraft unfortunately were not able to provide much information about the chemical composition of these clouds. However, if the Jovian atmosphere is even roughly similar to the solar photosphere in atomic composition, the most abundant condensible vapors ought to be water, ammonia, and hydrogen sulfide. In the cold uppermost reaches of the atmosphere, H_2O and NH_3 vapors should condense to form clouds composed of crystals of water ice and frozen ammonia. The hydrogen sulfide should react with ammonia to form solid crystals of ammonium hydrosulfide (NH_4SH).

Mathematical models of the Jovian atmosphere based on this presumed cloud composition predict that there should be an outer layer of ammonia-rich clouds about 30 km below the tropopause, where the temperature is $-120°C$. A layer of NH_4SH clouds is predicted at a level 60 km below the tropopause, at a temperature of $-75°C$. Finally, there should be a layer of water clouds 100 km below the tropopause, where the temperature is $+10°C$.

One problem with these cloud structure models is the lack of any concrete evidence for hydrogen sulfide in the Jovian atmosphere. It may, however, be located farther down in the atmosphere, hidden from view. Another problem is the apparent "dryness" of the Jovian atmosphere. One can occasionally see down through "holes" in the upper cloud deck to an atmospheric temperature of $+30°C$, where the water clouds should be, but infrared spectroscopy provides little evidence for their presence. Perhaps most of Jupiter's water got permanently trapped deep within its rocky core at the time of formation. Any reliable details about the Jovian cloud composition will probably have to await the arrival of *Galileo*.

The Jovian upper-atmospheric cloud deck exhibits a rich pattern of colors. The clouds have various hues of red, brown, and blue, as well as white. The colors vary roughly with height in the atmosphere. The highest clouds are reddish in color and lie about 10 to 30 km below the tropopause. Below these are the white clouds, lying at a depth of 20 to 50 km. Farther down are the brownish clouds, lying 40 to 80 km below the tropopause. The lowest layer of clouds that can be seen are blue in color and are at depths of 100 km or greater below the tropopause.

The substances in the clouds responsible for giving them their characteristic colors are uncertain, as all of the suspected cloud condensates (NH_3, H_2O, NH_4SH) are white. One possibility is that elemental sulfur is present. This substance is known to form several different phases that have widely different colors, depending in a complex manner on temperature. The sulfur could have been produced by the photodecomposition of H_2S, which has so far escaped detection in the Jovian atmosphere. Another candidate is phosphorus. Ultraviolet disruption of phosphine could have produced elemental phosphorus or various inorganic phosphates, which could give the clouds their characteristic colors. Organic molecules produced by the photodecomposition of ammonia, methane, or water vapor in the upper atmosphere are a third possibility. There may even be enough organic molecules in the upper Jovian atmosphere to sustain some sort of life there, although most workers believe this to be rather unlikely. The *Galileo* atmospheric-entry spaceprobe may provide the answer.

The Jovian atmosphere shows violent weather patterns, as revealed by the motions of the parallel bands that encircle the planet. The darker bands are called *belts,* whereas the lighter bands are called *zones.* They are sufficiently permanent to have been given names and extend as far as 60° north and south of the equator. The zones are regions of rising, warmer air, whereas the belts are made up of cooler, descending air. The light color of the zones is probably caused by clouds of frozen ammonia and water ice crystals that hover in the upper atmosphere. The belts often exhibit a reddish or brownish color. A schematic view of the pattern of Jovian belts and zones is shown in Figure 10.6.

The positions of the belts and zones correlate very well with a system of eastward- and westward-blowing winds in the upper atmosphere parallel to the equator. These winds can be tracked by carefully watching atmospheric features as they move across the planetary disk. Near the equator there is a band of strong winds blowing toward the east that reaches speeds as high as 100 m/sec. This band extends from the equator to points 15° north and south. Farther from the equator the winds are weaker (averaging about 20 m/sec) and break up into a pattern of alternating eastward- and westward-blowing parallel bands. The two Voyager spacecraft noted a pattern of lightning bolts on the dark side of the planet. The lightning appears in clusters of bolts that seem to be fairly uniformly distributed over the planet, indicative of a rather violent weather pattern within the Jovian atmosphere.

The belts and zones maintain a more or less constant appearance over time, but there are some transient phenomena. The best known of these is the

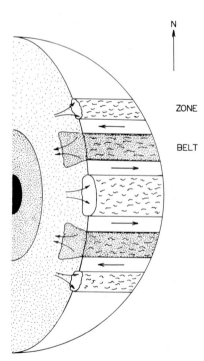

FIGURE 10.6 The belts and zones of the Jovian atmosphere. In this view, the depth of the upper atmosphere is grossly exaggerated in relation to the size of the planet as a whole.

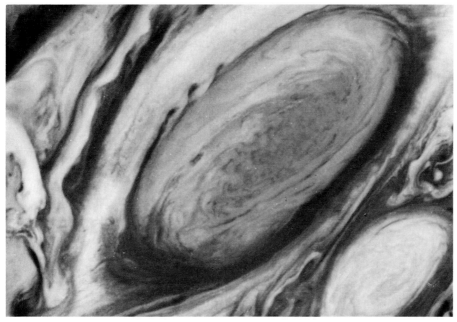

FIGURE 10.7 *Voyager 2* photograph of the Great Red Spot. (Photograph courtesy of NASA/JPL)

Great Red Spot, which is centered at about 20°S (Figure 10.7). It is about 50,000 km across and has been observed in the telescope for over 300 years. The Spot changes in color and in size with time and has drifted irregularly westward over the years. At one time it was imagined that the Great Red Spot was produced by some sort of volcanic disturbance, but its drift probably means that it is not connected with anything farther down in the Jovian interior. The Spot rotates counterclockwise with a period of about 6 days, and the temperature at its top is substantially colder than the surrounding regions. The top of the Great Red Spot exceeds the altitude of the highest clouds in the surrounding regions; it is probably a gigantic free-wheeling atmospheric disturbance that lies between the counterflowing north and south boundaries of a zone, more or less analogous to a terrestrial hurricane but much, much larger and longer lasting. Unless there is some sort of dissipative mechanism, the Great Red Spot could last forever.

Andrew Ingersoll and Friedrich Busse have proposed that the pattern of winds seen in the upper Jovian atmosphere actually extends 30,000 km into the interior of the planet, probably all the way to the outer boundary of the liquid-metallic-hydrogen core! This model is based on experiments performed on Earth with rapidly rotating fluids, where it was found that small-scale turbulent motions tend to align themselves in coaxial cylindrical shells parallel to the rotational axis, with each of the cylinders rotating at a slightly different velocity. The steady eastward and westward winds seen in the upper Jovian atmosphere may be nothing less than the outward signs of the rotational motion of giant coaxial cylinders of liquid hydrogen that extend virtually all the way through the interior (Figure 10.8). This motion is driven by the enormous flow of heat from the interior produced by the slow but steady gravitational compression of the planet. If this rotating-cylinder model is correct, there is unlikely to be any definite gas–liquid interface that could define a surface for Jupiter.

On the other hand, Gareth Williams has introduced a competing model that proposes that Jupiter's atmospheric flow pattern is a lot like Earth's, with a relatively thin (1000-km) hydrogen–helium atmosphere and a definite liquid-hydrogen surface. The pattern of belts and zones exists only in the very uppermost part of the atmosphere and is driven by the strong convective heat currents originating in the Jovian interior. When the rising hot air in the zones reaches the top of the clouds, it cools and spreads out toward the poles. However, the Coriolis force exerted on the air as a result of rapid rotation of the planet causes the poleward-flowing air to change course. In the northern hemisphere of Jupiter, northerly flows are turned toward the east and southerly flows to the west. The direction is reversed in the southern hemisphere. Without a solid surface, there is nothing to dissipate the winds, and the atmospheric circulation pattern in the upper Jovian air quickly spreads out to cover the entire circumference of the planet.

JUPITER'S ENORMOUS MAGNETOSPHERE

Jupiter has an enormous magnetic field, much larger and more intense than Earth's. The magnetic intensity at the cloud tops is about 4 gauss, 10 to 20 times stronger than the field at the surface of the Earth. Field lines extend outward as far as 2.0×10^7 km from the planet. Because of its extensive size, the Jovian magnetic field contains 400 million times as much energy as Earth's does. A schematic view of the Jovian magnetic field is shown in Figure 10.9.

Jupiter's magnetic field is probably produced by convective electrical currents within the liquid-metallic-hydrogen core. It has two distinct parts, an inner and an outer region. The inner field extends from the cloud tops out to

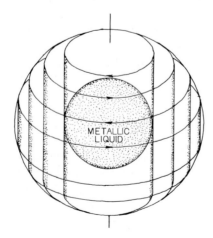

FIGURE 10.8 Rotating-cylinder model of the Jovian interior.

METALLIC LIQUID

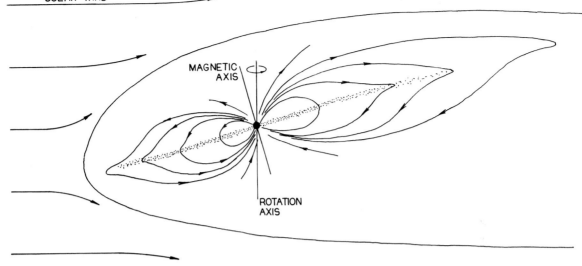

SOLAR WIND

MAGNETIC AXIS

ROTATION AXIS

FIGURE 10.9 Jupiter's magnetosphere.

about 2.3×10^6 km from the planet. The center of the field is offset from the center of Jupiter by 7000 km, and the dipole axis makes an angle of 11° with the equatorial axis. As a result, the field pattern appears to bob up and down by an angle of 22° in latitude as the planet rotates. The outer field is considerably weaker and is more nearly aligned with the rotational axis. Its extent varies with the strength of the solar wind. When the Sun is in the active part of its cycle, the solar wind is strong enough to push the outer boundary of the magnetosphere as close as 6×10^6 km from the planet. When the Sun is quiet, Jupiter's magnetosphere can extend as far as 2.0×10^7 km from the planet.

The Jovian magnetic field traps large numbers of charged particles in radiation belts that look much like Earth's Van Allen belts. The particle density pattern conforms to the geometry of the Jovian magnetic field, with the radiation intensity being strongest in the plane of the magnetic equator and declining rapidly at higher latitudes. Most of the particles are high-energy electrons and protons, although there are some helium, carbon, oxygen, and sulfur ions present as well. The radiation is at its peak intensity at a distance of 286,000 km from Jupiter, where the particle flux can be as high as 10^8 particles/cm² sec. This is many thousands of times greater than the radiation intensity in Earth's Van Allen belts.

During its flyby mission, the *Pioneer 11* spacecraft approached within 42,800 km of the Jovian clouds and received a radiation dose 100 times the amount that would have been fatal to a human being. The radiation was so intense that some of the more delicate electronic components aboard the craft were permanently damaged. It is virtually certain that the high radiation intensity will prevent astronauts from ever entering the inner regions of the Jovian magnetosphere.

Where do the particles in the Jovian radiation belts come from? The particles appear to come from three distinct locations. The carbon and helium ions were originally solar wind particles that were trapped in the Jovian magnetosphere. Most of the hydrogen ions probably came from hydrogen molecules in the Jovian upper atmosphere that were photoionized by the absorption of ultraviolet light from the Sun. The oxygen and sulfur ions probably came from the moon Io. The ions were perhaps stripped of the exposed surface of the moon by the intense radiation in the inner reaches of the Jovian magnetosphere.

Whatever their origin, once the ions enter the magnetosphere, they are constrained to spiral along the magnetic lines of force. Since the magnetic field rotates with the planet, the entire charge distribution rotates as well. There is thus an enormous "sheet" of electric current in the plane of the magnetic equator that corotates with Jupiter.

Where do the particles in the radiation belts get their enormous energies? It now appears that the agent responsible for the high particle energies is the solar wind. The Jovian magnetic field deflects the solar wind as it passes the planet. As the planet spins on its axis, the wobbling, off-center magnetosphere "rubs" up against the solar wind. The frictional energy thus generated is transferred to the ions spiraling along the magnetic field lines. In the outer magnetosphere, however, the magnetic field is sufficiently weak to permit some of the particles to spill out and flow away from the planet, flooding nearby space with radiation. Spacecraft have been able to detect radiation leaking from Jupiter at distances as large as 0.3 AU from the planet.

Jupiter is a strong source of radio emission. Radio waves coming from the planet were first detected in 1955 by Bernard F. Burke and Kenneth L. Franklin, although they had actually been observed for several years earlier than that without being recognized. The total amount of radio power emitted by Jupiter is about 40 GW.

Radio emissions that take place above a frequency of 10 GHz (1 GHz = 10^9 Hz) are produced by molecular motion in the outer Jovian atmosphere. Radio noise in the 0.4- to 10-GHz range comes from the radiation that is emitted by high-energy electrons as they spiral along the magnetic lines of force in the Jovian magnetosphere. The intensity of this radio emission varies as Jupiter rotates on its axis. In fact, the period with which this emission varies can be used as an accurate measure of the "true" rotational period of the planet's interior, where the magnetic field presumably originates. The "radio period" of Jupiter turns out to be 9 hours, 55.5 minutes, just slightly longer than the equatorial rotational period. Jupiter's outer atmosphere rotates just a bit faster than its interior does, producing a net eastward-moving circulation at the equator that could actually extend as far as 30,000 km into the interior. Perhaps the most interesting radio emission of all is in the 10- to 50-MHz range. This radiation is emitted in the form of massive bursts lasting an hour or more. It appears to be stimulated by electrical discharges that take place in the Jovian ionosphere and inner magnetosphere. Most surprising of all, the timing of the bursts seems to correlate with the motion of the large Jovian moon Io.

11

Saturn — the Jewel of the Solar System

Of all the planets in the Solar System, it can probably be said that the giant ringed planet Saturn (Figure 11.1) is the most beautiful.

Saturn is approximately twice as far (10 AU) from the Sun as is Jupiter. Apart from the rings, Saturn is similar in size and overall appearance to Jupiter. Like Jupiter, Saturn is noticeably flattened at its poles (the equatorial and polar diameters are 116,400 and 104,000 km, respectively; Figure 11.2). Saturn also rotates very rapidly about its axis, a point on the equator making a complete turn once every 10.2 hours.

There are, however, some significant differences. Saturn is only 30 percent as massive as Jupiter, so it has a much smaller density (0.71 g/cm^3). Like Jupiter, Saturn is composed of lighter elements such as hydrogen and helium, although there must be appreciable differences in the internal structures of the two planets to account for the marked density difference. Saturn also differs from Jupiter in having an appreciable equatorial inclination ($27°$) with the plane of its orbit. Seasonal effects may therefore be important on Saturn.

Until a few years ago, the facts and figures just listed were about all that could be stated with certainty about the planet. In September of 1979 the

Voyager 2 photomosaic of Saturn. Three inner satellites are visible as small dots of light just beneath the main disk of the planet. The shadow of a fourth is visible on the main disk of the planet just beneath the outer edge of the ring system. (Photograph courtesy of NASA/JPL)

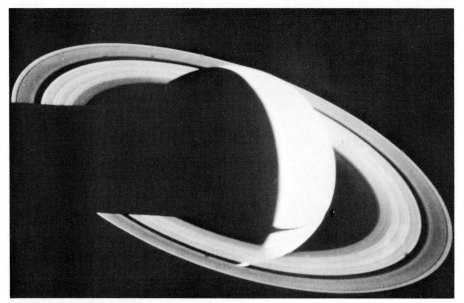

FIGURE 11.1 *Voyager 1* photograph of Saturn. This photograph was taken by the spacecraft after it had passed the planet. It shows a crescent Saturn casting a shadow on the rings. The C ring is quite transparent, since the outline of the planet can be seen through it. The Cassini division between the B and A rings is clearly visible. (Photograph courtesy of NASA/JPL)

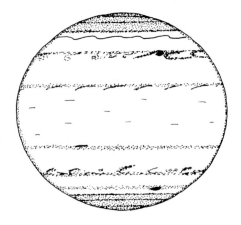

FIGURE 11.2 Relative sizes of Saturn and the Earth. In this view, the rings of Saturn are seen edge-on and are hence invisible.

spacecraft *Pioneer 11* flew past Saturn on its way out of the Solar System. It was followed by the much more sophisticated Voyager probes in 1980 and 1981 (see Figure 10.2). These missions increased knowledge about the Saturn system a millionfold.

THE INTERIOR OF SATURN

These spaceprobes provided enough data so that a realistic model of the internal structure of Saturn could be constructed. The interior is a lot like Jupiter's, with a hydrogen-rich mantle and a dense rocky core (Figure 11.3).

Saturn appears to have a dense central core of about 19 Earth masses, with a radius of 12,000 km. The average density of the core is about 19 g/cm^3. The composition of the core is uncertain but it is suspected to be approximately 25 percent metals and silicates and 75 percent water, methane, and ammonia. The pressure and temperature at the center of Saturn are estimated to be of the order of 5.0×10^7 atm and 15,000°C, respectively. Like Jupiter, the planet Saturn may be the product of a catastrophic collapse of large amounts of hydrogen and helium from the solar nebula induced by the initial accumulation of this massive core.

The core is surrounded by a much less dense 17,000-km-thick layer of liquid metallic hydrogen. The outer 30,000 km of Saturn is liquid molecular hydrogen. Saturn's liquid-metallic-hydrogen layer is appreciably thinner than Jupiter's, and the liquid molecular layer is proportionally thicker. The relative shortage of liquid metallic hydrogen within Saturn as compared to Jupiter is undoubtedly a result of the appreciably smaller mass and the correspondingly smaller amount of internal compression.

Like Jupiter, Saturn has an appreciable internal energy source and gives off to space 3 times as much energy as it receives from the Sun. The effective temperature of Saturn is −179°C, about 11°C warmer than it should be for a body at that particular location heated only by absorbed sunlight. The mass of Saturn is too small for Jovian-style gravitational contraction to be the major source of the internal energy. The most likely candidate for the source of Saturn's excess heat is the slow sinking of outer-mantle helium into the interior of the planet. Detailed calculations indicate that the temperature of the interior

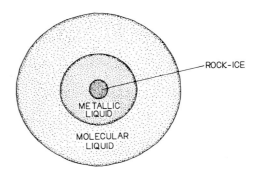

FIGURE 11.3 The interior of Saturn.

could actually be low enough to permit Saturn's hydrogen and helium to separate, with the helium falling inward toward the planet's center and releasing its gravitational potential energy in the form of heat.

THE AIR OF SATURN

Saturn's atmosphere is much like Jupiter's, with the dominant components being hydrogen and helium. There are alternate light and dark bands parallel to the equator that are produced by dense cloud layers high in the atmosphere. These bands are sufficiently permanent to have been given names, but they do not have as much fine structure and contrasting color differences as do those on Jupiter. Above the clouds the atmosphere is 11 percent (by mass) helium, compared with 19 percent on Jupiter. Most likely, Saturn does not contain proportionally less helium than does Jupiter; the helium depletion in the upper atmosphere may be the result of the presumed separation of hydrogen and helium within the Saturnian upper mantle, with much of Saturn's helium having sunk to the center of the planet.

Ammonia and methane gas are present, but their abundance is less than 1 part per 1000. As on Jupiter, the atmosphere of Saturn is enhanced in both nitrogen and carbon over the solar abundances of these elements. Oddly enough, water vapor has so far escaped detection in the Saturnian atmosphere. Like Jupiter, Saturn also seems to lack any measurable amount of hydrogen sulfide (H_2S). Both of these gases are almost certainly present on Saturn; perhaps they lie much farther down in the atmosphere, perpetually hidden below the dense upper clouds. Gases such as phosphine (PH_3), ethane (C_2H_6), and acetylene (C_2H_2) are present only in trace amounts of 1 ppm or less. Methylacetylene (C_3H_4) and propane (C_3H_8) may also be present.

Radio-occultation measurements by the Pioneer and Voyager spacecraft have made it possible to establish a temperature–pressure–depth profile for the upper atmosphere of Saturn (Figure 11.4). The atmospheric profile is quite similar to that of Jupiter. There is a troposphere and a stratosphere separated by a tropopause. The tropopause is at a temperature of −190°C and a pressure of 0.07 atm. Below the tropopause both the temperature and pressure steadily increase with depth, reaching a value of −130°C at the depth where the pressure is 1.2 atm (the lowest point reached by the radio-occultation measurements). The rate of temperature increase in the upper troposphere is 0.85°C for every kilometer of descent. The temperature of the tropopause in the southern hemisphere appears to be about 10°C warmer than it is in the northern, which may be a seasonal effect produced by Saturn's equatorial tilt.

The upper atmosphere of Saturn is significantly colder than that of Jupiter. On Saturn, a given troposphere temperature is reached at a higher pressure (greater depth in the atmosphere) than it is on Jupiter. For example, the −120°C level is at a pressure of 0.7 atm on Jupiter but at 1.4 atm on Saturn. As a result, the cloud layers lie deeper in the atmosphere on Saturn than they do on Jupiter and have a blander appearance when observed from above. Saturn's clouds lie

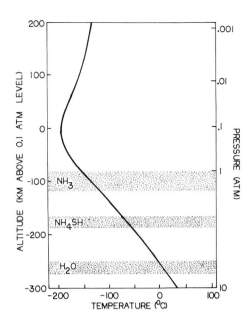

FIGURE 11.4 Vertical profile of Saturn's atmosphere.

deeper in the atmosphere at the poles than they do at the equator, apparently because of a lower temperature.

As with Jupiter, it is unclear whether or not Saturn has a definite liquid surface lying below its atmosphere. The atmosphere may simply blend smoothly and continuously into the liquid "mantle" of the planet as one descends, with no clear and distinct place where atmosphere ends and surface begins.

SATURN'S WEATHER

The composition of the Saturnian clouds is uncertain, but in all likelihood they have a chemical composition similar to that presumed for the Jovian clouds. Mathematical models of Saturn's atmosphere predict an ammonia cloud deck beginning 110 km below the tropopause, an intermediate NH_4SH cloud layer 180 km deep, and water clouds at a depth of 275 km. A significant flaw in the model is the current lack of evidence for the presence of either water or hydrogen sulfide in the Saturnian atmosphere.

The atmospheric weather patterns on Saturn are roughly similar to those on Jupiter, but there are significant differences. On Saturn there is a broad band of eastward-moving equatorial winds in the upper atmosphere that can reach speeds as high as 400 m/sec, 4 times faster than any winds on Jupiter. This band of winds extends as far from the equator as 40° north and south before the wind speed drops to 0. Unlike in the Jovian atmosphere, the pattern of equatorial winds appears to bear little relation to the banded cloud structure. The wind profile in fact resembles the flow patterns in the solar photosphere more than it does the pattern of belts and zones on Jupiter. The atmosphere does appear to break up into a pattern of alternatively eastward- and westward-blowing winds at latitudes higher than 40°, but features that appear along the boundaries between belts and zones on Jupiter are usually seen in the middle of zones on Saturn.

Saturn's atmosphere is much less colorful than Jupiter's, apparently because the clouds lie much farther down in the atmosphere. There is nothing nearly as dramatic as the Jovian Great Red Spot. There is a single 10,000-km-long oval spot with 100 m/sec circumferential rotation, but turbulent features any larger than 1000 km in size are rare.

These intense winds in the Saturnian atmosphere may be driven by condensing clouds of water and ammonia crystals high in the atmosphere. Perhaps convective heat currents originating from the interior also play a role. It is uncertain how far down into the atmosphere the pattern of strong winds extends. If the convective-rotation models proposed for Jupiter also apply to Saturn, the pattern of atmospheric circulation could extend all the way to the liquid metallic core of the planet.

SATURN'S MAGNETOSPHERE

Like Jupiter, Saturn has a magnetosphere, but it is not nearly as large and intense. The magnetic-field intensity at the cloud tops is about 0.2 gauss, about 20 times less than it is on Jupiter. Saturn's cloud-top magnetic-field intensity is slightly less than Earth's surface field, but, since Saturn's field occupies so much more space than Earth's, the total amount of stored magnetic energy is 1000 times greater. The sunward side of Saturn's magnetosphere extends as far outward as 1.3×10^6 km from the planet, making it only one-third the size of Jupiter's.

Saturn's magnetic field probably originates from electrical currents within the liquid-metallic-hydrogen interior, but it behaves as if it came from a giant bar magnet placed within 1200 km of the geometrical center of the planet. Curiously, the magnetic axis is almost exactly parallel (to within 0.7°) to the rotational axis. Most of the other planets with magnetospheres have significantly off-center and nonparallel magnetic fields. The significance of Saturn's departure from this norm is unclear.

There is a system of radiation belts associated with the magnetic field. However, Saturn's ring system acts to absorb a large fraction of the charged particles, so the radiation intensity is far less than it is in the Jovian magnetosphere. Most of this radiation is confined to the region outside of the ring system, and there are dips in the particle intensities at distances corresponding to the orbits of Saturn's moons. Even though the intensity of the radiation in Saturn's magnetosphere is much less than it is in Jupiter's, it is still strong enough to be fatal to any astronauts who might fly through it.

There is a large-scale azimuthal electrical-current system in the equatorial plane, which travels in an eastward direction. This pattern of electrical currents extends outward as far as 10^6 km from the planet. The current densities are, however, an order of magnitude smaller than those found within the Jovian magnetosphere. There is a plasma torus of hydrogen and oxygen ions lying between 350,000 and 400,000 km from Saturn. The ions in this plasma torus probably originated on the icy surfaces of the moons Dione and Tethys. These ions were stripped from the surfaces of those moons and ejected into space by collison with the intense radiation present in the Saturnian magnetosphere. There is also a torus of neutral hydrogen atoms lying between 5×10^5 and 1.5×10^6 km from Saturn that extends 4×10^5 km above and below the equatorial plane. The large moon Titan may be an important source of atoms and ions for this region. In addition, there is a plasma sheet rich in hydrogen (and possibly oxygen) ions that corotates with the magnetosphere. It extends outward as far as 10^6 km from the planet.

Like Jupiter, Saturn is a strong source of radio emission. There are recurrent bursts of strongly polarized radio waves, with frequencies in the 10-kHz range. These bursts of radio waves correlate strongly with the orientation of Saturn's magnetosphere. In fact, the correlation is so strong that the variation in the radio emission can be used as an accurate measurement of Saturn's internal rotation rate. An internal rotational period of 10 hours, 39.4 minutes is inferred, slightly longer than the observed equatorial rotational period of 10 hours, 14 minutes. The interior of the planet spins at a slightly slower rate than the outer atmosphere does.

There are also radio bursts that correlate with Saturn's specific orientation with respect to the Sun, and there is a modulation that correlates with the motion of the moon Dione. In addition, there are unpolarized radio bursts in the 10-kHz frequency range that appear to originate from electrical discharges within the ring system itself.

THE EXQUISITE RING SYSTEM

No discussion of Saturn would be complete without a description of its strikingly beautiful ring system. The rings were first noticed by Galileo in 1610. He described Saturn as appearing to have a set of "ears" attached to the sides of its disk.

For many years thereafter, early astronomers debated the true nature of what they saw in the telescope when they looked at Saturn. Some of the descriptions were quite imaginative but were unfortunately all incorrect. The rings were not correctly described until the work of Christian Huygens (1629–1695) in 1655. He reported that the ring system is a thin, flat disk that is coplanar with the Saturnian equator. It is nowhere actually attached to the planet.

The actual structure of the rings remained controversial for a long time. Many astronomers thought the rings to be a solid body. The seventeenth-century Italian astronomer Giovanni Cassini (1625–1712) seems to have been the first to suggest that Saturn's ring actually consists of a large number of smaller objects that cannot be individually resolved in the telescope.

The proof of this assertion was long in coming. In the nineteenth century, the famous Scottish physicist James Clerk Maxwell (1831–1879) proved mathematically that tidal forces exerted by Saturn would rapidly tear apart any large rigid body at that distance from the planet. A solid Saturnian ring is a theoretical impossibility. If the ring actually is a thin sheet of particles encircling the planet, then each particle should travel around Saturn in its own Keplerian orbit. The ring particles farthest from Saturn should then be traveling more slowly than those closer in. Different parts of the ring will appear in the telescope to be rotating at different rates. The differential rotation of Saturn's ring system was finally confirmed by James Keeler in 1895, during a study of the reflectance spectra of the light emitted by the rings.

A DISK OF SNOWBALLS

Spectroscopic observations of reflected light by Earth-based astronomers have established that the ring particles are primarily made of water ice. Voyager color photographs of the rings show a distinctly reddish tinge, indicating that other substances must also be present. Likely candidates are iron oxide or carbon-rich material.

Earth-based radar observations of the rings performed by the 210-ft radio antenna of the deep space network at Goldstone, California have established that typical ring particles are no larger than a few meters across. Although the Voyager spacecraft did not come nearly close enough to the ring system to photograph any individual ring particles, radio-occultation and optical-reflectance-spectroscopy studies by the spacecraft were able to establish that there is a continuous distribution in particle sizes, ranging from chunks a few meters across down to grains only 1 micron ($1 \mu m = 10^{-6}$m) in diameter. The temperature varies across the width of the ring system, being as warm as $-190°C$ on the sunlit side and as cold as $-220°C$ on the dark side.

The ring system has three primary components that can be seen in a telescope from Earth (Figure 11.5). These are labeled by the letters A, B, and C. The innermost component (C) begins about 14,000 km from the cloud tops and is a good deal fainter and more transparent than the other two. The brightest ring component is labeled B; it starts at a height of 31,600 km and extends outward to 57,100 km above the clouds. The outermost ring (or A) is separated from the B ring by a 4500-km gap. This gap is known as the *Cassini division,* named for the famous Italian astronomer who discovered it in 1675. It can be seen from Earth with relatively small and inexpensive telescopes. The A ring is somewhat fainter than the B component and extends outward to 76,300 km above the clouds. There is a much less pronounced (320-km) gap within the A ring known as the *Encke division.* It is at an altitude of 73,200 km, near the outer edge of the ring. The Encke division can be seen from Earth only in the largest telescopes and has so far proven impossible to photograph. It was not until the year 1977 that it was possible to confirm its existence, by photoelectrically observing the eclipsing of the moon Iapetus as it passed behind the ring system.

There is an "atmosphere" of neutral hydrogen atoms extending 60,000 km above and below the ring plane. It also extends somewhat beyond the outer edge of the A ring. This cloud may originate from photodissociation of the water molecules in the icy ring particles, or it could be produced by the sputtering of the ring particles by the intense radiation in Saturn's magnetosphere.

Because of Saturn's 27° axial tilt, the rings can be seen edge-on from Earth

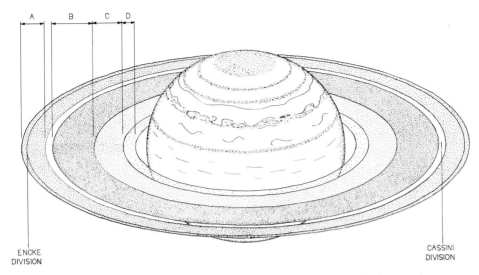

A B C D

ENCKE
DIVISION

CASSINI
DIVISION

FIGURE 11.5 Diagram of the main ring components of Saturn that are visible from Earth.

twice every Saturnian year. At that time, they are virtually invisible, indicating that they must be quite thin (less than 10 km thick). Occasionally, stars can be seen shining through them. Microwave measurements made by *Voyager 1* revealed that the ring system is actually even thinner than it appears from Earth. Ring A is 50m thick, whereas ring C is less than 10m thick! The discrepancy may be due to warps in the ring plane, or it may be due to the presence of tenuous material outside the primary ring system that make it appear from Earth to be much thicker than it really is.

RINGS WITHIN RINGS WITHIN RINGS

Spacecraft observations have been able to bring out much sharper detail in the ring structure. The three main rings A, B, and C appear rather uniform when observed from Earth, but upon closer examination, there is extensive structure. All three sets of rings are broken up into many hundreds of smaller ringlets, much like the grooves on a recording disk. Neither the Encke nor the Cassini division is entirely devoid of particles; they both have numerous narrow ringlets within them. Some of the narrow ringlets are clumped and discontinuous, and a few even have slightly eccentric shapes rather than being perfectly circular. The scale of ringlet structure extends downward to the limit of resolution of the Voyager measurements, which was of the order of a few hundred meters. The total number of discrete ring components may number in the hundreds of thousands (Figure 11.6).

FIGURE 11.6 *Voyager 1* photograph of the underside of Saturn's rings. On the dark side of the rings, diffuse regions of the ring system show up very brightly because of scattered light. Dark ring features are either areas of high optical density or nearly empty gaps. The thin outermost ring is the F ring. The broad dark band inside the F ring is the A ring. The Encke division is the thin dark line in the outer region of the A ring. The Cassini division is the bright band on the right. The B ring is the dark region on the far right. (Photograph courtesy of NASA/JPL)

The origin of the discrete ringlets is a controversial subject. It is commonly thought that gravitational forces exerted by the moons of Saturn have something to do with some of the features seen within the ring system. In particular, it is proposed that the motion of the moon Mimas is responsible for the Cassini division (Figure 11.7). A particle lying in an orbit at the outer edge of the Cassini division has a period precisely half that of Mimas. Any ring particle initially located in this part of the Cassini division experiences a perturbing force in the same direction every time Mimas is nearby, so that it is gradually nudged out of this orbit and into another one. In this way, the 2:1 resonance position is gradually swept clear of ring particles. Similar effects may explain the Kirkwood gaps in the Asteroid Belt.

However, resonance with the orbits of Saturn's moons cannot explain each and every one of the thousands of gaps actually found by the Voyager spacecraft. Careful measurement of the photographs did not turn up significant numbers of correlations between these moons' orbits and the many thousands of gaps present in the ring system. Some gaps did coincide with resonance points, but many other gaps did not. In search of a different explanation, it was then proposed that tiny "moonlets" embedded within the ring system were responsible for sweeping the gaps free of particles. *Voyager 2* (with a much more sensitive television camera) made an extensive search for such bodies when it flew past the planet a few months later. None were found, even though the spacecraft could have spotted any object within the ring system larger than 1 km across. The mystery continues.

MORE RING COMPONENTS

The Pioneer and Voyager spacecraft discovered additional smaller and fainter ring components that are invisible from Earth.

There is a very faint D ring lying between the inner edge of the C ring and the planet's atmosphere. Components of the D ring were seen as close as 6700 km to the cloud tops, but it is possible that some diffuse components could actually extend all the way to the top of the atmosphere. Even though several ground-based observers have claimed to have seen this ring in the telescope, it is probably unlikely that such a faint object could actually be seen from Earth up against the bright glare of Saturn's disk.

Pioneer 11 discovered a narrow ring only 500 km wide about 3700 km beyond the outer edge of the A ring. This narrow ring is designated by the letter

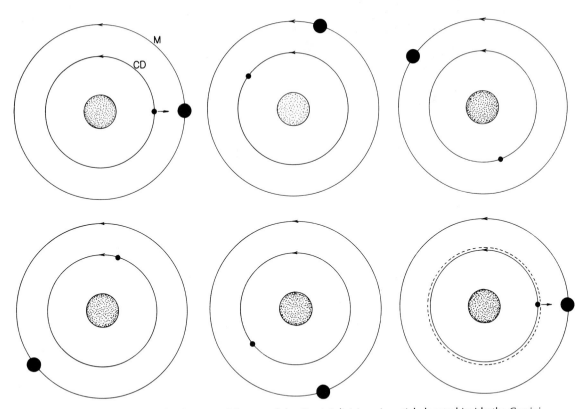

FIGURE 11.7 Resonance coupling between Mimas and the Cassini division. A particle located inside the Cassini division travels around Saturn exactly twice during the time that it takes for Mimas to go around once. This produces a periodic tug on the particle located within the gap, so that it is gradually nudged out of its original orbit.

F. The *Voyager 1* spacecraft's cameras were able to see much sharper detail, and found that the F ring actually consists of four or five distinct components, each about 70 to 100 km wide. Stellar-occultation measurements taken while the spacecraft flew past the planet showed that these F-ring components are further subdivided into many hundreds of thin strips, some as small as a few hundred meters across. Voyager photographs showed that the material within the F ring seemed to gather into distinct clumps at various locations around the circumference, at average spacings of 9000 km (Figure 11.8).

The structure of the F ring seems to vary with time. At the time of the *Voyager 1* encounter with Saturn, the outer pair of F-ring components appeared to be twisted or braided about each other. At the time of the *Voyager 2* flyby 9 months later, the braids had vanished and the number of distinct strands had changed. The cause of this transient behavior is uncertain.

There is a very faint ring component (designated by G) much farther out, at a distance of 110,000 km). It is so diffuse that it can be seen only in forward-scattered light. Finally, there is an even more diffuse ring component (termed E) ranging from about 150,000 km to about 240,000 km from the clouds. The E ring is centered near the orbit of the moon Enceladus and may actually have origi-

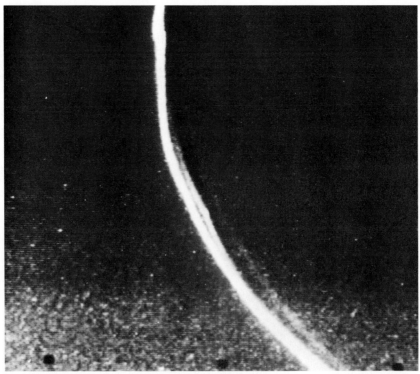

FIGURE 11.8 *Voyager 1* photograph of the F Ring. There are two components visible in this view, which appear to cross each other several times. Several bright clumps can also be seen. (Photograph courtesy of NASA/JPL)

TABLE 11.1 SATURN RING DATA

FEATURE	DISTANCE FROM CENTER OF SATURN (KM)	REMARKS
Cloud tops	60,330	0.1-atm level
D ring	67,000–73,200	Low optical density
C ring	73,200–92,200	
B ring	92,200–117,500	Spoke structure
Cassini division	117,500–121,000	Low optical density
A ring	121,000–136,200	
Encke division	133,500	200 km wide
F ring	140,600	Braided components
G ring	170,000	Very diffuse
E ring	210,000–300,000	Near orbit of Enceladus

SOURCE: Adapted from Stone and Miner (1981).

nated from particles of ice stripped off the surface of this moon. The various ring components are summarized in Table 11.1.

THE SATURNIAN WAGON WHEEL

The Voyager spacecraft found the central part of the B ring to have some rather odd transient spokelike markings. As the Voyager television cameras looked on, a radially oriented spoke would suddenly and mysteriously appear within the B ring, rapidly fan out and become tilted as the rings rotated, and then gradually fade away. A fresh spoke would appear within a time as short as 5 minutes and could be tracked for 360° or more around the planet before it finally disappeared. In some cases fresh radial spokes were found imprinted over older, more tilted ones. Oddly enough, the spokes arise without disturbing the underlying banded ringlet structure (Figure 11.9).

The B-ring spokes appear dark when seen in back-scattered light, bright when observed in forward-scattered light. The spokes must then be made up of particles no larger than 1 μm in size, since they preferentially scatter sunlight forward.

Although the true cause of the transient spokes is as yet unknown, most theories propose that they originate in some sort of interaction between the very smallest micron-sized ring particles and Saturn's rotating magnetic field. Saturn's magnetic field rotates in space at exactly the same rate as particles in the middle part of the B ring revolve around the planet. A newly formed spoke at first corotates with Saturn's magnetic field but it soon degrades into a more Keplerian orbital pattern as it becomes tilted and eventually fades out. Perhaps the finer particles in the B ring acquire an electric charge from the radiation in Saturn's magnetosphere, and a complex coupling between the charges and the magnetic field may be responsible for creating the transient spoke phenomena.

FIGURE 11.9 *Voyager 2* photograph of spokes in B ring. At the time that the photograph was taken, the spacecraft was 4×10^6 km from Saturn. A large number of spokes can be seen with different widths and tilts. The Cassini division is the dark band visible at the top. The bright band visible above the Cassini division is the A ring. The outer C ring is the dark area at the lower right. (Photograph courtesy of NASA/JPL)

Some radio bursts emitted by Saturn may actually originate from the spoke-forming regions of the B ring, indicating that some sort of electrostatic discharge is taking place there.

DISRUPTION, IMPACT, OR ACCRETION?

How did Saturn happen to acquire such an extensive ring system? There are three theories, which are roughly classified as *tidal disruption, meteoric collision,* or *gradual accretion.*

In 1848 the French mathematician Edouard A. Roche calculated that tidal forces exerted by Saturn would have disrupted any molten-liquid protomoon that happened to form any closer than 1.5 Saturnian radii from the surface of the planet. This critical distance, known as the *Roche limit,* is close to the outer edge of the present ring system. Roche proposed that a liquid moon initially located within this limit was torn apart by tidal forces to supply the ring particles. However, it is now suspected that Saturn never had a completely liquid moon at any time in its past, since all of its present moons are known to have been completely solid objects from the time of their formation.

Another possibility is that a large icy body from the outside happened to pass very close to Saturn and was tidally disrupted during the encounter. Several theoretical astronomers have performed calculations that demonstrated that no solid object smaller than 100 km in diameter could have been disrupted by Saturn's tidal forces, no matter how close it came to the planet. Even an appreciably larger body would have been tidally disrupted only if it came closer to Saturn than the very inner edge of the present ring system. So it is thought unlikely that the ring particles could have been produced by any sort of catastrophic disruption of a larger body that wandered into this region from the outside.

Meteoric impact theories have also been suggested. Perhaps one or more large icy moons formed in the region of space presently occupied by the ring system. These moons were subsequently fragmented into millions of tiny pieces by collisions with large meteoroids. The gravitational field of Saturn deflects the paths of passing meteoroids in such a way that there is a larger meteor flux within the inner parts of the Saturn system than in the outer. The meteor flux within the outer reaches of the Saturn system was sufficient only to scar the surfaces of the moons located there with numerous impact craters. However, the flux within the inner parts could have been high enough to shatter any moons initially present into millions of small chunks. Hence the ring system seen today.

Accretion theories are currently felt to have the best chance of giving a correct description of the origin of the rings. Saturn was initially a large spinning ball of hydrogen and helium gas intermixed with ice particles and silicate dust. At the time that proto-Saturn began to collapse to its present size, it separated into a roughly spherical inner part, which was destined to form the planet itself, and a flatter, disk-shaped outer part, which was eventually to form the system of rings and moons. As the outer parts of the disk cooled, the dust particles acted as nucleation centers for the condensation of water ice particles. Numerous collisions between particles as well as the viscous drag created by the hydrogen gas still present forced the disk to settle into a thin sheet of particles lying within the equatorial plane. Within the outer reaches of the disk the ice particles gathered into larger and larger clumps, eventually to form the large icy moons currently seen. Inside the Roche limit, however, tidal forces were strong enough to prevent the accretion of ice particles into larger bodies. Clumps of ice remained separate, forming the beautiful ring system seen today.

12

Jupiter's Moons — a Miniature Solar System

Jupiter has 16 known moons. The four largest (Io, Europa, Ganymede, and Callisto) can be seen from Earth by using a moderately good pair of binoculars. They could actually be seen by the naked eye were it not for the glare produced by nearby Jupiter. These four large Jovian moons were discovered by Galileo in 1610 and are commonly referred to as the Galilean satellites. Their unexpected discovery provided Galileo satellites. Their unexpected discovery provided Galileo with a miniature model of the heliocentric Solar System, with several smaller bodies orbiting about a single larger body. The four Galilean moons are large enough to be considered as worlds in their own right (Figure 12.1a and b). The other Jovian moons are much smaller and can be seen only through the largest telescopes.

The moons split quite naturally into groups of four apiece. The innermost four are quite small (three were discovered by the Voyager spacecraft) and lie in nearly circular orbits in the plane of Jupiter's equator. The next four are the large Galilean moons, also in nearly circular orbits in the plane of Jupiter's equator (Figure 12.2). The moons in the third group are all in rather eccentric orbits that make angles of about 30° with respect to the Jovian equator and move around Jupiter in the retrograde sense. Tidal coupling exerted by Jupiter's enormous

Voyager 1 photograph of Io, seen here up against the main disk of Jupiter. (Photograph courtesy of NASA/JPL)

(a)

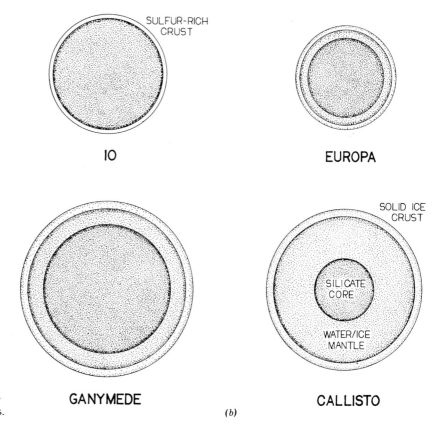

SULFUR-RICH
CRUST

IO

EUROPA

SOLID ICE
CRUST

SILICATE
CORE

WATER/ICE
MANTLE

GANYMEDE

CALLISTO

(b)

FIGURE 12.1 *(a) Voyager 1* composite photograph of the four Galilean moons of Jupiter, shown in their correct relative sizes. (Photograph courtesy of NASA/JPL) *(b)* A schematic view of the interiors of the four moons.

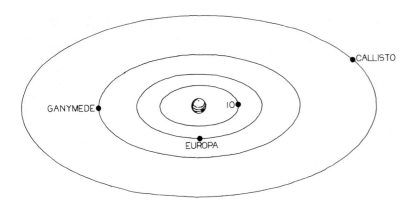

FIGURE 12.2 Orbits of the four Galilean moons of Jupiter.

gravitational field forces the inner moons to rotate in a synchronous manner, always turning the same faces toward the planet as they travel along their orbits. The outermost four moons are so tiny that virtually nothing is known about their rotational properties.

The Jovian moons have always been difficult to study in a telescope because of the interference of light from Jupiter itself. In 1979 the American planetary probes *Voyager 1* and *2* flew through the Jovian system, paying particular attention to the Galilean moons. Hundreds of detailed pictures of their surfaces were returned and many physical measurements were made. These worlds turned out to be far stranger than any science fiction novelist could have imagined.

IO—A WORLD TURNED INSIDE OUT

The innermost Galilean moon, Io (J1), is perhaps the single most interesting object in the entire Solar System. The diameter is 3630 km, which makes Io slightly larger than Earth's Moon. The mean density is 3.53 g/cm³. Io's orbital radius is 421,000 km, which places it in the more intense region of the Jovian radiation belts.

THE VOLCANOES OF IO

On its flight past Jupiter, *Voyager 1* paid especially close attention to Io, passing within 20,500 km of the surface. The photographs show Io to be a bright-orange-colored sphere, with irregular white splotches (Figure 12.3). The polar regions are a little darker. The average reflectivity is 60 percent. Surprisingly, the surface shows absolutely no impact craters. However, there are a large number of odd-looking sinkhole like depressions scattered over the surface. Over 100 of these sinkholes were spotted, a typical one being about 50 km across. Nothing similar had been found anywhere else in the Solar System, and their appearance on Io was at first entirely baffling.

Some weeks after the *Voyager 1* flyby, Linda Morabito, an engineer at the Jet Propulsion Laboratory in Pasadena, California, happened to be checking

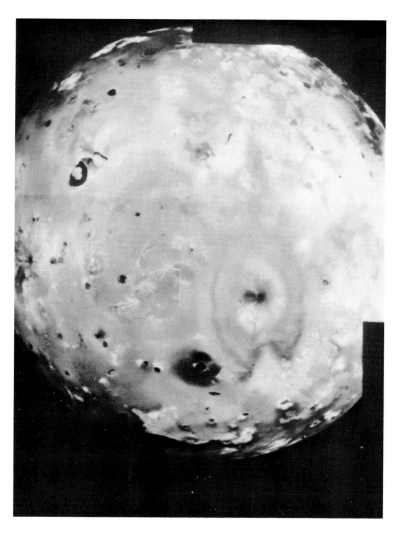

FIGURE 12.3 *Voyager 1* photomosaic of Io. North is at the top. This view shows the hemisphere that faces in a direction opposite to that of Io's motion along its orbit. The pear-shaped feature near the center is the top of the eruptive plume issuing from the active volcano Pele. The plume conceals the volcano and its related flow patterns from view. The dark spot seen southwest of Pele is Babbar Patera, the caldera of an inactive volcano. The dark, **D**-shaped feature in the northwest quadrant is Loki Patera, which is probably a lake of molten black sulfur. The bright region seen near the western limb is a region of extensive volcanic lava flows associated with Ra Patera. (Photograph courtesy of NASA/JPL)

some overexposed spacecraft photographs of Io to look for background stars when she spotted a large, bright form shaped like an umbrella at the moon's limb in one of the pictures. The umbrella was the plume of an active volcano! (Figure 12.4.)

The sinkholes on Io are the vents of volcanoes, of which at least eight are currently active. All but one of the active volcanoes lie within 30° of the equator. Eruptions take place on a continual basis and can last from a few months to a few years. Io is the only world in the Solar System besides the Earth known to have active volcanoes. The volcanoes have been given the names of deities associated with fire (Amirani, Loki, Marduk, Pele, Maui, Prometheus, and so on).

The umbrella-shaped plume is produced by the eruption of gaseous and particulate material from the volcano. The volcanic plume forms a gigantic

FIGURE 12.4 *Voyager 1* photograph in ultraviolet light of volcanic plume coming from Loki Patera. This photograph has been computer enhanced. The plume rises more than a 100 km above the surface. (Photograph courtesy of NASA/JPL)

fan-shaped cloud as it spreads out into the vacuum of space and falls back to the surface. Eruption velocities can be as great as 1 km/sec, and plumes as high as 270 km have been seen. It is thought that the gases emitted from the volcanic plumes are primarily sulfur dioxide (SO_2) and hydrogen sulfide (H_2S).

Beautifully colored flow patterns are seen radiating out from the volcanic sinkholes (Figure 12.5). The colors include black, yellow, red, orange, and brown. Much of the surface of Io is completely covered with this volcanic material. The reflectance spectrum of the volcanic lava quite closely matches the spectrum of elemental sulfur as well as the spectra of various sulfur-rich inorganic compounds. The sulfur in the lava may play the major role in producing the rich variations in color of the surface. Elemental sulfur can occur in several crystalline modifications that have strikingly different mechanical and optical properties, which vary in a complex manner with temperature.

There are many sinuous scarps and faults that have bluish-white patches lying among them. The bright patches may have been produced by volatile material leaking out of the interior and condensing. Some of this material may be SO_2 "frost" that settled back to the surface after the volcanic eruptions, but the rest may be solidified sulfur that was forced to the surface along cracks in the fault ridges. There are also extensive flat plains between the volcanic vents that may be composed of salts of potassium and sodium.

There are isolated mountains of up to 10 km in height in the polar regions, which suggest the presence of tectonic fracturing forces and the tilting of surface blocks. Some of the steepest relief present on Io may be evidence for an underlying silicate mantle, since a "mountain" made purely of sulfur probably could not support its own weight. These mountains may be the remnants of an ancient prevolcanic crust that is richer in silicate minerals than it is in sulfides.

The average daytime surface temperature on Io is $-145°C$, although temperatures as hot as $300°C$ have been measured near the active volcanoes. Nighttime temperatures get as low as $-190°C$. The active volcanoes have produced a thin SO_2 atmosphere, but the pressure is quite small, only about 1×10^{-7} atm.

Io does not appear to have any water ice on its surface, and there is probably not much water within the interior. The region of the Jovian system in

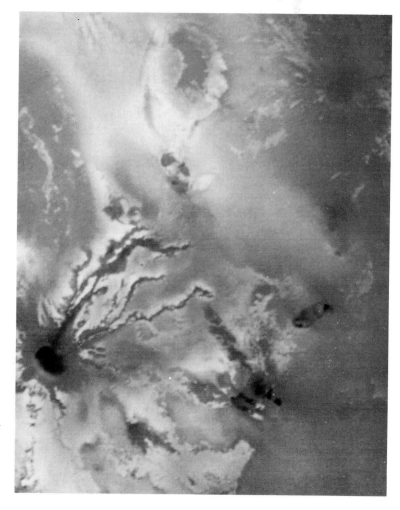

FIGURE 12.5 *Voyager 1* photograph of volcanic lava flows on Io. The dark spot on the left is Ra Patera, the caldera of a collapsed volcano. The total width of this view is approximately 1000 km. (Photograph courtesy of NASA/JPL)

which Io formed may have been too hot for any appreciable amount of water to condense, but not too hot for sulfur. Io's density is consistent with the presence of a rocky, metal-rich inner core and an outer crust of sodium and potassium salts that is rich in sulfur-containing compounds. Sulfur is actually rather common in the universe. In the Sun, sulfur is only a factor of 40 times less abundant than oxygen. As mentioned in Chapter 4, the relative scarcity of sulfur in Earth's crust is believed to be the result of the sinking of most terrestrial sulfur to the molten core very early in Earth's history.

The upper mantle of Io is probably almost entirely molten; it may have many pockets of liquid SO_2 as well as vast underground rivers of hot liquid sulfur. Heated by thermal energy originating from the interior, the liquid sulfur is forced upward towards the surface. When this molten sulfur comes into contact with liquid SO_2, the SO_2 is suddenly vaporized and forced upward through volcanic fissures until it vents to the surface. The SO_2 vapor quickly condenses into a solid when it is exposed to the cold of outer space, eventually falling to the ground and covering the surface with a thick blanket of frozen "snow." The excess molten sulfur from the interior then flows upward until it spills out of the volcano and covers the surrounding surface. A schematic view of the volcanic activity in Io's outer crust is shown in Figure 12.6.

Judging from the total absence of meteorite craters on Io, the volcanic activity is sufficiently intense to renew the surface once every few million years. The heat flow currently coming out of the Ionian interior is estimated to be 2 W/m^2, compared to only 0.06 W/m^2 for Earth. Most of the heat flow seems to come from the localized "hot spots" near the volcanic vents.

THE IO–EUROPA RESONANCE HEAT PUMP

What could be responsible for generating the heat required to drive such extensive volcanism? It is unlikely that a body of Io's low density (about that of Earth's Moon) could have had enough heavy radioactive elements in its interior to have produced such intense heat over so long a time. The answer may have been provided in a paper that appeared in *Science* magazine just a week before *Voyager 1* returned the first detailed photographs of Io.

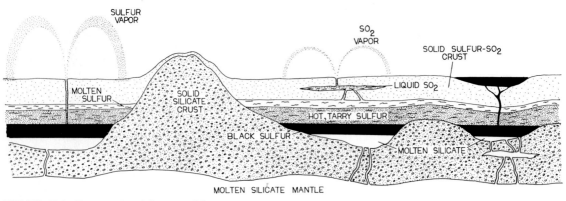

FIGURE 12.6 Cross-sectional diagram of the crust of Io.

In this paper, authors Stanton Peale, Patrick Cassen, and Ray Reynolds predicted that Io would be found to be volcanically active, the driving mechanism being the motion of the outer moon Europa. It had been known for a long time that the orbits of Io and Europa are locked into a perpetual 2 : 1 resonance. During the time it takes Europa to go around Jupiter once, Io goes around exactly twice (Figure 12.7). Over many years of time, the gravitational attraction between the two moons has forced both their orbits to depart significantly from perfect circles. Because of their elliptical orbits, the tidal forces exerted on these moons by Jupiter's enormous gravitational field vary in magnitude as they travel along their orbits. These time-varying tidal interactions alternatively force the moons to bulge outward toward Jupiter when at *perijove,* and then relax to a more nearly spherical configuration when at *apojove* (Figure 12.8). This continuous flexing heats up the interiors of both moons. Because of Io's closer proximity to Jupiter, its internal heating is much more intense than Europa's. If both of these moons continue in their current orbits, the energy source driving Io's volcanic activity could last forever.

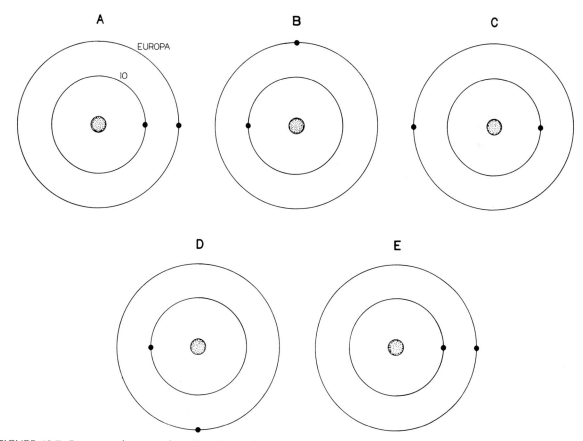

FIGURE 12.7 Resonance between the orbits of Io and Europa.

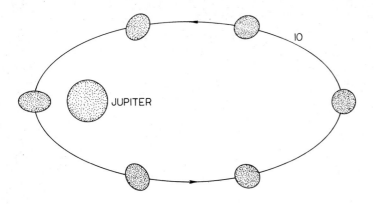

FIGURE 12.8 The elliptical orbit of Io causes its interior to be flexed periodically by the tidal forces exerted by Jupiter. This causes the interior of Io to get very hot, driving the intense volcanic activity.

THE IONIAN RADIO TRANSMITTER

There is a torus-shaped plasma region encircling Jupiter coincident with the orbit of Io. The Io plasma torus can be seen by observing the ultraviolet light that it emits, the total radiated power being about 2 TW (2×10^{12}W). The torus is about 120,000 km thick and is centered in Jupiter's magnetic equatorial plane rather than in the plane of the geographic equator. Since there is an appreciable angle (22°) between the Jovian magnetic and rotational axes, the plasma torus bobs up and down relative to the orbit of Io by an amount of 22° as Jupiter rotates. The plasma is rich in sulfur, oxygen, and sodium ions. These ions probably originated from the sputtering of the surface of Io by the intense radiation in Jupiter's magnetosphere. Once neutral atoms and molecules are stripped from Io's surface, they are immediately ionized by the high-energy radiation in the Jovian magnetosphere and a trail of plasma is left behind Io as it traverses its orbit (Figure 12.9).

The motion of Io is responsible for the powerful bursts of megahertz radio waves that originate from Jupiter. As Io moves through the Jovian magnetic field, a potential difference of 400,000 V is set up across the diameter of the moon. When Io reaches the right point in its orbit, an electrical discharge takes place and an enormous current flows between Io's ionosphere and Jupiter's upper atmosphere, causing an intense radio burst. The total power emitted during a typical burst is about 1 TW.

EUROPA—THE WATER WORLD?

The next Galilean satellite, Europa (J2), is in a circular orbit 670,900 km in radius. It is slightly smaller (diameter of 3130 km) than Io and is somewhat less dense (3.03 g/cm³). Europa is almost as interesting a world as is Io. There is a distinct possibility that Europa is the only other world in the Solar System besides Earth that has liquid water!

PLATE 1 *(top)* Photograph of Earth taken by the crew of *Apollo 11*.

PLATE 2 *(bottom)* *Apollo 15* lunar landing site. The site is located at the edge of Hadley Rille, a long channel cut into the lunar surface by an ancient lava flow.

(Photographs courtesy of NASA.)

PLATE 3 *(top) Mariner 10* photograph of Venus taken in ultraviolet light. The wave patterns that exist in the upper atmosphere can be clearly seen. In visible light, Venus is a completely featureless disk. *(Photograph courtesy of NASA.)*

PLATE 4 *(bottom)* Photographs taken on the surface of Venus by the Soviet landers *Venera 13 (top)* and *Venera 14 (bottom)*. *(Photograph courtesy of the USSR Academy of Sciences.)*

PLATE 5 *(top)* Photograph of Mars taken by the *Viking 1* orbiter. The giant basin Argyre can be seen at the terminator, and part of the enormous Valles Marineris canyon complex can be seen near the terminator at the top. The bright south polar cap can be seen at the bottom. *(Photograph courtesy of NASA.)*

PLATE 6 *(bottom)* Photomosaic of Valles Marineris made by the U.S. Geological Survey.

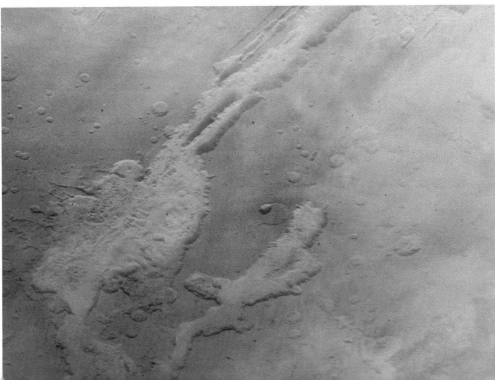

PLATE 7 Photograph taken on the surface of Mars by the *Viking 1* lander.
(Photograph courtesy of NASA.)

PLATE 8 *(top) Voyager 1* photograph of Jupiter.

PLATE 9 *(bottom) Voyager 1* photograph of Great Red Spot on Jupiter.

(Photographs courtesy of NASA/JPL.)

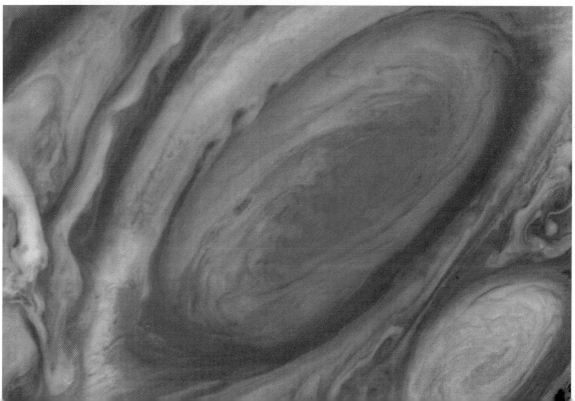

PLATE 10 *Voyager 1* photograph of Saturn. It shows a crescent Saturn casting a shadow on the ring system. *(Photograph courtesy of NASA/JPL.)*

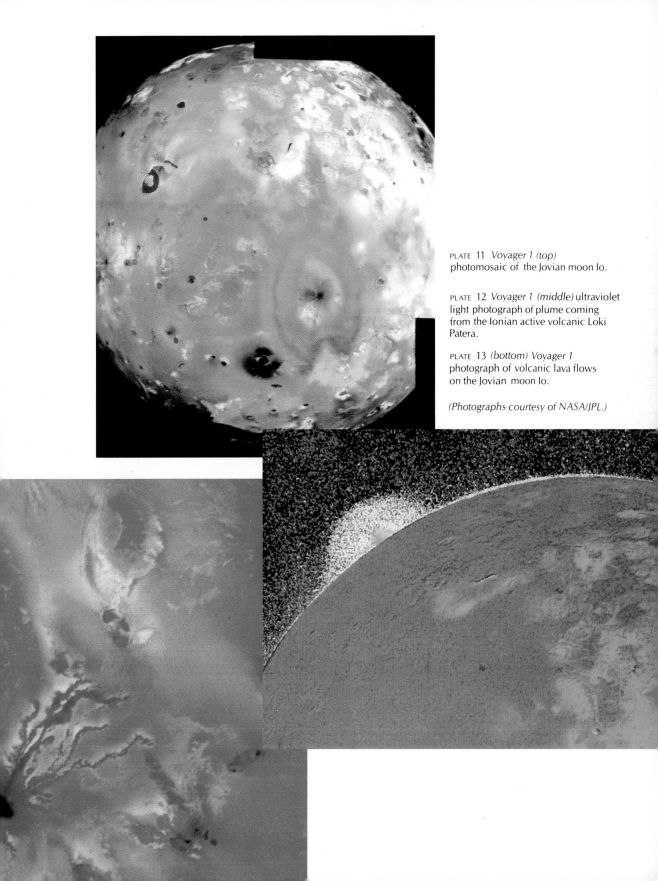

PLATE 11 *Voyager 1 (top)* photomosaic of the Jovian moon Io.

PLATE 12 *Voyager 1 (middle)* ultraviolet light photograph of plume coming from the Ionian active volcanic Loki Patera.

PLATE 13 *(bottom) Voyager 1* photograph of volcanic lava flows on the Jovian moon Io.

(Photographs courtesy of NASA/JPL.)

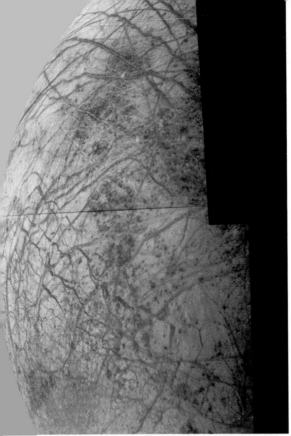

PLATE 14 *Voyager 2* photograph of Europa.

PLATE 16 *Voyager 1* photograph of Callisto.

PLATE 15 *Voyager 2* photograph of Ganymede.

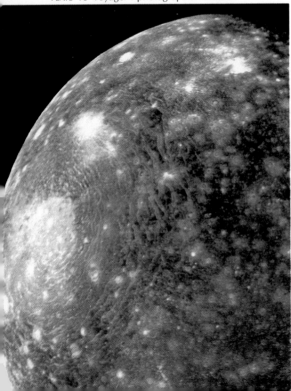

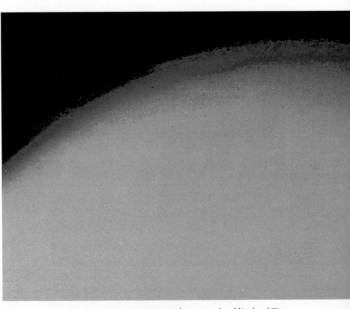

PLATE 17 *Voyager 1* photograph of limb of Titan.

(Photographs courtesy of NASA/JPL.)

THE FROZEN WASTELAND

Voyager 2 came within 206,000 km of the surface of Europa. The photographs show that Europa is much different from Io, with a light brown surface interspersed with white patches. Much of the terrain consists of interlocked depressions and mesas a few kilometers in size. The world is exceedingly smooth, with the surface relief being less than 1 km. There are few signs of any impact craters (only three probable craters 18–25 km across were seen) and no obvious basins. Unlike on Io, there are no sinkholes and no signs of volcanism, either current or in the distant past. Europa has absolutely no atmosphere, and the maximum daytime surface temperature is −180°C. A *Voyager 2* photograph of Europa is shown in Figure 12.10.

Europa is an extremely bright world, with a reflectivity of 70 percent. The reflectance spectrum of Europa shows signs of the presence of much water ice at the surface. The density of Europa is consistent with up to 20 percent of its bulk actually being made of water, with the rest being metals and silicates. Most of Europa's water seems to be either on or near the surface in the form of solid ice, with the denser metals and silicates being concentrated farther down in the interior.

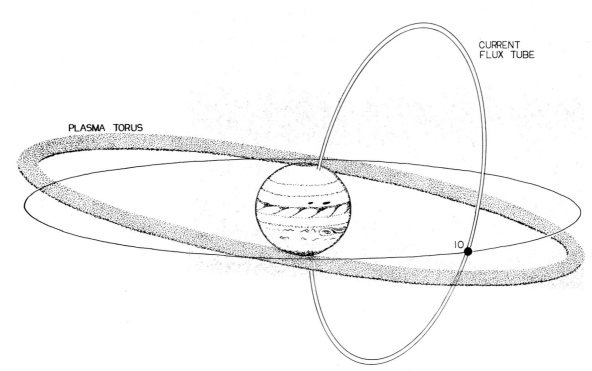

FIGURE 12.9 The Io plasma torus. There is a donut-shaped region of plasma coincident with the orbit of Io. It is tilted with respect to the plane of Io's orbit because of the influence of Jupiter's off-axis magnetic field. The plasma torus is not apparent in visible light but can be seen in the ultraviolet. There is an intense electrical current passing between Io and the Jovian ionosphere. The current follows the lines of flux in the Jovian magnetic field. The current in the flux tube can reach values as high as 5×10^6 A and is the source of the radio emissions that correlate with the motion of Io.

FIGURE 12.10 *Voyager 2* photograph of Europa. Features as small as 5 km across can be seen. (Photograph courtesy of NASA/JPL)

Europa is unique in having a surface criss-crossed with a vast network of narrow dark brown stripes tens of kilometers in width. In some cases, these stripes extend for thousands of kilometers. The lines look a lot like some of the drawings of the notorious Martian canals published during the early decades of the twentieth century. They have been given the names of some of the mythological figures that appear in the legend of Europa, one of the many paramours of the god Jupiter.

A SUBTERRANEAN OCEAN?

The absence of any large impact craters must mean that the present icy surface of Europa formed long after the era of intense meteoric bombardment came to an end. Europa probably formed from an initial accumulation of rocky debris that was intermixed with frozen water ice crystals. When Europa first coalesced as a solid body, it must have been far more homogeneous in its interior than it is now, with water ice and rocky debris being intermixed more or less uniformly throughout. However, the internal heating produced by the combined effects of radioactive decay and resonance coupling with Io gradually forced most of the interior water out of the rocks and upward to the surface. Quite early in its history, Europa must have acquired a deep ocean of liquid water that completely covered its surface. This Europan ocean may have lasted for a considerable length of time. However, as the years passed, the surface water gradually cooled and finally froze solid to form an outer ice crust perhaps 100 to 150 km thick.

The extensive network of narrow lines may represent fractures that formed in the ice crust as it froze and expanded. Many of the Europan lines appear to be radial and concentric around a point in the middle of the side of the moon that faces away from Jupiter. This may mean that the lines are fractures in the icy crust that were forced open by stresses induced by the tidal distortion that caused the despinning and synchronization of the moon's rotation.

Theoretical calculations of the current thermal state of Europa have raised an intriguing possibility. The results suggest that some liquid water may still be present within Europa, hidden deep underneath the icy crust. The Io–Europa orbital resonance may generate enough heat within Europa to maintain a perpetual 75- to 100-km-thick "mantle" of liquid water on top of the silicates, starting at a depth of 10 km below the icy surface (Figure 12.11). It has been suggested that the Europan water mantle may actually be warm enough to support some form of life!

Evidence for a much more fluid interior below the outer crust may be found in the absence of impact craters. The fact that no craters can be found on Europa must mean that the mean retention time for 10-km-diameter craters is

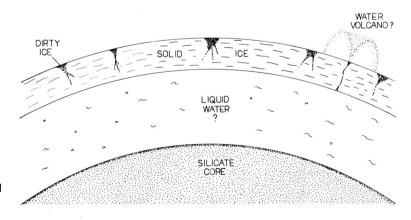

FIGURE 12.11 Cross-sectional diagram of the crust of Europa.

only 30 million years, an incredibly short time on the scale of the history of the Solar System. In order to maintain a near-surface viscosity small enough to allow the surface relief to relax so rapidly, the underlying icy crust of Europa must be significantly warmer than the surface.

WATER VOLCANOES?

The reflectance spectrum of Europa indicates that the surface ice has been contaminated with sulfur, probably originally coming from Io. The observed distribution of sulfur implanted from the Jovian magnetosphere into the icy surface of Europa can be explained by assuming that it was implanted contemporaneous with the deposition of water frost on the surface. Something is continually coating the surface of Europa with frost, about 0.1 μm every year.

Water frost condenses from the vapor phase rather than from the liquid. Where did the water vapor come from? Perhaps it came from the subterranean Europan ocean. It is possible that cracks in the outer ice crust periodically bring some liquid water to the surface, exposing it to the vacuum of outer space. Water so exposed will rapidly evaporate, forming a thin Europan atmosphere of water vapor. However, at the exceedingly cold temperatures that exist on this moon, any water vapor atmosphere will almost immediately condense into solid ice. This icy material then falls back onto the surface, covering the entire Europan landscape with a layer of fine frost.

GANYMEDE—A WORLD OF DRIFTING CONTINENTS

Ganymede (J3) is the largest of the Jovian moons (5260 km in diameter) and is possibly the largest moon in the entire Solar System. It is actually 10 percent larger than Mercury. It is in a nearly circular orbit about 10^6 km from Jupiter. The density is 1.93 g/cm³, far less than that of the inner Galilean moons Europa and Io. Ganymede is significant in that it may be the only other world in the Solar System besides the Earth that has a surface largely shaped by continental drift.

The average reflectivity of Ganymede is 50 percent. Evidence for the presence of water ice has been detected in the spectrum of the light reflected from the surface. The low density is consistent with Ganymede being at least 50 percent water ice in composition. There is probably a silicate-rich core and perhaps a mantle of liquid water or a layer of warm, convecting ice near the surface. Some telescopic observations seemed to indicate that Ganymede might have a tenuous atmosphere, but the Voyager missions turned up no evidence for its existence.

THE ICY, CRATERED SURFACE

Voyager 2 came within 62,000 km of Ganymede. The photographs show a surface superficially resembling that of Earth's Moon (Figure 12.12). There are two distinctly different types of terrain, one dark and the other bright. The darker regions are roughly polygonal in shape and are separated from each other by broad stripes of brighter terrain. Water ice is the dominant component

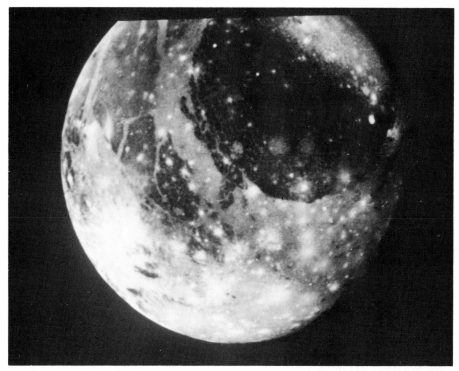

FIGURE 12.12 *Voyager 2* photograph of Ganymede, showing the hemisphere that faces away from Jupiter. The large dark area is Galileo Regio (3200 km across). The white splotch on the lower left near the limb is ejecta from the crater Osiris. (Photograph courtesy of NASA/JPL)

of Ganymede's crust. The darker regions may have a reasonable amount of silicate dust and rocky debris intermixed with the ice, but the brighter terrain is probably nearly pure water ice.

The darker regions on Ganymede are named after astronomers responsible for the discovery of Jupiter's moons (Barnard, Perrine, Michelson, and so on). The largest of these is, quite naturally, named after Galileo (Figure 12.13). Unlike the dark maria on Earth's Moon, these regions are quite heavily cratered, being nearly saturated with shallow craters 10 to 50 km across. Many of these craters have sharp rims and convex bowl-shaped floors, a form rarely seen on terrestrial planets. Some of the impact craters appear to be relatively fresh, whereas others are ancient circular forms that are little more than ghosts. Bright rays can be seen streaming outward from some craters, suggesting a much brighter material below the dark crust that was excavated by the crater impact. The high density of craters on the dark plains of Ganymede suggests that this part of the surface is quite old. It must date back to the period of heavy meteoric bombardment that ended 4 billion years ago.

There are no large-scale mountainous landforms, and the surface relief on Ganymede is not much greater than 1 km. There appear to be no large basins,

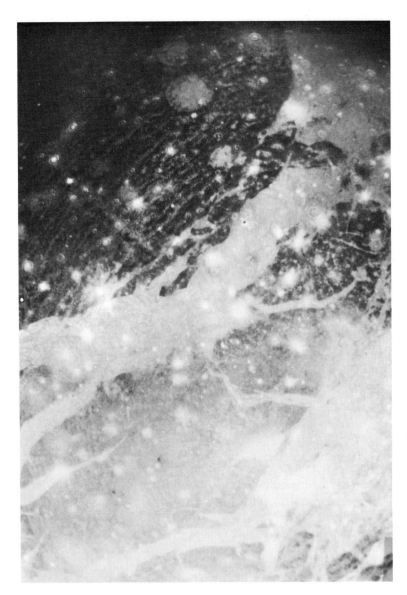

FIGURE 12.13 Voyager close-up view of Galileo Regio on Ganymede. North is at the top. Galileo is the dark region at the top of the photograph. The brighter landform below Galileo is Uruk Sulcus, an area of complex grooved terrain. Note the sharp boundary between the two landforms. The bright spots scattered across the surface are the remnants of ancient impact craters that have been nearly obliterated by crustal relaxation or by the flows of near-surface ice. (Photograph courtesy of NASA/JPL)

although Galileo does have a series of curved, bright parallel streaks that look like the bits and pieces of a ringed basin similar to Mare Orientale on the Moon or the Caloris Basin on Mercury. However, there is no visible center, and it appears that the Galileo impact basin—if that is indeed what it was—subsequently underwent drastic alterations by some sort of intense geological activity.

The brighter terrain differs from the darker polygonal regions in having an appreciably lower density of craters. Undoubtedly, this means that it is considerably younger, perhaps only 3.5 billion years old. Craters larger than 50 km in radius are virtually absent. Most of the impact craters in the brighter terrain also have bright ray systems radiating outward from their centers. This must mean that these particular regions are rich in bright material to depths of at least several kilometers. However, there are a few craters in the grooved terrain that are surrounded by broad, diffuse dark "halos." These dark halos were probably produced by meteoroid impacts that were sufficiently energetic to punch through the bright surface ice layer and throw out darker material lying underneath.

The main distinguishing feature of the bright terrain is the presence of large numbers of parallel grooves arranged in intricate patterns (Figure 12.14). The grooved patterns are as much as several hundred kilometers wide and are often hundreds of kilometers long. The grooves are so regular and so evenly spaced that many regions of the bright terrain look as if a gigantic rake had been passed over them.

The grooved "stripes" neatly separate the darker polygonal "islands" from each other, the boundary between light and dark regions being quite sharp. Features within the brighter regions generally lie somewhat below the level of the darker terrain. Features found within the brighter terrain are named after mythological figures prominent in ancient Egyptian and Middle Eastern legends (Osiris, Gilgamesh, Anu, Isis, and so on).

THE MOVING CONTINENTS

The grooved terrain is evidence for some sort of geological activity having taken place on Ganymede in the distant past. In most cases, the grooved patterns lie parallel to the boundary of a dark, heavily cratered region. Grooves are typically 3 to 10 km apart and the topographical relief is of the order of 300 to 400 m. There are no grooved features with relief any greater than 700 meters. There are occasional places where bands of grooved terrain are cut obliquely and offset by faults. Many regions are complex mosaics of discrete patches of grooves, with one system of grooves ending abruptly and another beginning. There are even some spots where several groove patterns cross each other. However, there are a few bright regions that are entirely groove-free. This grooved terrain is unique to Ganymede and seems to be indicative of a complex pattern of tectonic activity in the past.

The grooved features seem to have been the result of tensional stresses acting at right angles to the grooves. There is no convincing evidence for any sort of crustal compression such as is responsible for mountain building on Earth nor is there any evidence for subduction. Theorists now believe that the unusual rheological properties of ice are responsible for the crustal evolution that has taken place on Ganymede. The bright, grooved features were probably formed by a sudden expansion of the entire body of Ganymede, which took place during its initial cooling phase. Immediately after its formation, Ganymede must have been like a "dirty snowball," with ice and rock intermixed more or less uniformly throughout. The outer crust was undoubtedly the familiar ice-I present on Earth's surface, but high-pressure polymorphs of ice (ice-II,-IV,

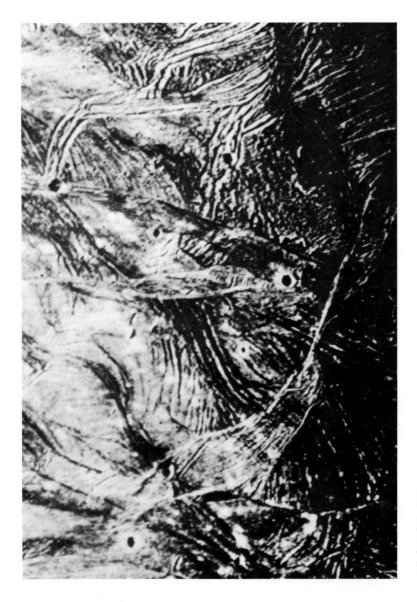

FIGURE 12.14 *Voyager 1* photograph of grooved terrain on Ganymede. The grooves are typically 10 to 15 km apart. Many groove patterns can be seen crossing each other, indicating a complex geological history in this region. (Photograph courtesy of NASA/JPL)

-VI,-VII, and -VIII) must have been present in the interior, where the pressure was much higher. All of these high-pressure forms of ice have a higher density than does ice-I.

Natural radioactivity in the silicate material partially melted the interior of Ganymede, allowing the rocky debris slowly to sink to the center of the moon. The sinking, rocky material gradually displaced the high-pressure ice originally located in the interior of Ganymede, forcing it upward toward the surface, where the pressure was smaller. As the pressure on this rising ice was slowly relieved, it converted back into the more familiar low-density ice-I. This conversion in turn forced the entire body of Ganymede to undergo a slow expansion, producing perhaps a 6 percent increase in overall outer surface area (Figure 12.15).

Since the outer crust of Ganymede probably remained largely solid all throughout this period of internal readjustment, it was subjected to enormously high tensile stresses. The crust must have cracked in many places, producing isolated "continents," which slowly drifted away from each other as the expansion continued. As the cracks formed and expanded, they were filled up by warmer ice flowing upward from the interior. The expanding cracks must have frozen solid almost as soon as they were formed, creating fresh uncratered terrain. The expansion of Ganymede was probably completed approximately 3.5 billion years ago. After that all geological activity ceased, leaving Ganymede a cold, dead world forever after.

CALLISTO—THE RING OF VALHALLA

Callisto, the outermost of the Galilean moons, is 1.88×10^6 km from Jupiter and has a diameter of 4900 km. It has the lowest density (1.79 g/cm^3) as well as the darkest surface (reflectivity of only 20 percent) of any of the four Galilean moons.

Callisto has no atmosphere to speak of, and the surface pressure cannot be much greater than 10^{-11} atm. It has the highest noonday temperature of any of the Galilean moons ($-110°C$), due possibly to the darker color.

The infrared reflectance spectrum of Callisto shows extensive signs of water ice, and Callisto's low density probably means that it is over 50 percent water ice in composition. Because Callisto is less dense than Ganymede its silicate content must be considerably smaller. Callisto probably has a dense silicate core with a crust of solid ice mixed with large amounts of dirt and rocky debris.

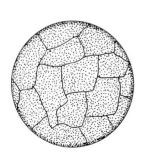 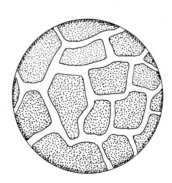 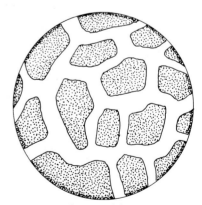

FIGURE 12.15 The grooved terrain on Ganymede may be caused by the slow expansion in the size of the entire moon. This expansion produces cracks in the solid outer crust. The darkly colored islands are parts of the original crust that gradually separate from each other as the expansion proceeds. The lighter regions are made of fresh water ice that is forced upward from the interior of the expanding moon.

Voyager 1 passed within 126,000 km of the surface of Callisto. A *Voyager 1* photograph of Callisto is shown in Figure 12.16. The photographs show the surface to be somewhat similar to the dark cratered terrain on Ganymede. However, Callisto is even darker in color and more heavily cratered. The density of craters on Callisto is about the same as that in the lunar highlands, but there is a relative shortage of craters larger than 100 km across. There are many craters with bright floors, but there are no extensive bright ray systems. This undoubtedly means that the surface is rich in darker material to substantial depths. The craters are all named for gods and heroes who appear in Scandinavian mythology.

FIGURE 12.16 *Voyager 1* photograph of Callisto. North is at the top. The bright feature on the left is the giant basin Valhalla, which lies in the forward-facing hemisphere of Callisto. Features as small as 10 km across can be seen. (Photograph courtesy of NASA/JPL)

The most prominent feature on Callisto is a large multiringed structure known as Valhalla. It is nearly 3000 km across, and is reminiscent of Mare Orientale on the Moon and the Caloris Basin on Mercury. Valhalla has a bright central patch 300 km in diameter that is surrounded by 10 to 15 discontinuous ringlike ridges. The spacing between rings is quite even, with the gentler slopes facing radially inward and the steeper slopes, radially outward. Oddly enough, there does not appear to be any sort of impact basin at the center of Valhalla and there are no radial ejecta or ring mountains. In addition, there is a rather strange variation in crater density across the width of Valhalla, with only about one-third the number of craters per unit area near the center as there are near the periphery.

There is no evidence that Callisto was ever volcanically active. Unlike Ganymede, Callisto shows no evidence of any past geological activity. There are no surface cracks, no bright resurfaced areas, and no systems of grooves. Callisto's low density probably means that there were only relatively small amounts of heavier radioactive elements incorporated into the body of the moon, certainly not enough to generate the intense heating required to drive any sort of tectonic activity. The surface of Callisto is quite old and must date back to the period of early meteoric bombardment that took place over 4 billion years ago.

However, the primitive icy crust must have been sufficiently soft so that a considerable amount of surface flowing and relaxation could take place. Some of the craters on Callisto have very subdued topographies, indicating that some sort of viscous relaxation has taken place. The Valhalla ring system was probably created by the impact of an asteroid-sized body after most of the craters had already been formed. The impact erased any previously existing craters within 350 km of the center, but craters farther out were preserved. At the time of the Valhalla impact, the icy crust was probably still not rigid enough to support any steep topographical relief. Any ring mountains that initially formed quickly collapsed, and the impact basin rapidly disappeared. Shortly thereafter, the cooling surface became more rigid, and the surface features became frozen into their final form.

INNER MOONS

There are four small moons lying in orbits close to Jupiter (Figure 12.17), all inside the orbit of Io.

The largest of these is Amalthea (J5), which orbits 89,540 km above Jupiter's cloud tops. It was discovered by the American astronomer Edwin E. Barnard in 1892. Because of its dark color and close proximity to Jupiter, Amalthea is very difficult to see from Earth. Voyager photographs show Amalthea to be an irregular-shaped object, 270 km long and 150 km wide. It is 10 times larger than the Martian moon Phobos and is approximately the size of Long Island (Figure 12.18). As expected Amalthea always points its long axis toward Jupiter as it revolves.

Amalthea has a dark reddish color, indicating that its composition is quite different from that of the Galilean moons. It is redder than Mars but not as red as

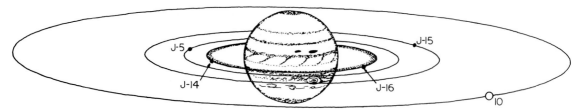

FIGURE 12.17 Orbits of the four inner moons of Jupiter. The dark band lying inside the orbits of J14 and J16 is the Jovian ring.

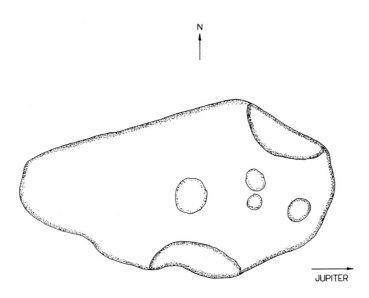

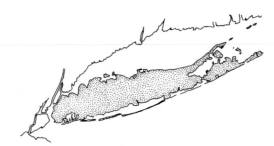

FIGURE 12.18 Amalthea is approximately the size of Long Island.

Io. The *Voyager 1* spacecraft came no closer than 420,000 km to Amalthea, so the photographs are quite indistinct. The pictures do show two large circular depressions (named Pan and Gaea) that are probably impact craters. These craters are much deeper than comparable-sized craters on the Moon. There is probably a substantial regolith (perhaps up to 1 km deep) with perhaps an extensive fracture zone beneath.

The Voyager spacecraft did not come close enough to Amalthea to measure its density, so its composition is largely a matter of conjecture. If this moon formed in place during the superluminous phase of Jupiter, it is probably a rather dense object made up of refractory oxides, nickel–iron metals, and silicates. If it was captured from the Asteroid Belt, it is more likely to be made up of dark, low-density, volatile-rich material. The optical-reflectance spectrum of Amalthea is consistent with at least the surface being made of carbonaceous material, perhaps contaminated with sulfur that managed to escape from Io during volcanic eruptions. Amalthea orbits in a particularly intense region of Jupiter's radiation belt, and prolonged exposure to this radiation may play a role in establishing the dark reddish color of the surface.

The other three moons are a lot smaller and were seen for the first time by the Voyager spacecraft. All three of these tiny moons have reflectance spectra like that of Amalthea, suggesting that their compositions are similar. The closest is Metis (J16), orbiting 56,700 km from Jupiter's cloud tops. Metis is quite small, with a diameter of approximately 40 km. Next out is Adrastea (J14). It has an orbit only 500 km outside that of Metis and has a diameter of the order of 25 km. The last of these small moons is Thebe (J15). It orbits 150,800 km above the Jovian cloud tops, well outside the orbit of Amalthea. The diameter is approximately 80 km. Thebe was subsequently spotted from Earth by using sensitive charge-coupled optical-detection devices placed at the focus of the Mt. Palomar 5-m telescope.

OUTER MOONS

Leda, Himalia, Lysithea, and Elara are all in elliptical orbits that have average distances of 1.1×10^7 to 1.2×10^7 km from Jupiter. These orbits are inclined at angles of between 26° and 29° with respect to the Jovian equator. Himalia (J6) and Elara (J7) were discovered by Charles Perrine in 1904 and 1905, respectively. Lysithea (J10) was discovered by Seth Nicholson in 1938. Leda (J13) was first spotted by Charles Kowal in 1974.

These moons range in diameter from 20 to 200 km. They are much darker than any of the Galilean moons, and no evidence for the presence of water ice has been found in any of their reflectance spectra.

The four outermost moons are Anake, Carme, Pasiphae, and Sinope. Sinope (J9), Carme (J11), and Anake (J12) were discovered by Seth Nicholson in the years 1914, 1938, and 1951, respectively. Pasiphae (J8) was discovered by P. Mellote in 1908. These moons are all small, of the order of 10 to 20 km in diameter. All are in elliptical orbits averaging between 2.0×10^7 and 2.5×10^7 km from Jupiter. Their orbits are so steeply inclined with respect to the Jovian equator that they are virtually polar in orientation. These four moons revolve around Jupiter in the retrograde sense. They are dark in color and do not seem to have any water ice.

The dark colors and inclined, elliptical orbits of these outer moons have led many astronomers to speculate that these objects are not indigenous to the Jovian system. Perhaps they were originally asteroids that happened to wander too close to Jupiter and were captured in highly elliptical orbits. These small outer Jovian moons may have originally come from the very outermost regions of the Asteroid Belt or from the Trojan population of asteroids coorbital with Jupiter. Most of these asteroids are of the D variety. However, none of the outer Jovian moons has a reflectance spectrum that is similar to any of the D-type asteroids. Their spectra, in fact, more closely resemble those of the C-type asteroids, which are usually found much closer to the Sun. Consequently, the origin of the outermost moons of Jupiter is an open question.

JUPITER'S RING

One of the more surprising results of the Voyager missions to Jupiter was the discovery of a ring system similar to Saturn's but far less extensive. No telescopic observations had ever hinted at the existence of a Jovian ring system, but *Pioneer 11* may have detected some indication of its presence when it flew past the planet in 1974. This spacecraft's particle counters noted a sudden decrease in the intensity of charged-particle radiation at an altitude of 50,000 to 55,000 km above the Jovian cloud tops. Mario Acuña and Norman Ness of NASA's Goddard Space Flight Center proposed that a previously unknown moon or ring system located at that particular position in space could be responsible for absorbing some of the particles in the radiation belts. When *Voyager 1* flew past Jupiter in 1979, its television cameras made a special search of this region. The ring system was found (Figure 12.19).

Jupiter's ring lies in the plane of the planet's equator but is much closer to Jupiter (its "primary") than the familiar ring system of Saturn is to that planet. The main part of the ring system is only 10,000 km wide and begins about 48,000 km above the cloud tops. It shows up best when viewed from the far side of Jupiter, backlighted by the Sun.

The ring has four main components, termed *bright, diffuse disk, gossamer,* and *halo*. The bright part is the one most readily seen and has a width of 6000 km. It has a sharp outer boundary at an altitude of 58,000 km above the Jovian cloud tops. The outer boundary of the bright ring is just below the orbit of Metis, Jupiter's innermost moon. A much fainter and more diffuse disk component lies inside the bright ring. It seems to be much wider than the bright ring, and dispersed components of the diffuse disk may actually extend nearly all the way to the cloud tops. An extremely faint gossamer ring component has been detected in a single Voyager photograph that was specially enhanced to bring out contrasting features. This component extends beyond the edge of the bright ring, perhaps reaching out as far as 140,000 km from the Jovian cloud tops. The halo ring component is a diffuse band that extends 20,000 km above and below

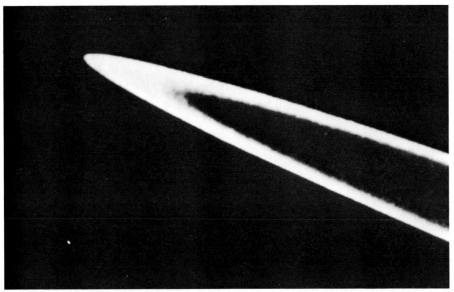

FIGURE 12.19 *Voyager 2* photograph of Jupiter's ring. This picture was taken when the spacecraft was on the far side of Jupiter, with the ring system being backlighted by the Sun. (Photograph courtesy of NASA/JPL)

the main ring plane. This out-of-plane component may have been formed by electromagnetic interactions between charged ring particles and Jupiter's inclined, off-axis magnetic field.

Like Saturn's rings, the Jovian ring is a sheet of many millions or even billions of small particles, each in an independent orbit around the planet. The particles in the Jovian ring seem to have a characteristic size of a few microns, although there are probably a few particles at least 1 cm across. The reflectance spectrum of the ring system seems to indicate that the particles within it are not made up of ices of either ammonia, methane, or water. They are probably made up instead of silicates, sulfates, or carbon-rich materials.

The ring appears to be intimately related to the innermost small moons Metis and Adrastea. Perhaps the particles in the ring are debris produced by the meteoric degradation of these moons. Alternatively, the ring may have been formed by an accumulation of volcanic ejecta and debris that managed to escape from the surface of Io.

The ring particles all appear to be slowly spiraling inward toward Jupiter under the combined influences of solar radiation pressure and magnetic drag produced by Jupiter's enormous magnetic field. A micron-sized particle will move from the outer edge of the bright ring to its inner edge in a time as short as 20 years. Two hundred more years will bring the particle through the diffuse disk and into the Jovian atmosphere. As a result, Jupiter's ring may have a short lifetime. However, the ring may be undergoing continuous replenishment by the steady accumulation of debris originating farther out in the Jovian satellite system, so that a steady-state ring particle population density is maintained.

THE FORMATION OF JUPITER'S MOONS

Jupiter and its moons represent a sort of Solar System in miniature. The most striking similarity is the manner in which the Galilean moons happen to be placed according to their densities, with the denser, rocky moons Io and Europa orbiting close to the planet and the less dense icy moons Ganymede and Callisto lying farther out. At the time of the formation of the Solar System, Jupiter must have been a giant ball of gas made up of hydrogen, ammonia, methane, and water vapors interspersed with grains of metal-rich silicate minerals. It was probably several hundred times its eventual size. Proto-Jupiter soon began to shrink under the influence of its own gravity and became very hot in its interior. Had it been only 50 times more massive, Jupiter would have become hot enough at its center to initiate hydrogen fusion, becoming a small sun.

Even though Jupiter was far too small ever to fuse hydrogen into helium, models developed by James B. Pollack indicate that the contracting Jovian gas ball actually got hot enough to become at that early time a stronger local energy source than the Sun itself. An object located at the present orbit of Io probably got more energy from Jupiter at that time than Earth now does from the Sun. It became so hot in the inner parts of the Jovian system that virtually all of the water located there remained in the vapor phase. The outer regions of the primordial Jovian system, however, remained cool enough for water vapor to condense into ice crystals as the Jovian gas ball continued to contract. This ice intermixed with rocky dust and debris and eventually coalesced to form the outermost icy Jovian moons. The densities of Ganymede and Callisto are about what one might expect if they are made up from a gas of essentially solar composition that condensed at the freezing temperature of water ice. The denser material that did not coalesce into larger bodies in the outer regions of the Jovian system fell inward as the rapidly forming planet continued to shrink. This material condensed at a somewhat later time to form the largely rocky inner moons Io and Europa.

13

Saturn's Moons — Worlds of Ice, Worlds of Mystery

Saturn has 17 known moons, more than any other planet. Some of these moons are rather substantial bodies, with diameters of 400 km or more (Figure 13.1). The largest of them, Titan, is bigger than Earth's Moon and is nearly as large as Jupiter's giant moon Ganymede. All of the moons except Phoebe (and perhaps Hyperion) appear to have rotational rates that are locked by tidal forces into synchronization with their orbital periods so that they perpetually turn the same faces toward Saturn. All but the very outermost of the moons have nearly circular orbits that lie very close to the plane of Saturn's equator and rings (Figures 13.2 *a* and *b*).

Many of the moons show marked differences in surface features between one hemisphere and the other, with the hemisphere that is turned toward the direction of the orbital motion being quite different from the trailing side. All but one of the moons are completely without atmospheres. The lone exception, Titan, has an atmosphere denser than that of Earth and may actually be of substantial biological interest.

In November of 1980, the *Voyager 1* spacecraft passed through Saturn's system of moons, providing the first close-up views of these extremely interesting bodies. *Voyager 2* followed in August of 1981. The spacecraft photographed

Voyager 2 photograph of Titan. In this view, the moon is backlit by the Sun, and the atmosphere is clearly visible as a crescent. (Photograph courtesy of NASA/JPL)

245

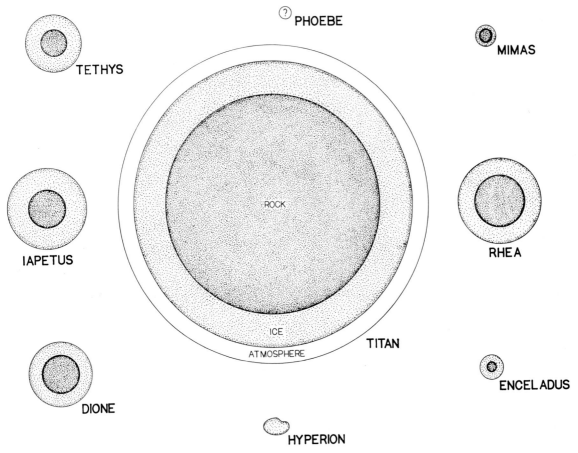

FIGURE 13.1 Relative sizes and interiors of the moons of Saturn.

the surfaces of many of the moons and sent valuable data back to Earth about their densities and surface compositions. *Voyager 1* came within 6490 km of the giant moon Titan, which corresponds to a major planetary encounter in its own right.

MIMAS—"THE DEATH STAR"

The innermost large moon of Saturn is Mimas (S1), discovered by the English astronomer William Herschel in 1789. The radius of this moon's orbit is 185,540 km.

Mimas is a small spherical world with a diameter of 390 km. The average density is quite low, only 1.44 g/cm³. Spectrographic analysis of the light reflected from Mimas indicates that its surface is covered with a thick layer of water ice. In fact, the low density probably means that as much as 60 percent of the mass of Mimas is water ice, with the rest being a rocky rubble rich in metals and silicate minerals.

Voyager 1 passed within 88,400 km of Mimas. The photographs of the surface show a densely cratered terrain that looks very much like the lunar highlands (Figure 13.3). Many of the large craters have rudimentary central peaks, but most of the more ancient craters are severely degraded. Apparently no ray craters are present. Most of the large craters on Mimas are bowl shaped and are much deeper than craters of comparable size on Earth's Moon or on Jupiter's Galilean satellites. Most of the prominent features on Mimas are named after people and places that appear in the stories surrounding the legendary King Arthur of Britain.

The most prominent crater on Mimas is Herschel, a feature 130 km in diameter centered in the leading hemisphere (the side that always faces in the direction of the moon's travel along its orbit around Saturn). The walls of Herschel are 5 km high, and some points on the floor are as deep as 10 km below the surrounding terrain. There is a central peak 20 to 30 km wide and 6 km high. Herschel is nearly one-third of the diameter of Mimas itself. If the meteoroid that excavated the crater had been even slightly larger, the moon itself probably would have shattered under the impact! At the time of the Voyager flyby, some wit pointed out that this giant crater makes Mimas resemble the "Death Star" in the motion picture *Star Wars*.

There are several grooves on Mimas that are as long as 90 km and up to 10 km wide. Some grooves are as much as 2 km deep. The most conspicuous grooves trend northwest and west-northwest and may have been produced by the impact that formed the giant leading-hemisphere crater Herschel. Or perhaps the cracks developed as a result of tidal interactions shortly after the moon's formation. There is a local cluster of hills in the trailing hemisphere that may be ejecta from the Herschel impact.

Even though Mimas is an exceedingly tiny world, there is evidence for some sort of surface-altering geological activity having taken place in the distant past. The best indication for this is the widely varying crater densities that occur at different locations on the surface. The leading hemisphere west of Herschel is covered with many craters greater than 40 km in size, but the south polar areas generally lack craters any larger than 20 km across. Furthermore, there is a relative shortage of small craters all over the surface of Mimas. Some mechanism has selectively erased many of the small craters, while leaving the larger ones intact. Intense internal heating has obviously taken place within Mimas, but the mechanism responsible is obscure. Radioactive decay within the interior of so small a body should be insufficient.

ENCELADUS— THE FURROWS OF SAMARKAND

The next large moon, Enceladus (S2), was discovered by William Herschel in the same year that he found Mimas. The orbit of this moon has an average radius of 238,040 km. Enceladus is nearly as intriguing an object as the Jovian moon Io. In

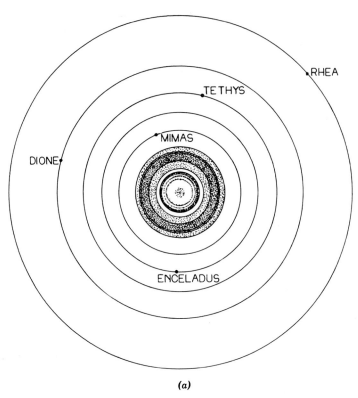

FIGURE 13.2 Orbits of the large moons of Saturn. (a) The inner moons Mimas, Enceladus, Tethys, Dione, and Rhea. (b) The outer moons Titan, Hyperion, and Iapetus.

spite of its tiny size, Enceladus appears to be almost as geologically active as the Earth!

The diameter of Enceladus is 500 km and the mean density is 1.2 g/cm³. The low density of Enceladus is consistent with its being at least 60 percent water ice in composition. This moon is the most reflective body in the entire Solar System. The bright surface must be virtually pure water ice, intermixed with only relatively minor amounts of darker-colored contaminants. Enceladus is much too small to retain any sort of atmosphere. The surface temperature during daytime is only −220°C, owing to the high reflectivity.

A *Voyager 2* photograph of Enceladus is shown in Figure 13.4. The prominent features on Enceladus have been named after people and places appearing in the *Tales of the Arabian Nights*.

The icy surface of Enceladus is quite a bit smoother than that of any of Saturn's other moons. It also has far fewer numbers of impact craters. There are no craters any larger than 70 km wide, most craters being less than 10 km across. In some regions, the craters are relatively well preserved, with sharp rims and deep, bowl-shaped interiors. In other places, the craters are in an advanced

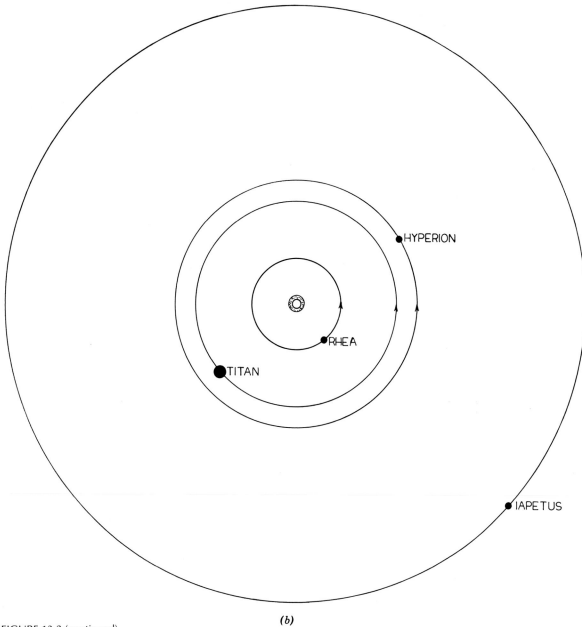

HYPERION

RHEA

TITAN

IAPETUS

(b)

FIGURE 13.2 (continued)

state of collapse, with sunken rims and degraded central peaks. Some plains are entirely crater-free; a few have margins marked by braided ridges.

Some of the smooth plains on Enceladus have a pattern of rectilinear faults and ridges that is reminiscent of the grooved terrain on Ganymede. There are also some long, narrow curvilinear grooves. Grooves and furrows tend to con-

FIGURE 13.3 *Voyager 1* photograph of Mimas, taken at a range of 425,000 km. Here the equatorial region is shown. The large crater Herschel with its prominent central mountain peak is located at the terminator. (Photograph courtesy of NASA/JPL)

gregate in regions where there are relatively few craters. There are no grooves at all in the most densely cratered terrain. There is a particularly long series of furrows and ridges that extends from southern latitudes almost to the north pole. This feature has been named Samarkand.

The most densely cratered regions on Enceladus are about as heavily cratered as the lunar highlands. This probably indicates that they are at least 3.8 to 4.0 billion years in age. The less densely cratered terrain must be appreciably younger, the very smoothest crater-free terrain being perhaps less than 1 billion years old. Enceladus's complex topology must mean that extensive geological resurfacing has taken place over much of its history.

The distinctly different types of terrain present on Enceladus seem to indicate that there were several discrete episodes of intense geological activity separated by relatively quiet periods, rather than one continuous resurfacing process. During active episodes in the past, heating within Enceladus melted icy material deep within the interior. This molten material migrated upward and flowed out over the surface. It quickly froze into solid ice, covering up the older terrain and creating a fresh, uncratered surface. Elsewhere, the heating sped the

FIGURE 13.4 *Voyager 2* mosaic of Enceladus. The enormous trench complex Samarkand can be seen near the center. The north pole of Enceladus is located at the left end of Samarkand, near the terminator. An older cratered terrain can be seen to the left of Samarkand near the terminator. (Photograph courtesy of NASA/JPL)

viscous relaxation of crater topologies. The internal heating also produced a net thermal expansion of the entire moon, generating tensile stresses strong enough to form the rectilinear fractures seen on some plains. During inactive periods, the interior cooled and the resurfacing abated. This cooling caused the moon to contract, generating compressive stresses that formed the ridges currently seen. Since there are some regions that are completely free of craters, the most recent episode of resurfacing may have taken place only a few hundred million years ago. There is a distinct possibility that some geological activity may still be taking place on Enceladus even at present!

Some evidence for currently active water volcanism on Enceladus may be found in Saturn's tenuous E ring. The brightness profile of the E ring peaks at the orbit of Enceladus, and it seems likely that the particles presently found within this ring originally came from this moon. If Enceladus were not actually the source of the particles, the moon's gravitational field should long ago have cleared a wide path through this ring. The E ring has to be a relatively young feature, since the intense radiation in Saturn's magnetosphere is capable of destroying all of the tiny ice particles presently located in this ring in a time as short as a few hundred years.

The interior of Enceladus may be warm enough to sustain an extensive "mantle" of liquid water under the outer crust of solid ice. Perhaps a recent volcanic episode brought some of this interior liquid water to the surface, exposing it to the vacuum of outer space. At least some of this exposed water would have evaporated and escaped into outer space, where it would instantly condense and form solid ice particles. Meteor impacts may have also played a role, punching through the icy crust to release some of the subsurface water.

Where does the energy that drives the intense geological activity come from? A body of such small size and low density should not have enough heavy radioactive elements to provide any significant amount of radiogenic internal heating. It has been proposed that resonant gravitational coupling with the nearby moon Dione might be responsible. The orbital period of Enceladus is precisely one-half that of Dione, and the mutual gravitational attraction between these two moons forces both of them into more highly elliptical orbits. As Enceladus travels along its eccentric orbit, it is subjected to continually varying tidal forces exerted by Saturn. These variable tidal forces periodically stretch the body of Enceladus away from a perfectly spherical shape, causing its interior to get hotter. A similar mechanism is thought to be responsible for the volcanic activity on the Jovian moon Io.

TETHYS—THE CANYON OF ITHACA

Next out from Saturn is Tethys (S3), discovered by Giovanni Cassini in 1684. It orbits 294,670 km from the center of Saturn. Tethys is a spherical body 1060 km in diameter and has an average density of only 1.2 g/cm³. This moon is probably 60 percent water ice, with the remainder being rocky material perhaps intermixed with a few percent of solid ammonia. Tethys is much too small to have any sort of atmosphere, and its surface daytime temperature is −185°C.

Voyager 2 came within 93,000 km of the surface, and the photographs show a rather dense population of deep craters. The prominent features on Tethys have been named after gods, heroes, and places mentioned in Homer's *Odyssey*. The region north of the equator on Tethys is rough, hilly, and rather densely cratered, but most of the craters there are rather heavily degraded. This terrain is probably at least 4 billion years old.

There is a particularly large crater named Odysseus in the leading part of the outward-facing hemisphere (Figure 13.5). Odysseus is 400 km in diameter and is 15 km deep. The width of the crater is over two-fifths the diameter of Tethys itself. It is the largest crater yet found in the Solar System that has a well-developed central peak. The floor of Odysseus has rebounded to match

FIGURE 13.5 *Voyager 2* photograph of leading hemisphere of Tethys, taken at a range of 830,000 km. The large crater near the terminator is Odysseus. (Photograph courtesy of NASA/JPL)

the spherical contour of the moon, and both the central peak and the rim have undergone extensive collapse. At the time that Odysseus was formed, Tethys was probably still largely liquid in its interior. If Tethys had actually been fully solid at this time, the Odysseus impact almost certainly would have shattered the moon into millions of fragments.

The trailing hemisphere of Tethys (Figure 13.6) is generally less rugged than the forward-facing side. Some sort of resurfacing episode seems to have taken place there. This face of Tethys has only a few large craters, most of which are surrounded by large plains containing relatively small numbers of craters. These plains may have been produced by a flood of icy slush flowing upward from the interior, which covered up and obliterated any preexisting craters when it froze. The intercrater plains on the trailing side of Tethys are nevertheless relatively old, having formed perhaps as long ago as 3.8 to 4.0 billion years, immediately after the era of the most intense meteoric bombardment had come to an end.

There is a huge trench system, named Ithaca Chasma, centered in the Saturn-facing hemisphere. The canyon complex extends nearly three-quarters of the way around the moon. In places, the trench is over 100 km wide and up to

FIGURE 13.6 *Voyager 2* photograph of trailing hemisphere of Tethys, taken at a range of 280,000 km. The north pole of Tethys is at the upper left-hand corner in this photograph. Ithaca Chasma can be seen in the lower central part of the visible disk. The prominent crater located at the top of Ithaca is Telemachus. The craters Phemius, Polyphemus, and Ajax are lined up along the terminator on the right. (Photograph courtesy of NASA/JPL)

5 km deep. The crater density in the region surrounding Ithaca Chasma is about the same as it is in other heavily cratered regions of Tethys, suggesting that the formation of this trench complex is an ancient event. The surface within the canyon itself is the least cratered of any region on Tethys, which probably means that it is the youngest terrain on the moon. Ithaca Chasma seems to be geometrically related to the giant crater Odysseus. The trench complex follows a rough great circle centered on Odysseus for at least 270°, suggesting that the Odysseus impact event may have had something to do with its formation.

Shortly after the formation of Tethys, the interior of the moon must have completely melted, leaving a sphere of liquid water covered with a thin solid ice crust. When the interior cooled and froze solid, it forced an expansion of the surface. Calculations indicate that the amount of surface expansion was just enough to have produced a crack of about the same total area presently occupied by Ithaca Chasma. It is not clear, however, why only a single crack should develop rather than many smaller ones distributed more or less uniformly over the surface. The era of tectonic activity responsible for Ithaca Chasma was probably quite short-lived, lasting only a few hundred million years before the surface of the entire moon froze solid into its final configuration.

DIONE — THE WISPS OF AMATA

Next out from Saturn is Dione (S4), discovered by Cassini in 1684. Dione is in an orbit 377, 420 km in radius. It is nearly the same size as Tethys (with a diameter of 1120 km), but may be slightly more dense (1.4 g/cm³). Dione is probably made up of a 60:40 mixture of water ice and rock, with perhaps a few percent solid ammonia.

Voyager 1 came as close as 161,500 km to the surface of Dione and provided extensive photographic coverage. The more prominent features on this moon have been named after people and places described in Vergil's *Aeneid*.

Striking surface asymmetries were found on Dione, with the forward face being quite different from the trailing side. The forward hemisphere is dark and is covered with a dense population of impact craters. However, there are only a few craters larger than 30 to 40 km across and almost no ray craters. The trailing hemisphere is completely different. It consists of a complex network of bright, wispy markings superimposed over a darker surface scarred with numerous large craters.

The forward face of Dione (Figure 13.7) looks superficially like the lunar highlands. There are two distinctly different types of terrain: a rough unit with many craters larger than 20 km across and a smooth plains unit with relatively few craters. Many of the larger craters on Dione exhibit central peaks. Unlike the lunar craters none of the craters on Dione exhibit any easily identified associated ray patterns or ejecta, nor are there any obvious secondary impact craters.

Scattered among the craters are numerous linear features such as scarps, ridges, and troughs. Oddly enough, most of these features seem to be oriented either northwest or northeast. This may mean that they were formed by internal stresses associated with tidal despinning and orbital recession at the time that Dione's rotation became coupled with its revolution.

The rearward-facing hemisphere of Dione (Figure 13.8) was relatively poorly imaged during the Voyager flybys, so not much is known about it. It consists of a relatively dark terrain overlain by a complex series of bright, wispy streaks. The dark terrain is probably composed largely of water ice intermixed with some sort of dark-colored contaminant of uncertain composition. The wisps are probably nearly pure water ice in composition. The Voyager photographs show the presence of numerous quasi-circular features that are presumed to be impact craters. However, most of these craters are rather subdued and have dark centers and bright rims. The wisps are clearly younger than the darker terrain. Many of the bright, wispy streaks are geometrically related to a large circular feature named Amata. It is presumably the remains of a gigantic

FIGURE 13.7 *Voyager 2* mosaic of the leading face of Dione. North is at the top. Much of this side of Dione is quite heavily cratered, but some of the wispy steaks prominent on the trailing face can be seen near the eastern limb. The prominent crater in the northern hemisphere is Aeneas; the large crater in the southern hemisphere is Dido. To the west of Aeneas is a large curvilinear trough known as Latium Chasma. (Photograph courtesy of NASA/JPL)

impact crater. It is possible that the wisps were formed by the flow of fresh, uncontaminated ice forced upward along geological fracture lines created by the Amata impact.

The complex surface of Dione shows unmistakable signs of a period of substantial crustal evolution and extensive resurfacing. Like Earth's Moon, the earliest period of Dione's history appears to have been dominated by the impacts of large asteroid-sized bodies left over from the initial accretion of the major planets. After this episode of intense bombardment had passed, the surface of Dione suddenly became active. Much of the terrain was inundated with successive floods of water driven upward from the interior by the intense heat. When this flood of water cooled and solidified into solid ice, it covered up many of the deep craters and erased most of the record of the early period of intense meteor bombardment. Eventually, the interior of Dione cooled off and the volcanic activity subsided. At a somewhat later time, meteoric impacts from smaller icy fragments left over from the initial accumulation of Saturn's moons began to scar the newly formed frozen plains of Dione. Since these loose objects originated within the Saturnian system itself, the meteoric impacts took place preferentially on the forward-facing side of Dione. Hence the two-faced world seen today.

Some sort of heat source had to have been present within Dione to have driven the extensive resurfacing episodes of the past. The high density of Dione (as compared to Saturn's other icy moons) may mean that it had sufficient heavier radioactive material to provide an appreciable internal energy source. Another possibility is orbital coupling with the moon Enceladus. The orbit of Dione is locked in a 2 : 1 resonance with that of Enceladus. During the time that it takes Dione to go around Saturn once, Enceladus goes around exactly twice. This coupling may produce sufficient gravitational flexing of the interior of Dione to provide the internal heating necessary to drive the intense surface-altering events that have taken place in the past. Dione may be the Saturnian equivalent of Europa.

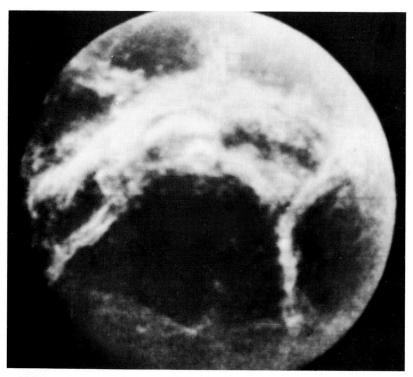

FIGURE 13.8 *Voyager 2* mosaic of the trailing face of Dione, taken at a range of 790,000 km. The north pole is at the top. The complex pattern of wisps appears to radiate outward from a giant impact crater named Amata. (Photograph courtesy of NASA/JPL)

RHEA—THE DIONE TWIN

Rhea (S5) is the largest airless moon in the entire Saturnian system and was discovered by Cassini in 1672. It is 527,100 km from the center of Saturn and has a diameter of 1530 km. The average density is 1.3 g/cm³, similar to that of Dione and Tethys. The internal composition is probably approximately 60 percent water ice, 40 percent rock and/or metal. Rhea has absolutely no atmosphere, and the surface temperature is −175°C on the sunlit side.

Voyager 1 passed within 59,000 km of Rhea. The photographs show a surface much like that of Dione, with a bright, heavily cratered leading hemisphere and a darker trailing hemisphere covered with bright, wispy features. The more prominent features identified on Rhea have been given the names of gods or supernatural beings that appear in creation myths and legends from at least 30 different terrestrial cultures, both ancient and modern.

The bright forward face of Rhea (Figure 13.9) is probably covered with nearly pure water ice. Like that of Dione, this hemisphere of Rhea can be subdivided into two distinctly different terrain types.

The first of these is a terrain very heavily scarred with large craters, ones with diameters ranging between 40 and 130 km. Regions of this type are found in the vicinity of the north pole of Rhea, as well as near the equator between 0°

FIGURE 13.9 *Voyager 1* mosaic of heavily cratered north polar region of Rhea. The north pole is located at the top, near the terminator. The region to the left is quite heavily scarred by large craters, but the region to the right is totally lacking in large craters. (Photograph courtesy of NASA/JPL)

and 150°W longitude. These densely cratered landforms superficially resemble the highlands on the Moon and Mercury. However, compared to the Moon and Mercury there is a relative shortage of small craters, that is, ones having diameters less than 20 km. As yet no continuous ejecta deposits or ray craters have been identified. Many of the larger craters are relatively fresh, but others are so badly degraded that they are almost indistinguishable from the surrounding terrain. Some of the large craters have an irregular or polygonal shape, suggesting the presence of a deep rubble zone below the surface.

The second major forward-face landform consists of terrain heavily scarred with numerous small, relatively fresh craters but with almost no craters any larger than 40 km across. This type of terrain extends south from the north polar region as far as 60°S and covers virtually all of the eastern half of the Saturn-facing hemisphere of Rhea.

The remnants of three large multiringed basins have been identified on the heavily cratered leading hemisphere of Rhea. These multiringed basins resemble the Valhalla feature on Callisto and the Caloris Basin on Mercury, but they are not nearly as prominent. In fact, they were completely missed in the early examinations of the Voyager photographs. A relatively fresh basin is centered at 35°N, 150°W. Its most prominent ring is 450 km in diameter. A second, more degraded basin has been identified at 30°N, 320°W. Its total radius is 700 km. What rings that can be seen are little more than indistinct curvilinear features that rarely extend any more than 90° around the basin center. The most prominent of these ring fragments is a series of broad ridges and coalescing pits approximately 275 km in radius. A third basin appears at 10°S, 310°W. Although this basin is quite badly degraded, two continuous rings can be identified. These two rings have diameters of 350 and 700 km, respectively. As yet, none of these basins have been assigned a name.

There are numerous linear features found on the forward face of Rhea. Many of the regions that have many large craters also have numerous parallel linear troughs and coalescing pit chains, generally with a north-northeast orientation. Most of these troughs and pit chains are superimposed upon large craters but are overlain by small craters. This must mean that their formation largely postdates the era of intense asteroid bombardment but probably precedes the episode of ice volcanism that resurfaced much of Rhea. The troughs and pit chains seem to have been formed during an episode of global expansion caused by the slow cooling and eventual freezing of the Rhean interior. Scarps and ridges with relatively high relief can be seen in numerous places, especially inside lightly cratered regions. These features clearly postdate the troughs and ridges. They seem to have been produced during a time of global contraction that began shortly after the era of intense water volcanism had come to an end.

The trailing hemisphere (Figure 13.10) of Rhea looks a lot like that of Dione, with a complex pattern of bright wisps overlying an ancient darkly colored terrain. These markings were observed only at relatively poor resolution during the Voyager 1 encounter, so their nature is uncertain. Perhaps the darker regions present on this side were produced by the extrusion of massive amounts of dirty ice from the interior of Rhea. This extruded ice covered up and obliter-

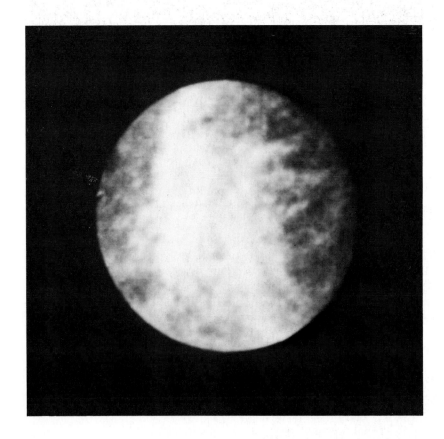

FIGURE 13.10 *Voyager 1* photograph of complex pattern of wisps on trailing face of Rhea. North is at the top. The pattern of wisps does not appear to follow any pattern laid down by craters. Some heavily cratered terrain can be seen near the western limb. (Photograph courtesy of NASA/JPL)

ated any craters that might have previously been present. The brighter wisps seem to be geologically younger than the darker terrain. They may have been produced by fresh, relatively uncontaminated ice flowing upward from the interior along geological fault lines produced by the impacts of large meteoroids. Although no meteoric impact craters have been positively identified on the rear face of Rhea, some of the dark features do have a roughly circular outline, suggestive of an impact origin. We cannot be certain until more detailed photographs become available.

The widely differing terrain types suggest that Rhea, like Dione, underwent a period of extensive water volcanism in the past. The terrain now heavily scarred with so many large craters was probably formed by the impacts of asteroid-size bodies soon after Rhea first formed a solid surface. This bombardment came to an end approximately 4 billion years ago. By that time, most of the large asteroid-sized objects in the Solar System had been swept up by a collision with one or another of the new planets. Shortly thereafter the icy surface of Rhea became geologically active. Hot liquid water was forced upward from the interior, flowed out over the surface, and froze solid. The volcanic activity was so intense that much of Rhea was completely resurfaced. In some places, the layer of fresh ice was so deep that it erased all of the preexisting craters, leaving behind a pristine, unscarred surface. In other locations it covered only the smaller craters, leaving the larger ones intact. The era of Rhean water volcanism came to an end shortly thereafter. At a somewhat later time smaller objects originating from within the Saturn system itself rained down upon the frozen surface of Rhea, excavating the smaller craters seen today in numerous plains units.

TITAN—AN ANCIENT, PREBIOLOGICAL EARTH?

Saturn's largest moon, Titan (S6), lies in an orbit 1,221,860 km in radius. It was discovered by the Dutch astronomer Christian Huygens in 1655. Titan's diameter is 5150 km, which makes it just slightly smaller than Jupiter's moon Ganymede. Its density is 1.9 g/cm³, the largest of any of Saturn's moons. The density is consistent with Titan being a 50:50 mixture of rock and water ice. The most interesting feature of Titan is that it possesses a significant atmosphere!

THE PRE-VOYAGER ATMOSPHERE

The existence of Titan's atmosphere has been known since 1944, when the astronomer Gerard P. Kuiper discovered the absorption spectrum of gaseous methane in the light reflected from the moon. This finding caused a sensation, since it was the first evidence for the presence of an atmosphere on a satellite. However, in the pre-Voyager era specific details about the composition, temperature, and pressure of Titan's atmosphere were largely a matter of conjecture.

Titan is extremely difficult to study from Earth because of the interference produced by light reflected from nearby Saturn. Nevertheless, terrestrial astronomers were able to obtain some excellent infrared spectra during the 1960s. Detailed study of these spectra turned up some rather odd anomalies. For one, the spectra seemed to indicate that Titan's atmosphere might be much denser than would be expected for a world of such low mass. In addition, the amount of infrared light emitted by the moon is unusually high, which might indicate that Titan could be much warmer than would be anticipated for a body so distant from the Sun.

Several models of the Titan atmosphere were proposed to explain the data. The models were in sharp conflict with each other, some predicting a thin atmosphere with cold surface temperatures and others proposing a dense, thick atmosphere with surface temperatures approaching those of Earth. Some theorists proposed that the unusually intense infrared radiation actually comes from a strong temperature inversion layer high in the upper atmosphere, and that the temperature at the surface is actually rather cold, about −190°C. They estimated the surface pressure to be only 0.02 atm. Other theorists proposed that the data can be best explained by the presence of a massive amount of nitrogen in Titan's atmosphere. There could be enough nitrogen in the air to produce a surface pressure of 20 atm and a greenhouse-enhanced surface temperature as high as −70°C. This would make Titan's surface nearly as warm as some of the colder regions on Earth! The surface of Titan could conceivably be sufficiently benign to allow some sort of life to be present there.

THE POST-VOYAGER ATMOSPHERE

It was not until 1980 (with the flight of *Voyager 1*) that enough detailed information could be obtained so that some sort of rational choice could be made between the various conflicting models of Titan's atmosphere. In that year, *Voyager 1* passed within 6490 km of Titan, flying behind the moon so that the radio signal transmitted back to Earth would momentarily pass through the densest parts of the atmosphere near the surface. The resulting radio-occultation pattern provided the first measurement of the temperature–density–depth profile of Titan's atmosphere (Figure 13.11).

Titan's atmosphere turned out to be rather denser and thicker than most people had expected. The atmosphere is actually thicker than Earth's and extends 10 times farther above the surface. The surface pressure is 1.5 atm, although the temperature there is quite cold, about −180°C. There is no sign of any atmospheric heating originating from the interior, although there may be a moderate greenhouse effect that keeps the surface slightly warmer than would be expected for a body at Titan's distance from the Sun. There is a strong inversion layer at an altitude of 200 km, where the temperature is as high as −70°C. This warm layer of air was undoubtedly the source of the excess infrared light emanating from Titan.

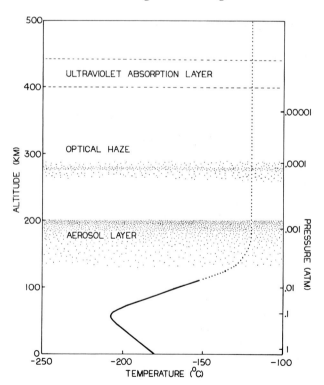

FIGURE 13.11 The vertical profile of the atmosphere of Titan.

Voyager measurements indicate that the dominant component (82–94 percent) of the atmosphere is nitrogen gas (N_2). The methane concentration is only about 0.6 percent, although it is possible that the methane concentration is as high as 10 percent near the ground. The presence of argon is suspected, but it was not detected by the instruments aboard the spacecraft. The argon concentration is probably less than 6 percent, although there could be as much as 12 percent argon closer to the surface. No water vapor was detected; indeed, none would be expected at such cold temperatures. Any water present on Titan must be in the solid state, permanently frozen to the surface. There does not appear to be any oxygen or neon.

The instruments aboard *Voyager 1* detected trace amounts (in the part per million range) of such gases as ethane (C_2H_6), acetylene (C_2H_2), propane (C_3H_8), ethylene (C_2H_4), diacetylene (C_4H_2), methylacetlyene (C_3H_4), cyanoacetylene (HC_3N), cyanogen (C_2N_2), and hydrogen cyanide (HCN). Ground-based astronomers using the 4-m telescope of the Kitt Peak National Observatory have detected 50 to 150 ppm carbon monoxide (CO) in the Titan atmosphere. There are even trace amounts of carbon dioxide in the parts per billion range.

The hydrocarbon gases such as ethane and acetylene are almost certainly not indigenous to Titan, but were instead formed quite recently by the ultraviolet disruption of methane in the upper atmosphere. Photochemically excited molecules of methane are highly reactive and readily combine with other methane molecules to form higher-molecular-weight gases such as ethane, acetylene, and ethylene. The hydrogen cyanide in the Titanian atmosphere almost certainly was also created photochemically via reactions between excited methane and nitrogen species. The presence of even a trace amount of carbon dioxide in the atmosphere is somewhat of a mystery, as this gas should long ago have condensed out as a solid at the exceedingly cold temperature present on Titan. The CO_2 in the atmosphere is possibly a product of steady-state reactions between carbon monoxide (which does not condense) and the by-products of the photochemical destruction of water. However, since there is currently no detectable amount of water vapor in Titan's atmosphere, this hypothesis is uncertain.

The earlier Saturn probe *Pioneer 11* noted a cloud of hydrogen trailing behind Titan in its orbit. This cloud extends from the orbit of Rhea outward to nearly 1.5×10^6 km from Saturn. The hydrogen in this cloud probably originated in the ultraviolet destruction of methane high in the Titanian atmosphere, the methane molecule being split up into hydrogen atoms and various carbon-containing radical species. The atmosphere of Titan is cold enough to trap the carbon-rich radicals, but the hydrogen is able to escape into space.

Titan has no detectable neon gas in its atmosphere. This is rather puzzling, since neon has a high cosmic abundance and must have been available in ample supply in the primitive solar nebula from which the planets and moons condensed. The Sun actually has almost as much neon as it does nitrogen. Perhaps the gravitational pull exerted by the accumulating proto-Titan was not strong enough to trap neon directly from the solar nebula.

The isotopic composition of the argon presumed to be in the atmosphere is unknown. However, most of the argon present is likely to be primordial, as there is only enough rocky material in Titan's interior to give the atmosphere at most 70 ppm radiogenic argon via the decay of ^{40}K.

Like Earth's atmosphere, Titan's atmosphere must be secondary. It accumulated only after the moon itself had fully formed. The presumed predominance of primordial argon suggests a model in which Titan's entire atmosphere was produced by the outgassing of primordial volatiles accreted from the solar nebula that had become trapped within clathrate compounds in Titan's crust and interior. A clathrate compound is one that is able to hold a chemically inert species such as argon within a sort of internal "cage" without actually forming a chemical bond. Titan's original atmosphere was probably a highly reducing mixture of ammonia and methane intermixed with inert gases such as argon. At this time there could have been enough ammonia in the atmosphere to have produced an appreciable greenhouse effect, and the initial surface temperature could have been as high as −120°C. However, the ammonia underwent rapid ultraviolet photodecomposition, forming N_2 and H_2. The hydrogen gas was so light that it quickly escaped into space, but the heavier nitrogen gas remained trapped on Titan. In this way, a much less reducing atmosphere rich in nitrogen was created. It is believed that Earth's atmosphere underwent a similar transition during the first billion years of its existence. Titan is of such intense current interest precisely because its present state may shed much light on what happened on our world during its early prebiological existence.

THE ORANGE CLOUDS OF TITAN

Voyager 1 found Titan's surface to be completely concealed by a dense orange-colored aerosol layer hovering about 200 km above the surface (Figure 13.12). In addition, there is a thin layer of haze lying about 100 km above the top of the aerosol layer. The aerosol layer is a suspension of many billions of tiny solid dust particles averaging about 0.5 μm in size. The composition of the aerosol particles themselves is uncertain, but the orange color of the layer indicates that they are probably rich in organic polymers. The best candidates for the organic material in the aerosol layer are long-chain hydrocarbons produced by the photochemistry of methane in the upper atmosphere.

This aerosol layer appears to be the cause of the temperature inversion in the upper atmosphere. The particles absorb a good deal of the incident sunlight, heating the upper reaches of the atmosphere to temperatures as high as −70°C. So little sunlight reaches the surface of Titan that it is probably never any brighter there than it is on a moonlit night on Earth.

THE SURFACE—A GLOBAL OCEAN?

Because of the dense aerosol layer, next to nothing is known about the surface of Titan. It cannot even be verified that Titan's rotation rate is synchronous with its orbit around Saturn, but this is almost certainly the case. Because of our ignorance, speculation about Titan's surface can roam fairly freely. In spite of its cold temperature, there is a distinct possibility that Titan possesses an ocean!

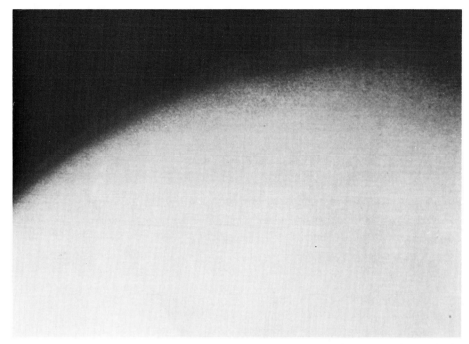

FIGURE 13.12 *Voyager 1* photograph of the limb of Titan. (Photograph courtesy of NASA/JPL)

Calculations indicate that the current rate of ultraviolet decomposition of Titan's atmospheric methane is sufficiently high so that all of the methane now found in the Titanian atmosphere should have been destroyed a long time ago. Why, then, is the methane still there? The only possible answer is that there must be an internal source that has continuously replenished the methane in Titan's atmosphere over the years. Perhaps the source is a reservoir of liquid or solid methane on the surface that vaporizes and replaces the methane in the air as rapidly as it is destroyed.

The temperature at the surface of Titan is quite near the *triple point* of methane, the temperature at which methane coexists in the solid, liquid, and vapor phases. However, it is probably not cold enough for nitrogen to liquify. Some have proposed that Titan has a methane ocean as deep as 0.5 km that completely covers the surface, leaving no land areas exposed. On the other hand, the methane concentration in the atmosphere just above the surface at the two points where the Voyager radio-occultation measurements were made seems to be somewhat less than the saturation value that should exist over any sort of global methane ocean. This may mean that the liquid methane (if it actually exists) does not cover the entire surface. At least some "dry" land is exposed to the air. Alternatively, some of this methane may exist in the solid state as part of an outer crust.

Other liquids have been considered for the Titanian ocean. It is possible that the ocean is made up largely of liquid ethane, originally created by the photochemistry of methane in the atmosphere. Alternatively, there might be an ocean of liquid ammonia at the surface that is as deep as 100 km. We can only speculate until more knowledge becomes available.

What might the solid surface of Titan be like? The surface may consist of a mantle of ammonia–methane clathrates that supports an asthenosphere made up of a liquid-water–ammonia slurry that is topped off by a lithosphere rich in water ice. There could conceivably be a thick blanket of hydrocarbons, amino acids, nitrogenous bases, and other such organic molecules covering the entire surface of Titan. These might be formed via a steady rain of heavy molecules that are created photochemically in the upper atmosphere. There could be enough surface hydrocarbons present to form a thick organic crust. The landscape of Titan may be a surrealistic image of methane rivers slicing their way through hydrocarbon canyons. Titan's surface is probably a rich organic chemistry laboratory, where the ingredients necessary for the creation and maintenance of life are in great abundance. However, the surface temperature is probably much too cold for life itself to have originated there. Further evidence must await the arrival of sophisticated landing probes that can explore the surface.

NASA and the European Space Agency are currently exploring the possibility of a joint mission to the Saturn system, one which could provide some of the answers to the questions about the surface of Titan. The proposed mission, named *Cassini*, will be launched in 1993-94 and will arrive in the Saturn system some 6 or 7 years later. The craft will perform a tour of the Saturnian system of moons, using Titan for gravity-assisted maneuvers. During one of the Titan encounters, the probe will release a lander that will parachute through the atmosphere of this moon and land on its surface. The landing probe will carry instruments to study the structure and composition of the Titan atmosphere on its way down to the surface. As yet, no funding has been provided for this mission.

HYPERION—THE TUMBLING WORLD

The odd moon Hyperion (S7) was independently discovered by the American astronomer George Bond (1825–1865) and the English astronomer William Lassell (1799–1880) within a few months of each other in 1848. Both are commonly given credit for the discovery.

Hyperion is a small moon about the size of Mimas that lies in a rather eccentric ($e = 0.1$) orbit averaging 1,481,000 km distance from Saturn. *Voyager 2* came no closer than 471,000 km to this moon, but the photographs are sufficiently detailed to show Hyperion to be the most irregular of Saturn's larger satellites. It is three-fifths as wide as it is long, its longest dimension being 450 km. The moon has been described as looking like a gigantic hamburger with a bite taken out of it (Figure 13.13). Hyperion may be a fragment of a larger object that was shattered by a giant impact in the distant past.

Hyperion is one of the darkest objects in the Solar System, with a reflectivity of only 20 to 30 percent. Craters 20 to 50 km in diameter are common, and there is no evidence for any internally driven resurfacing such as is found on

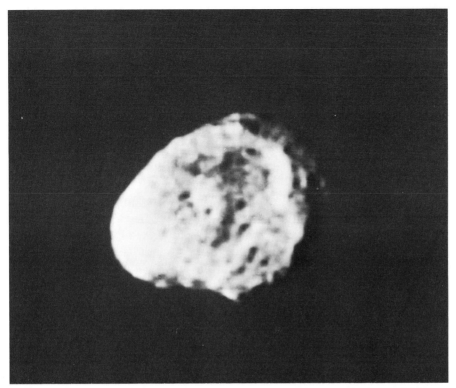

FIGURE 13.13 *Voyager 2* photograph of Hyperion. (Photograph courtesy of NASA/JPL)

some of the inner moons of Saturn. Hyperion may have the oldest surface in the Saturn system, having remained relatively inactive ever since the initial fragmentation event that formed it. The largest crater on Hyperion is 120 km across and about 10 km deep. There is a series of long scarps linked together into one single giant sinuous scarp 300 km long. It may be the boundary of an ancient crater 200 km in diameter. This ridged feature is named Bond – Lassell, after the moon's codiscoverers.

The density of Hyperion is unknown. However, the spectrum of water frost has been detected in the light reflected from the moon, and it can be presumed that the interior is largely water ice. The origin and composition of the dark surface is a mystery. It could be a relatively thin layer of extrinsic material that originally came from farther out in the Saturn system. Or it could be a dark, carbonaceous material that extends throughout the entire volume of the moon. The answer awaits further exploration.

Hyperion's orbital period is nearly 22 days long, but the sequence of images taken by *Voyager 2* covers such a short arc of its orbit that it cannot be definitely established that the moon is in synchronous rotation or even if the moon is rotating at all. An additional problem is that the long axis of Hyperion does not point toward Saturn, as would be expected for an oblong body locked into synchronous rotation. It is, in fact, tilted at an angle of approximately 45° with respect to the plane of its orbit. Recent ground-based observations seem to suggest an average rotational period of only 13 days rather than the 21 days that would be required for synchronous rotation.

The cause of Hyperion's anomalous rotational behavior is as yet unknown. It is close enough to Saturn that it should long ago have become permanently locked into synchronous rotation. Perhaps Hyperion was recently struck by a large meteoroid and knocked out of synchronous rotation and has not had sufficient time to lock back in. However, no scars of a recent meteoroid strike are observable.

Several researchers have proposed that the odd shape and eccentric orbit of Hyperion force the moon to tumble in space in a chaotic fashion as it travels around Saturn. Hyperion's orbital eccentricity is kept high by periodic gravitational tugs exerted by nearby Titan. As Hyperion moves along its eccentric orbit, the tidal forces exerted by Saturn attempt to lock the moon into synchronous rotation. However, these forces vary by such a large amount over the extent of the eccentric orbit that the lock is periodically broken and Hyperion is sent erratically tumbling. Many more studies of Hyperion are needed before this model can be verified.

IAPETUS—THE TWO-FACED WORLD

Iapetus (S8) is one of the strangest objects in the Solar System. It is about 3,560,800 km from Saturn, with a diameter of 1460 km and a density of 1.16 g/cm^3. It is slightly less dense than the inner moons, which may mean that its relative proportion of water ice is correspondingly larger. What makes this particular moon so interesting is that it is 6 times more reflective on one side than on the other!

Iapetus was discovered by Giovanni Cassini in 1671. During his early observations, he noted that the moon mysteriously vanished during part of its orbit! Later telescopic observations found that the reason for this odd behavior was that the trailing side of Iapetus has the reflectivity of snow (50 percent) whereas the leading side is no more reflective than asphalt (4 percent; see Figure 13.14).

Voyager 2 passed within 900,000 km of Iapetus in 1981, and the photographs that were sent back only deepened the mystery. Voyager photographs indicate that virtually the entire leading hemisphere is covered with an enormous dark splotch (named Cassini Regio) that has no apparent markings of any sort (Figure 13.15). It has a reddish color, indicating that this region may be rich in carbonaceous material. The daytime temperature of the dark zone is a rather high −160°C, owing to the extraordinarily low reflectivity. The trailing hemisphere is much brighter and is probably largely covered with water ice. The boundary between the two hemispheres is not sharp, but rather gradational and meandering.

Much of the bright face of Iapetus is heavily cratered, but there is no evidence there for any sort of resurfacing similar to that which has taken place

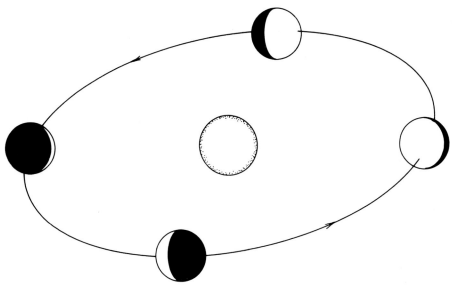

FIGURE 13.14 Iapetus is in synchronous orbit around Saturn. The hemisphere of Iapetus that faces in the direction of its orbital motion is much darker than the trailing face.

on some of the inner moons of Saturn. The prominent features on the moon are named for characters and places appearing in the *Song of Roland,* a French epic describing the defense of Europe against Saracen invaders during the early Middle Ages. Many of the craters near the boundary of Cassini Regio seem to have rather dark floors, reminiscent of craters on the Moon that were filled up by lava when volcanic material flowed over them. This suggests that the dark side of Iapetus might be a good deal younger than the bright side.

The origin of Iapetus's asymmetric surface is a mystery. It is unlikely that the asymmetry reflects any difference in bulk interior properties between the two hemispheres. If this were the case, Iapetus would probably have hemispheres of different densities, which would tend to align in the direction of Saturn rather than in a forward–trailing orientation. An alternative explanation is that water ice was preferentially eroded off the leading surface by meteoroid impacts, exposing a much darker surface underneath. Another possibility is that water ice was preferentially deposited on the trailing surface by some as-yet-unknown mechanism.

Perhaps Iapetus was preferentially coated with dark material originating elsewhere within the Saturn system. A model originally proposed in 1974 by Cornell University researcher Steven Soter suggested that the outer moon Phoebe is the source. According to the model, dark material eroded from the surface of Phoebe spiraled inward toward Saturn under the influence of solar : radiation pressure. As it fell inward, some of this material was swept up by the forward-facing hemisphere of Iapetus.

Phoebe is indeed rather dark, but telescopic observations indicate that its

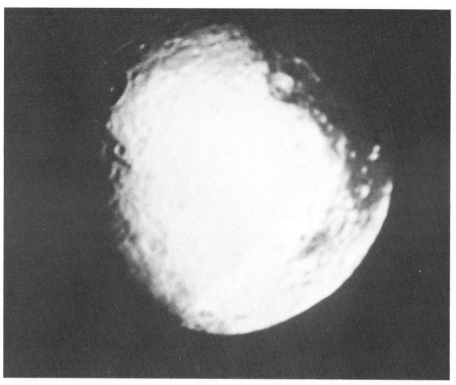

FIGURE 13.15 *Voyager 2* photograph of Iapetus, taken at a range of 10^6 km. North is at the top. The dark Cassini Regio can be seen on the eastern limb in this view. The bright region near the center is Roncevaux Terra. The craters near the terminator are Charlemagne and Osier. (Photograph courtesy of NASA/JPL)

reflectance spectrum is somewhat different from that of the dark face of Iapetus. So it is probably unlikely that the dark-side material on Iapetus originated there. If Phoebe were actually the source of the dark material, the rate of cratering on that particular moon would have to have been quite high. Such a high cratering rate could have taken place only very early in the history of the Saturn system, at the time of the most intense meteoric bombardment over 4 billion years ago. This requires an ancient age for Cassini Regio. If this were true, then it seems likely that we should see several bright craters within this dark region that were excavated by later meteoric impacts that penetrated the dark blanket to expose the bright material underneath. None are found, which means either that the dark material is several kilometers deep or that it was deposited in the relatively recent past.

The most currently favored model is that the dark material in Cassini Regio was extruded from the interior of Iapetus, since it is unlikely that particulate material falling onto the surface from the outside would preferentially cover only the bottoms of craters. This dark material is almost certainly not ordinary lava like that present on the Moon. It is more likely to be a slurry of ammonia, water ice, and carbon-rich compounds — perhaps not all that different in composition from frozen coal tar. The rather low density of Iapetus is consistent with its being 40 percent water ice, 25 percent rock, 5 percent ammonia ice, and 30 percent hydrocarbons. The hydrocarbons could have formed from methane buried in the interior being subjected to high temperatures and pressures early in Iapetus's history. There is no evidence for any appreciable amount of methane on the small inner moons of Saturn, but it may well have been cold enough in the outer regions of the proto-Saturn nebula for appreciable amounts of methane to condense.

PHOEBE — WORLD AT THE EDGE

The outermost Saturn moon, Phoebe (S9), was discovered by the American astronomer William Pickering in 1898. Phoebe is in a highly inclined elliptical retrograde orbit averaging 12,954,000 km from Saturn. Its discovery caused a sensation, because it was the first retrograde satellite to be found.

Phoebe is quite small (only 220 km in diameter) but is nearly spherical in shape. *Voyager 2* came no closer than 1.5×10^6 km, so not much is known about the moon. Voyager photographs did not show any distinct impact craters, but the distance was so great that no craters smaller than 50 km across could have been resolved. The general spherical shape of Phoebe does, however, suggest that it has not been subjected to the same amount of catastrophic meteoric bombardment that the inner Saturn moons have undergone.

Earth-based observations in the early 1970s seemed to indicate that Phoebe is not in synchronous rotation. Voyager photographs confirmed that such is the case: the rotational period is only 9 hours (the period of revolution is 1.5 Earth years) and the rotation is prograde rather than retrograde. The rotational axis is within 35° from the perpendicular to its orbital plane. Phoebe is evidently too far from Saturn for tidal forces to have played any role in establishing its rotation rate.

The density of Phoebe is unknown, so any details about its composition and internal structure are uncertain. The surface is quite dark, with a reflectivity of only about 5 percent. Phoebe is about as dark as the dark material on Iapetus, but is somewhat less reddish. In contrast to the other airless moons of Saturn, the infrared reflectivity spectrum of Phoebe shows no evidence for the presence of water ice. The spectrum more closely matches that of some of the darker asteroids orbiting close to the Sun. Perhaps Phoebe originally formed in the inner part of the Solar System and was accidentally thrown out to Saturn by a close encounter with Jupiter. The plane of Phoebe's elliptical orbit is much closer to the ecliptic plane than it is to Saturn's equatorial plane, which could be evidence that this outermost moon was originally a small asteroid that approached too close to Saturn and was captured.

In 1977, California Institute of Technology astronomer Charles Kowal discovered a small, asteroid-sized object he named Chiron orbiting in the region between Saturn and Uranus. It has a diameter estimated to lie somewhere between 50 and 320 km and is in a highly elliptical orbit (eccentricity of 0.38) with a large inclination (7°) with respect to the ecliptic. Chiron's orbit crosses that of Saturn, and it is possible that this object may have originally been an outer moon of Saturn that managed to escape. Or perhaps it is an errant asteroid that has so far managed to avoid capture by the giant planet. Other escaped moons of Saturn may lie in this region waiting to be discovered.

SATURN'S SMALL MOONS

In the pre-Voyager era, many Earth-bound astronomers made careful searches of the region near Saturn for additional moons. Dozens of sightings were reported, and numerous claims of discovery were made. However, sharp controversy surrounded each and every report. The observation of small moons orbiting close to such a bright source as Saturn and its rings is a notoriously difficult enterprise. It is not easy to be certain whether an indistinct spot on a photographic plate is really a previously unknown moon or simply an artifact of the imaging process.

When a new moon of Saturn is reported, it is first assigned a number. The numbering scheme is rather complicated and cumbersome. For example, the number 1980S27 designates the sighting of the 27th previously unknown object within the Saturn system in the year 1980. A single moon could in principle have several different numbers! Only when there is general agreement that the newly reported body actually exists and its orbit is well determined is it assigned an official name by the International Astronomical Union.

A resolution of the controversy came only with the passage of the Voyager spacecraft through the Saturn system. The photographs sent back to Earth were carefully scrutinized for the presence of previously undiscovered moons. A total of eight were found (Figure 13.16). Some claims of discovery from Earth were confirmed. Others were disproven. Some objects were found whose existence was entirely unexpected. The small moons of Saturn turned out to have some of the strangest orbits yet seen anywhere in the Solar System.

COORBITALS AND SHEPHERDS

In 1967 Audouin Dollfus, working at the Pic du Midi Observatory in France, announced the discovery of a small moon he named Janus orbiting at a distance of 159,300 km from Saturn's center. Texereau and Walker independently confirmed the existence of a small moon in this region. In 1978 Fountain and Larson of the University of Arizona noted that there appeared to be two small moons orbiting in this region with a hint of a possible third. However, Aksnes and Franklin of the Harvard–Smithsonian Observatory reexamined the relevant photographs and concluded that the orbital parameters that had been reported for these new moons should be regarded as questionable at best and that some of these objects may not actually exist.

An unambiguous answer had to await the arrival of the Pioneer and Voyager spacecraft. It now appears that there are a total of five moons orbiting just outside the ring system (Figure 13.17).

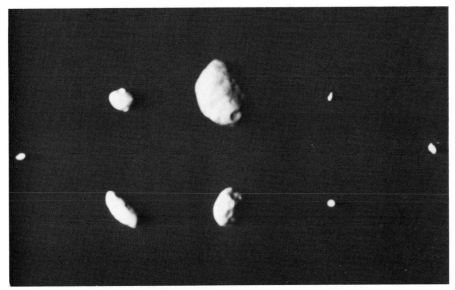

FIGURE 13.16 Montage of the small satellites of Saturn. Starting from the left, the moons are: the A-ring shepherd 1980S28, the F-ring shepherds 1980S27 and 1980S26, the coorbitals 1980S1 and 1980S3, the Tethys Lagrangians 1980S13 and 1980S25, and finally the moon Dione B (or 1980S6) on the far right. The moons are shown in their correct relative sizes. (Photograph courtesy of NASA/JPL)

The outermost pair of moons may be the same objects that were seen by Dollfus and by Fountain and Larson. *Pioneer 11* had actually spotted a small moon in this general region during its 1979 flyby of Saturn; it was nicknamed "Pioneer rock" by the imaging team. The spacecraft seems to have flown through the magnetic wake of one or the other of these moons, perhaps only narrowly avoiding a collision. *Voyager 1* spotted both moons during its flypast in 1980. These objects were initially assigned the numbers 1980S1 and 1980S3. A couple of years later, it was decided to name the first of these moons Janus (S10) in acknowledgment of Dollfus's claim. The other was named Epimetheus (S11). It is not clear which of these two objects was the one that Dollfus saw. The two moons have rather irregular shapes, with Epimetheus being 140 km long and 100 km wide and Janus being 220 km long and 160 km wide.

Careful Voyager measurements of the orbits of Janus and Epimetheus turned up an unexpected surprise. These two objects pass so close to each other that they periodically exchange orbits! This exchange process is shown schematically in Figure 13.18. The two moons lie about 151,450 km from Saturn in orbits that are only 50 km apart. At the time of the *Voyager 1* encounter in November 1980, Epimetheus was closer to Saturn and was slowly catching up to Janus in its orbit. The moons passed each other quite shortly thereafter but they did not actually collide. Instead the mutual gravitational attraction between the moons when they neared each other caused them to exchange orbits, so that Janus became the closer moon. By the time of the *Voyager 2* encounter in

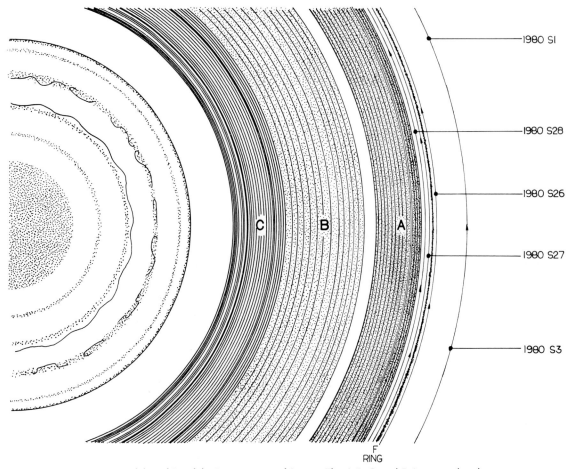

FIGURE 13.17 Diagram of the orbits of the inner moons of Saturn. The A,B, C, and F rings are also shown.

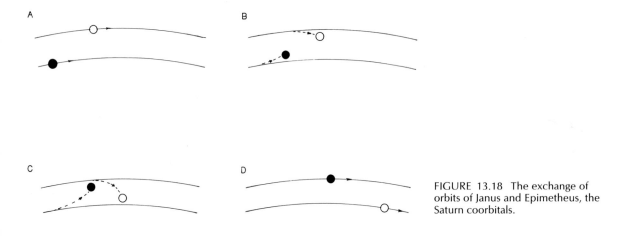

FIGURE 13.18 The exchange of orbits of Janus and Epimetheus, the Saturn coorbitals.

August 1981, this exchange had already taken place. This celestial dance takes place about once every 4 years, when the two partners reexchange orbits. The fact that Janus and Epimetheus are in such close orbits may mean that they were originally part of a larger body that was broken in two by a severe impact. The two moons are sometimes referred to as the "Saturn coorbitals."

Voyager 1 also discovered three new moons orbiting even closer to Saturn. None of them had been previously reported by Earth-based astronomers.

The nearest moon is 1980S28, at a distance of 137,670 km from Saturn's center. It has a diameter of only 30 km and just skims the outer edge of the A ring. It has now been assigned the official name Atlas. The motion of Atlas seems to "sharpen" the outer edge of the A ring. It does so by preventing the particles near the outer edge of the ring from wandering further out into space under the influence of gravitational forces exerted by other moons. Consequently, Atlas is sometimes called the "A-ring shepherd."

Next out are 1980S27 and 1980S26. These moons were assigned the names Prometheus and Pandora, respectively. They lie on opposite sides of the narrow F ring in nearly circular orbits only 2300 km apart (Figure 13.19). Neither of these moons is larger than 150 km in diameter. It seems that these two nearly coorbital moons act to keep the particles in the F ring confined to a narrow strip. The periodically varying structure of clumps and braids exhibited by the F ring may be a result of the complex motions of these two moons. Since these moons act to herd the F-ring particles into a narrow strip, these objects are sometimes referred to as the "F-ring shepherds." Prometheus and Pandora (as well as the F ring itself) may be the debris left over from a larger moon that was split in half by an ancient meteoroid impact.

The five tiny inner moons were poorly imaged by Voyager, so any surface or compositional details are uncertain. They all seem to have irregular shapes and rough, cratered surfaces, which indicate that they underwent intense meteoric bombardment in the distant past. The masses and densities of these moons are unknown. The spectrum of water ice has yet to be detected in the reflectance spectra of any of these moons, so its presence cannot be positively confirmed. But it is almost certainly present; the reflectivities are consistent with a surface of water frost contaminated by small amounts of dark material.

THE SATURNIAN LAGRANGIANS

The Saturnian system is unique in having several large moons with tiny companions that travel along with them in their orbits.

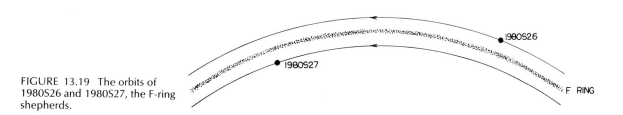

FIGURE 13.19 The orbits of 1980S26 and 1980S27, the F-ring shepherds.

Dione has a small partner only about 160 km in diameter. Known as yet only by the number 1980S6 (or S12), or sometimes simply as Dione-B, it occupies the L-4 Lagrange point 60° ahead of Dione in its orbit (Figure 13.20). It was first reported in 1980 by French astronomers just a few months before the arrival of *Voyager 1*. Its presence was confirmed in photographs taken by the spacecraft. There is even some evidence for the presence of a Dione-C (more formally designated by 1981S7) 60° behind Dione in its orbit. So far, it has not shown up in any of the Voyager photographs.

Tethys has two smaller companions located at the Lagrange points of its orbit (Figure 13.21). The object 1980S25 (S14) lies 60° ahead of Tethys, whereas 1980S13 (S13) is 60° behind Tethys. They have now been formally named Calypso and Telesto, respectively. These two small moons were both sighted by Earth-based astronomers just a few months before the arrival of the Voyager spacecraft in the Saturn system. Both were subsequently identified in Voyager photographs. They are quite small, being no larger than 35 km across. The trailing Tethys Lagrangian seems to be brighter than the leading object.

There are some reports of the existence of a third and even a fourth

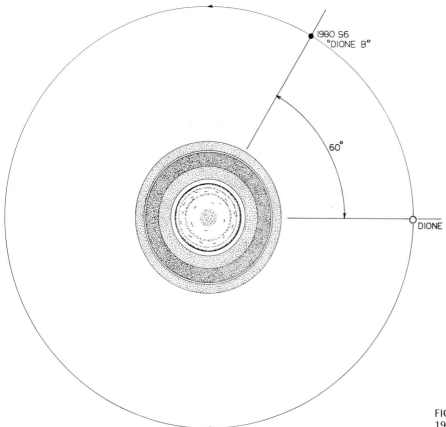

FIGURE 13.20 The orbit of 1980S6, the Dione Lagrangian.

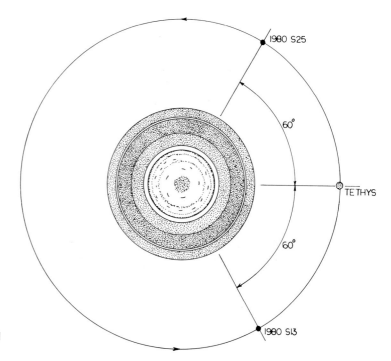

FIGURE 13.21 The orbits of 1980S25 and 1980S13, the Tethys Lagrangians.

companion of Tethys. These as-yet-unconfirmed objects have been designated 1980S34 and 1981S6. It is not yet clear whether these reports represent two separate moons or just two different sightings of the same object. Neither one of them has shown up in any of the Voyager photographs, so their existence is questionable at best.

Some charged-particle measurements made by the Voyager spaceprobes seem to indicate that another small moon may exist in approximately the same orbit as Mimas. It is designated 1980S12. Its presence has not yet been confirmed in any of the Voyager photographs. If this moon actually exists, it can be no larger than 5 to 10 km across.

Some Earth-based telescopic observations seem to indicate that there might be a bunching of E-ring particles at the trailing Lagrange point of the orbit of Enceladus. Perhaps a new moon is in the process of forming there. If so, this event would be exciting to watch in the years to come.

ODDS AND ENDS

There have been occasional reports throughout the years of tiny objects orbiting between the major moons of Saturn. In 1904 William Pickering reported the discovery of a moon he named Themis orbiting in the space between Titan and Hyperion. This object has never been spotted by any other ground-based astronomers nor has it appeared in any of the Voyager photographs. Even though Themis probably does not exist, it nevertheless often appears in tables and catalogs of Solar System data.

There is some indication for the presence of a small moon between the orbits of Tethys and Dione at a distance of 350,000 km from Saturn. This object has been designated 1981S10. If it exists, it cannot be much larger than 15 to 20 km across. There have also been reports of another tiny moon at a distance of 470,000 km from Saturn, between the orbits of Dione and Rhea. It has been designated 1981S9. Neither of these moons has yet appeared in any of the Voyager photographs.

DESTRUCTION AND REBIRTH?

The surfaces of Saturn's icy moons show evidence of two distinct eras of meteoroid bombardment. The earliest craters were created by the impacts of large numbers of asteroid-sized bodies still orbiting the Sun after the initial formation of the planets, whereas the later craters were a result of impacts of smaller particles originating from within the Saturn system itself. The initial asteroid bombardment of the moons of Saturn was probably contemporaneous with the formation of the lunar highlands. The impact craters left behind by these bodies are large, ranging in size from 20 to 100 km in diameter. This bombardment ended rather abruptly about 4 billion years ago, as the number of free asteroids orbiting the Sun rapidly became depleted. Subsequently, some of the larger moons such as Dione and Rhea underwent extensive resurfacing, which obliterated many of these large early craters. At a later time, impacts by smaller bodies orbiting within the Saturn system (perhaps of a size similar to the particles still lying in the ring system) scarred the surfaces of the moons with large numbers of much smaller craters. These impacts appear to have occurred preferentially on the forward-facing sides of the large moons such as Dione and Rhea, with the trailing faces being comparatively well protected.

The gravitational field of Saturn acts to focus meteoroids entering the system from the outside. This causes the average meteoroid flux to be significantly higher in the inner parts of the Saturn system than in the outer. The meteoroid flux is as much as two orders of magnitude larger at the distances of the innermost moons Mimas and Enceladus than it is at the position of the outer moons Iapetus and Phoebe. Estimates of the rates at which large meteoroids entered the Saturn system at the time of the intense bombardment epoch over 4 billion years ago have led to a startling conclusion. All of the moons initially located inside the present orbit of Dione must have been struck at least once by a body with enough energy to completely fragment them! After disruption, the bits and pieces of the fragmented moons entered separate orbits around Saturn, forming ringlike sheets of particles. Eventually, the fragments reassembled to form large moons once again. This probably happened at least once for objects that were initially located at the present position of Dione and as many as five times for objects that were located at the present position of Mimas. The numerous Lagrangian objects reported in the Saturn system may be fragments left over from the last disruption episode that became trapped in stable orbits and never recombined.

As in the case of the Jovian satellite system, there appears to be a general trend toward smaller densities as one travels outward from Saturn. The trend is, however, not nearly as marked as it is for the Jovian system. There are also significant exceptions to the rule, for example, the high density of Titan. It is possible that Saturn's moons were originally distributed in a manner similar to that now seen for the Galilean moons of Jupiter, with the inner moons being significantly denser than the outer moons. However, the numerous fragmentation events taking place in the inner reaches of the Saturn system in the distant past acted to mix the parts of different moons together, so that any initial differences in density between the moons located there were almost entirely eliminated.

PART FOUR

WORLDS OF ICE:
The outer solar system

Telescopic photograph of Comet West.

14

Uranus — a Planet Turned Sideways

All five of the inner planets were well known to the ancients and had been plotted on star charts almost since the invention of writing. In 1781 an additional planet was discovered purely by accident by William Herschel. Uranus is actually visible to the naked eye and had been plotted on star charts as early as the 1690s. However, its movement across the sky had gone undetected; Uranus was dismissed as nothing more than just another faint star of no particular significance. Herschel was the first to notice its movement against the stellar background.

Herschel originally proposed to name his new planet after his royal patron, King George III of England. After much persuasion he decided instead to continue with the ancient tradition of naming planets after deities. He chose the name Uranus, who was the father of Cronus (Saturn) in Greek mythology.

URANUS FROM THE EARTH

Uranus is in a slightly elliptic orbit with an average distance from the Sun of 19.2 AU. This places it almost twice as far out as Saturn, and it takes 84 Earth years for Uranus to make a full circuit of the Sun.

Voyager 2 view of a crescent Uranus taken at a distance of one million km. (Photograph courtesy of NASA/JPL)

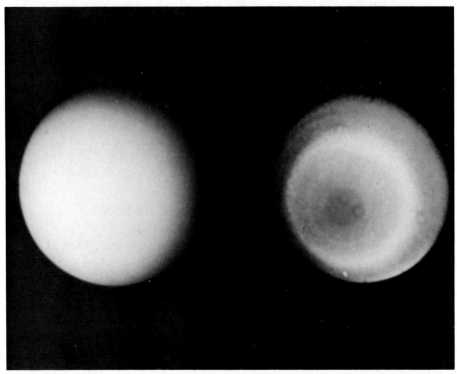

FIGURE 14.1 Full-face view of Uranus taken by *Voyager 2* during its initial approach to the planet. The south pole of the planet is near the center of the disk. An unprocessed photograph of the planet is on the left, showing a rather featureless disk. A computer-enhanced version of the left-hand view is shown on the right. This view brings out some detail in the outer atmosphere of the planet, showing a cap of haze over the south pole. (Photograph courtesy of NASA/JPL)

Because of its extreme distance, Earth-based telescopic measurements of the properties of Uranus are exceedingly difficult to make. Uranus appears in the telescope as a pale blue-green disk with no discernible surface detail (Figure 14.1). The planet is 26,145 km in radius at its equator, which places it intermediate in size between the small inner terrestrial planets and the large outer giant planets (Figure 14.2). Uranus is slightly flattened at the poles, the oblateness being about 2.4 percent. This oblateness is appreciably smaller than that of Jupiter or Saturn.

Uranus has five moons observable by telescope from Earth. Consequently, the mass of the planet can readily be determined, being equal to 8.67×10^{28} kg (14.5 Earth masses). The average density of the interior is 1.24 g/cm³, so Uranus (like Jupiter and Saturn) must be made up largely of lighter elements. However, the density is sufficiently high so that the planet cannot be entirely or even largely made up of hydrogen and helium. The dominant components are probably water, ammonia, and methane, since these substances are the most abundant chemicals likely to have condensed out of the solar nebula at Uranus's distance from the Sun.

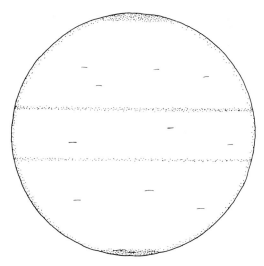

FIGURE 14.2 The relative sizes of Uranus and the Earth.

Because of the lack of apparent surface detail, the rotational rate of Uranus (like that of Venus) is exceedingly difficult to measure. One must rely on various indirect techniques (such as observations of the Doppler shifts of sunlight reflected from various parts of the planet's visible disk) to estimate the spin rate. The American astronomers Moore and Menzel used this technique during the 1930s and reported a rotational period of 10.6 hours.

The early measurements of Uranus's rotation turned up an unexpected surprise: the equator of the planet lies nearly perpendicular to its orbital plane (Figure 14.3). The planet is turned over on its side! All the other planets (with the possible exception of Pluto) have equators that lie approximately parallel to the planes of their orbits around the Sun. The north pole of Uranus is tilted slightly below its orbital plane, so the planet is actually classified as spinning in the retrograde sense.

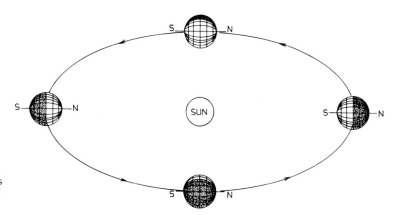

FIGURE 14.3 Diagram of Uranus's orbit around the Sun. The equator of Uranus is very nearly perpendicular to the plane of its orbit.

As Uranus travels along its orbit, the Sun is directly over the north pole of Uranus at one time in its year and directly over the south pole at another. Since a Uranian year equals 84 Earth years, each season is approximately 21 Earth years long. At present, Uranus's south pole is pointed almost directly toward the Sun, so the northern hemisphere is in the middle of a very long winter of utter darkness.

The sideways orientation of Uranus is somewhat of a puzzle, since most theories of planetary formation predict that all planets form in such a way that their equators lie approximately parallel to their orbital planes. One possible explanation is that Uranus did indeed originally condense from the primordial nebula with its equator oriented much closer to the plane of its orbit. At some time in the distant past, a large object collided with Uranus and knocked its spin axis over onto its side. Some of the debris from this catastrophic collision became trapped in orbit around the planet, ultimately to form the presently seen ring system and moons.

URANUS FROM SPACE

On January 24, 1986 the probe *Voyager 2* passed close to Uranus, approaching to within 81,600 km of the cloud tops. The craft sent back pictures of the outer atmosphere of the planet, studied the spectrum of the light that it emitted, and made careful measurements of the magnetic fields and the radiation density in the immediate vicinity of the planet.

A CORE OF SLUSH

The *Voyager 2* flyby provided enough information so that a realistic model of the interior of the planet could be constructed. The model predicts that the interior of the planet has three components: an outer layer of liquid and/or gaseous hydrogen, an inner mantle of water, and a dense rocky core (Figure 14.4).

The outermost part of Uranus visible from space consists of a dense atmo-

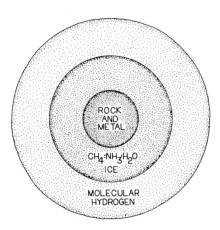

FIGURE 14.4 The interior of Uranus.

sphere of hydrogen and helium gases. The temperature and pressure of the atmosphere both increase with depth. The outer gas layer eventually merges with a far denser and hotter layer of liquid hydrogen. The liquid hydrogen ocean may extend as far as 11,000 km below the visible surface. Unlike Jupiter or Saturn, the mass of Uranus does not appear to be large enough to sustain any liquid metallic hydrogen in the interior.

Even though Uranus certainly does have massive amounts of hydrogen and helium, it has proportionally less of these light elements than does Jupiter or Saturn. Hydrogen and helium make up 90 percent of the mass of these giant worlds, but only about 10 percent of the mass of Uranus. The dominant component of Uranus seems to be a "mantle" of water lying beneath the outer hydrogen-rich layer. The mantle is approximately 8000 km thick and comprises approximately two-thirds of the matter in the planet. Significant amounts of ammonia and methane are probably also present within the mantle. Ammonia, water, and methane can exist only in the vapor or liquid phases at the moderate temperatures and pressures typically found in the Earth's mantle. However, the interior of Uranus is at such high pressures and temperatures that these volatile chemicals can exist only in the solid phase!

The temperature of the uppermost layers of the mantle (at the bottom of the liquid hydrogen ocean) may be as high as 2500°C, the pressure being as large as 200,000 atm. At a depth of 15,000 km into the planet the temperature is up to 7000°C and the pressure as high as 6×10^6 atm. Under these extreme conditions, water and ammonia are probably ionized. The Uranian mantle may be an excellent conductor of electricity. It is even possible that the methane in the lower mantle has been entirely stripped of its hydrogen by the high temperatures and pressures, leaving behind elemental carbon. The elemental carbon could be under such high pressure and temperature that the interior of Uranus is rich in diamonds!

There is probably a small, dense core at the very center of Uranus. The Uranian core is probably about 3.5 times as massive as the Earth and is 6500 km in radius. The matter within the core is probably largely metals and silicates. The pressure and temperature at the center of the core are estimated to be 1.7×10^7 atm and 7000°C, respectively. This core is undoubtedly hot enough to be largely molten.

THE TILTED MAGNETIC FIELD

The presence (or absence) of a magnetic field can tell scientists much about the structure and composition of the interior of a planet. Long before the Voyager encounter, many workers speculated on whether or not Uranus would be found to possess a magnetic field. Voyager confirmed that a field was indeed present. However, this field turned out to be perhaps the strangest yet encountered in the Solar System.

Voyager 2 first encountered the Uranian magnetic field at a distance of 470,000 km from the planet, on the sunward side. Scientists found to their complete astonishment that the magnetic axis of the Uranian magnetic field makes an angle of approximately 55° with respect to its rotation axis. (Figure

14.5). Most of the planets that have magnetospheres also have off-axis magnetic fields, but Uranus is the most extreme example found to date. In addition, the center of the Uranian magnetic field is offset from the geometrical center of the planet by 7500 km in the general direction of the north geographic pole.

Because of the extreme angle of inclination and the large amount of offset in the Uranian magnetic field, the intensity of the field at the cloud tops ranges from as small as 0.1 to as large as 1.1 gauss. These field intensities are comparable to magnetic fields found at the surface of the Earth. However, the Uranian field is much larger in volume than the terrestrial field and contains 50 times as much stored energy. Large as the Uranian magnetic field is, it still contains only 0.25 percent of the energy stored in the intense Jovian field.

The origin of the Uranian magnetic field, like that of any other world, is largely unknown. The dense metal- and silicate-rich core of Uranus is presumably molten. However, theorists believe that the Uranian field is much too strong to originate solely from convective currents within a molten core, as the terrestrial field presumably does. Perhaps the dominant component of the field is produced by ionic currents within Uranus's hot, compressed mantle of water, driven by the thermal energy flowing outward from the molten core.

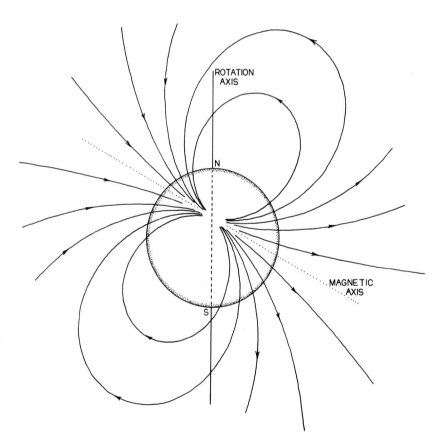

FIGURE 14.5 The tilted magentic field of Uranus.

The causes of the Uranian magnetic field's extreme angular inclination and large offset distance are even more uncertain. Perhaps these oddities are related in some way to the planet's unusual axial tilt. It is possible that the sideways orientation of Uranus is a consequence of a catastrophic event in the distant past, perhaps a collision with another body. Any process sufficiently violent to tilt the axis of so massive a planet by such a large amount was undoubtedly also strong enough to disrupt large volumes of the interior of the planet as well, producing the highly asymmetric field currently seen. Another possibility is that the Uranian magnetic field is currently in the act of reversing its polarity, just as the terrestrial field has done many, many times in the geological past. Whether or not this speculative theory corresponds to reality must await the advent of better models of the origin of planetary magnetism.

THE URANIAN MAGNETOSPHERE

As the solar wind flows past the planet, it is deflected by the magnetic field. The force exerted by the solar wind compresses the field lines in the Sun-facing direction and stretches them out in the downwind direction. The bow shock was encountered by *Voyager 2* at a distance of 600,000 km from the planet. Because of the high angular inclination of the magnetic field, the rotation of the planet causes the magnetic field lines in the downwind direction to twist around each other in a corkscrew fashion.

The magnetic field of Uranus entraps a large number of charged particles, forming a magnetosphere of high-temperature plasma and intense particle radiation. The charged particles within the magnetosphere appear to be exclusively protons and electrons. No heavier elements have yet been found. There are two principal plasma ion populations: a "warm" component (temperature of 100,000°C) confined largely within a distance of 175,000 km from the center of Uranus and a "hot" component (10^6°C) confined outside a radius of 125,000 km. The average density of the warm component is roughly 2 protons/cm^3, but the density of the hot component is about 20 times less. The warm component corotates with Uranus, but the ions in the hot component are slowly drifting inward toward the planet.

The high-energy trapped-ion population beyond a distance of 130,000 km from the planet is dominated by protons, but energetic electrons are also found. Proton fluxes are comparable to those measured in the outer magnetosphere of Saturn. The radial distribution of high-energy electrons exhibits dips in intensity that correlate with the orbits of the innermost three large Uranian moons Miranda, Ariel, and Umbriel. In fact, the radiation intensity is sufficiently high throughout the entire Uranian system that it is likely to have played an important role in the chemical and physical evolution of the material exposed on the surfaces of these moons.

Where do the particles in the Uranian magnetosphere come from? Possible sources of magnetospheric particles are the solar wind, the outermost hydrogen atmosphere of Uranus, the surfaces of Uranus's moons, or perhaps even a "corona" of neutral hydrogen that probably surround the planet out to considerable distances. However, if Uranus's moons are a significant source of energetic particles, it appears odd that no oxygen ions or heavier particles are present in the high-energy radiation pattern. The true source is as yet uncertain; perhaps all four mechanisms play a role.

THE ROTATION REMEASURED

The Moore–Menzel measurement of the rotation rate of Uranus was universally accepted virtually without question for many years. However, such a rapid rotation seems to be inconsistent with Uranus's relatively small oblateness. Measurements made during the 1970s seemed to indicate an appreciably slower rate of spin. Astronomers at the Kitt Peak National Observatory repeated the Moore–Menzel measurement in 1975. They obtained a rotational period of about 24 hours. University of Texas workers performed an identical experiment and concluded that the period probably ranged somewhere between 19 and 23 hours in length. Caltech researchers obtained an estimate of 12.9 hours. Workers at Harvard University even reported some evidence for a differential rotation rate, with higher latitudes having a slower rate of rotation than equatorial regions.

The controversy was settled during the Voyager mission. The magnetosphere of Uranus was found to emit a pattern of weak radio waves. As the planet rotates, the intensity of the radio emission varies in a periodic fashion. The period of radio intensity variations can be used to measure the rate at which the interior of the planet is spinning. The radio period of Uranus is 17.24 hours, appreciably longer than the Moore–Menzel measurement of the 1930s and more nearly consistent with the more recent Earth-based measurements.

THE BLAND ATMOSPHERE

Like Jupiter and Saturn, the surface of Uranus that can be seen from space is a dense and thick atmosphere. As expected, the dominant gases present in the upper atmosphere are hydrogen and helium. The infrared-emission and radio-science measurements made by the *Voyager 2* spacecraft determined that the mass fraction of helium in the upper Uranian atmosphere is 26 percent, with an uncertainty of about 8 percent.

The Uranian atmospheric helium mass fraction is unexpectedly high, much higher than that of Saturn (11 percent). It is probably even larger than that of Jupiter (19 percent). It is, in fact, quite close to theoretical estimates for the helium abundance within the protosolar nebula at a time before the Sun and the planets condensed. If so, then the interior of Uranus (unlike that of Saturn) has undergone little if any differentiation of helium from hydrogen. The atmosphere of Uranus remains much as it was immediately after the planet condensed from the primeval solar nebula. If the Uranian helium abundance is truly identical to that which existed in the protosolar nebula, then not only has differentiation of helium from metallic hydrogen taken place in Saturn, it has probably also begun within the interior of Jupiter.

The atmosphere also contains an appreciable amount of methane, which accounts for Uranus's characteristic green color. Ethane (C_2H_6), acetylene (C_2H_2), and some heavier hydrocarbons have been detected. These hydrocarbons may have been produced by ultraviolet photochemistry of methane in the upper Uranian atmosphere. The Voyager measurements found no evidence for the presence of ammonia or water vapor in the atmosphere. However, because of the exceedingly cold temperatures in the upper Uranian atmosphere, much of the ammonia and water in the air may have condensed into clouds that lie so deep in the atmosphere that they cannot be seen from above.

As the Voyager probe flew past the planet, it momentarily passed behind the Uranian disk, allowing the radio transmissions from the craft to skim past the top of the atmosphere as they went toward Earth. This provided a measurement of the pressure–density–depth profile of the outermost layers of the atmosphere (Figure 14.6). The atmosphere is so thick that only the outer few dozen kilometers could be probed.

The atmosphere of Uranus is quite similar to that of Jupiter and Saturn. However, its upper layers are appreciably colder, due to the fact that Uranus is almost twice as far from the Sun as is Saturn. There is a troposphere and a stratosphere separated by a tropopause. The Uranian tropopause has a temperature of −222°C and lies at the point where the pressure is 0.1 atm. Below the tropopause the temperature of the air gets about 1°C warmer for every kilometer of descent, reaching a temperature of −190°C at the level where the air pressure is 1.6 atm.

The sideways tilt of Uranus produces a slightly greater average solar energy input to the poles than to the equator. Consequently, one would expect that the

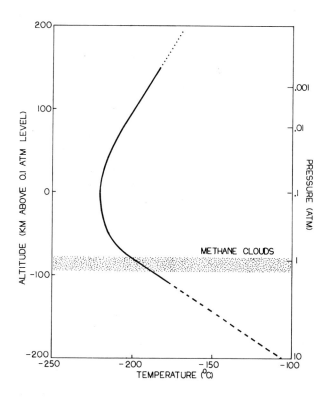

FIGURE 14.6 Vertical profile of the upper atmosphere of Uranus.

temperature of the air over the poles should be slightly warmer than that over the equatorial regions. The Voyager measurements found instead that the temperature of the upper atmosphere was quite uniform over the entire planet. No significant equatorial–polar temperature gradient was found. Some means must exist for rapid redistribution of the excess solar energy in the polar regions.

The steady increase in the temperature of the troposphere with depth is indicative of an appreciable energy outflow from the interior. Like Jupiter and Saturn, Uranus probably radiates more energy to space than it receives from the Sun. However, the Voyager infrared instruments established that the upper limit for the effective temperature of Uranus is only −213°C, not appreciably different from that expected for a body of Uranus's albedo at that distance from the Sun. Any Uranian internal energy source must be exceedingly small, outside the limits of measurement for the Voyager instrumentation.

In the stratosphere, the temperature of the air steadily rises until one reaches an extremely hot thermosphere composed largely of hydrogen atoms and molecules. At an altitude of 1500 km above the tropopause (where the pressure is only 10^{-9} atm) the temperature is as high as 1000°C. The energy source responsible for the high temperature of the thermosphere appears to be the absorption of ultraviolet sunlight. The high temperature and the high number density of hydrogen atoms within the thermosphere implies that there is an extensive thermal corona of hydrogen atoms that extends outward a considerable distance from Uranus, perhaps out as far as 30,000 km above the cloud tops. This corona may be an important source for the ions of the magnetosphere.

The upper atmosphere above the level where the pressure is 10^{-6} atm emits a glow of ultraviolet light. This light comes from hydrogen molecules in highly excited states. This ultraviolet emission was seen only on the day side of the planet, implying that the absorption of sunlight is somehow responsible. However, the amount of solar energy deposited in the upper atmosphere is far too small to be solely responsible for the intensity of the ultraviolet light emitted. Some other mechanism must be operating. The nature of this energy source is mysterious. Because of the uncertain origin of this ultraviolet light, the Voyager team of scientists suggested a new name for the phenomenon — *electroglow*. Although the cause of electroglow is largely unknown, a detailed analysis of the spectrum of the electroglow light has led to the conclusion that collisions with low-energy electrons may be responsible for the excitation of the hydrogen. In these collisions, some of the molecules of hydrogen are torn apart, producing hydrogen atoms that slowly escape outward to populate the diffuse corona.

THE WEATHER ON URANUS

The Voyager radio-occultation experiment detected a layer of methane-ice-crystal clouds in the upper troposphere. The cloud bottoms were at a temperature of −190°C, where the pressure is approximately 0.9 to 1.3 atm. Ammonia and water clouds may lie much farther down in the atmosphere, so deep that they cannot be seen. There is a layer of hydrocarbon "smog" above the methane clouds. The hydrocarbons in the smog layer were probably created by the ultraviolet photochemistry of methane in the upper atmosphere. Much of the hydrocarbon smog lies at a level where the pressure is 0.13 atm and extends 40 km above the tops of the methane clouds. This hydrocarbon haze is so dense that it completely obscures everything below it from view. An even denser "hood" of hydrocarbon covered the sunlit south pole.

The hydrocarbon haze in the upper atmosphere causes Uranus to have an extremely low contrast when viewed from space. All that is seen is a featureless blue-green disk. However, when the Voyager photographs were computer processed to bring out subtle detail, a banded structure reminiscent of that on Saturn emerged. Like Jupiter and Saturn, the bands in the upper Uranian atmosphere are parallel to its equator. The weather in the upper atmosphere is controlled by the rotation of the planet rather than by solar heating.

The bands are produced by strong zonal upper-atmospheric winds that blow predominantly in the east–west direction. Observations of the few discrete cloud patterns that could be identified suggest that the wind velocity increases with altitude. The zonal wind velocity also varies with latitude. As measured relative to the 17.24-hour radio period, the upper atmosphere poleward of $-20°$ latitude rotates slightly faster than the planet's interior does. This corresponds to a wind blowing from west to east. The equatorial regions of Uranus were poorly imaged by Voyager, but it is possible that the winds there blow in the opposite direction, from east to west.

THE MOONS OF URANUS

Uranus has five moons visible in the telescope from Earth: Ariel, Umbriel, Titania, Oberon, and Miranda. The moons all bear the names of characters that appear in the works of the playwright William Shakespeare or the satirist Alexander Pope. All five occupy nearly circular orbits lying in the plane of Uranus's equator. Their orbits thus share the unusual axial inclination of the planet itself. Uranus is the third world in the Solar System (besides Jupiter and Saturn) to have a regular system of large satellites (Figure 14.7).

Ariel, Umbriel, Titania and Oberon are quite similar in size, with diameters of the order of 1100 to 1600 km. This makes them approximately the size of the intermediate Saturnian moons Tethys, Dione, and Rhea. Miranda is considerably smaller (with a diameter of the order of 500 km). It is approximately the size of the Saturnian moon Mimas. All five moons have reflectance spectra that show evidence for the presence of water ice on their surfaces. Their densities are uniformly low, ranging between 1.3 and 1.6 g/cm³. Like the moons of Saturn, the Uranian moons are probably half silicate rock, half ice in internal composition. There does not seem to be any obvious trend in reflectivity or in average density with orbital distance from Uranus. A schematic view of the interiors of these moons is shown in Figure 14.8.

There is no evidence that any of the moons of Uranus has any sort of an atmosphere. The daytime temperatures on the moons are exceedingly cold, on the order of $-190°C$.

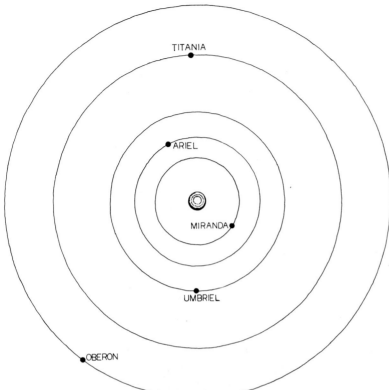

FIGURE 14.7 Orbits of the five
large moons of Uranus.

In January 1986, the *Voyager 2* spacecraft passed through Uranus's system
of moons, returning numerous photographs and other data. As expected, all five
of Uranus's large moons are in synchronous rotation. Unfortunately, this meant
that the south poles of all five of these worlds were facing the Sun at the time of
the Voyager encounter. The northern hemispheres of all five moons remained
in total darkness. Therefore, only half of their surfaces could be photographed
by the spacecraft's television cameras.

Before the arrival of Voyager, it had been generally assumed that the
moons of Uranus would turn out to be just another set of dull and uninteresting
balls of ice. However, the results surprised almost everyone. These moons
turned out to be some of the most interesting and puzzling worlds yet encountered in the Solar System.

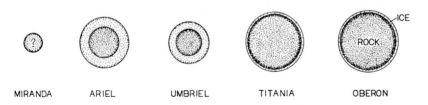

FIGURE 14.8 Interiors of the
five large moons of Uranus.

OBERON

The outermost moon of Uranus is Oberon, orbiting 580,000 km from the planet (Figure 14.9). It was discovered by William Herschel in 1787, only 6 years after his discovery of Uranus itself.

Oberon has a diameter of 1550 km. The density is approximately 1.5 g/cm³. The entire surface is rather dark, with an average reflectivity (or albedo) of only 24 percent. The surface has a gray color, similar to the surface of the Saturn moon Phoebe as well as the surfaces of several of the C-type asteroids. Although the spectrum of water ice has been identified in the light reflected from the moon, the low albedo indicates that other, more darkly colored materials must be present on the surface in significant amounts.

The surface of Oberon has a high density of large impact craters. The density of these large craters is similar to that found in the lunar highlands, suggesting that the surface of Oberon is quite ancient. A single 20-km-high mountain was seen near the limb in one of the Voyager photographs. It may be the central peak of a gigantic impact structure, hundreds of kilometers in diameter. Some linear and curved scarps are faintly visible.

Some of the craters have extremely dark floors, suggesting that volcanic activity may have filled up the interiors of these craters with some sort of darkly colored lava. The most conspicuous deposits of dark material are found inside a

FIGURE 14.9 *Voyager 2* photograph of Oberon, taken at a distance of 660,000 km. A large mountain peak can be seen at the moon's lower limb. (Photograph courtesy of NASA/JPL)

couple of particularly large craters that have bright ray patterns extending outward from them. This suggests that meteoroid impacts occurring near the end of the period of the most intense bombardment may have triggered some sort of volcanic activity that flooded some low-lying areas with a darkly colored lava. Other than the dark pools of material seen on the floors of some craters, the surface of Oberon seems to be relatively unmodified by geological activity.

UMBRIEL

Umbriel is a dull-gray world 1190 km in diameter, located 266,000 km from Uranus (Figure 14.10). It was discovered by the English astronomer William Lassell in the year 1851.

Umbriel is the darkest of the large Uranian moons, having an average reflectivity of only 19 percent. It also has a weaker water-ice signature in its reflectance spectrum, probably indicative of the presence of relatively larger amounts of darker material on its surface. Umbriel has a density of roughly 1.4 g/cm³.

The Voyager photographs show a surface with a high density of muted, overlapping craters and impact basins. There is a dense population of large impact craters, especially those between 50 and 100 km in diameter. The distribution of large craters across the surface of Umbriel closely resembles that of the

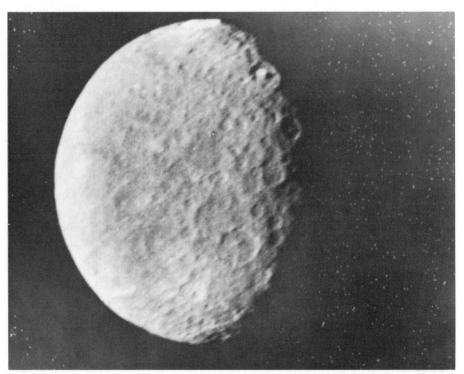

FIGURE 14.10 *Voyager 2* photograph of Umbriel, taken at a range of 560,000 km. A ring of bright material can be seen at the limb. (Photograph courtesy of NASA/JPL)

lunar highlands, as well as that of many of the most ancient, heavily cratered bodies in the Solar System. There are no crater rays visible.

Only a few bright features could be seen on the entire visible surface of Umbriel. The most prominent of these was a ring of bright material about 80 km across. This ring seems to lie on the floor of a large crater. There is little evidence for any significant past geological activity. The surface of Umbriel is probably the oldest of any of the moons in the Uranian system, having remained relatively unaltered ever since it first solidified over 4 billion years ago.

It is odd that Umbriel's surface is so uniformly dark, with almost no bright features. Numerous hypotheses have been proposed to explain this uniformity, none of them very convincing. Perhaps Umbriel is rich in dark material to a considerable depth on a global scale. However, this hypothesis is inconsistent with the presence of the bright ring. Alternatively, it is possible that the dark material is only an outer coating, with much brighter material lying just beneath the surface. Perhaps some sort of catastrophic event coated the entire surface with a uniform blanket of dark material that has completely covered any previous surface markings. Umbriel itself would have to be the source of the dark material, since neither of the nearby moons Titania nor Ariel show any signs of the presence of such deposits. Perhaps a large meteor happened by chance to strike a spot on Umbriel that was especially rich in dark material, the debris from the impact coating the entire surface of the moon with a dark layer of dust. However, it does seem rather unlikely that no smaller meteoroids came along at a later time to penetrate this thin layer of dust and expose brighter underlying material. A third possibility is a recent explosive eruption on the surface, perhaps a violent release of methane gas or the chemical dissociation of carbon monoxide compounds. Perhaps the bright ring is somehow associated with this volcanic event. However, any sort of recent internal volcanism seems inconsistent with Umbriel's ancient, densely cratered surface. The mystery remains.

ARIEL

Ariel is a spherical object 1170 km in diameter orbiting 191,000 km from Uranus (Figure 14.11). Ariel is the brightest of Uranus's moons (the albedo is 40 percent) and has an average density of 1.7 g/cm³. It was discovered by William Lassell in the same year that he discovered Umbriel.

The Voyager photographs show a surface with large variations in brightness. Much of the terrain is heavily scarred with the remains of ancient impact craters. However, there are relatively few craters in the 50- to 100-km size range and a great abundance of craters smaller than 20 km across. Many large craters must have been erased either by viscous relaxation or by the flow of melting ice. The largest identifiable crater remaining on Ariel is severely flattened and has a domed floor partly encompassed by a shallow trough.

The cratered plains unit is broken and fractured by a global system of faults. The most prominent of these is a spectacular system of narrow, deep valleys lying at mid-latitudes in the leading hemisphere. Some of the older fault scarps have many small craters superimposed upon them, whereas the younger ones are entirely free of craters. The entire system of faults may have developed

FIGURE 14.11 *Voyager 2* photograph of Ariel, taken at a distance of 130,000 km. Numerous faults and valleys can be seen. (Photograph courtesy of NASA/JPL)

in response to crustal extension that took place at the time that the interior of Ariel froze solid.

There are a few places where long faults are interrupted when they cross valley floors, presumably because these valleys were filled up by some sort of material that covered over and obliterated the faults. There are extensive regions of sparsely cratered terrain located at high latitudes. These plains seem to have been formed by a flood of icy material that covered over and buried most of the older craters. Occasionally the "ghosts" of underlying craters can be seen protruding out of these plains.

TITANIA

Titania is a world 1590 km in diameter that orbits 436,000 km from Uranus (Figure 14.12). It was discovered by Herschel in the same year that he discovered Oberon. The average density of Titania is roughly 1.6 g/cm³. The albedo of the surface of the moon is 28 percent, somewhat less than that of Ariel. The surface is a neutral gray in color.

The surface of Titania is heavily scarred with numerous bright ray craters.

There are also a few large impact basins, with diameters ranging between 100 and 200 km. The density of small craters (those having diameters of 20 km or less) is about the same as it is on the Moon. However, there is a relative shortage of large impact craters, especially those having diameters between 50 and 100 km. Some sort of geological process has selectively erased the larger craters, while leaving the smaller ones intact. Furthermore, there are several patches of smoother terrain with a much lower crater density, suggesting a prolonged early period of resurfacing. The surface of the entire moon has a relatively low relief, indicative of an ancient episode of internal heating that collapsed some of the steeper terrain.

There is a complex system of trenches that is part of a massive system of geological faults and fractures that spans half the diameter of the moon. Most faults occur in a branching, partly intersecting network. The vertical relief of the scarps ranges from 2 to 5 km. Along the lengths of some of the scarps, brighter material has been exposed to the surface. The faults cut through the larger craters and do not seem to be strongly modified by the smaller craters. These faults must be among the youngest features on Titania.

The geological evolution of the surface of Titania was undoubtedly quite complex. Titania must have been bombarded by the same population of large

FIGURE 14.12 *Voyager 2* photograph of Titania, taken at a range of 480,000 km. (Photograph courtesy of NASA/JPL)

meteoroids that scarred the surfaces of Umbriel and Oberon. However, most of the large impact craters excavated on the surface of Titania had to have been obliterated by the extrusion of icy material onto the surface very early in the history of the moon. Viscous relaxation resulting from internal heating may also have played a role. A later bombardment by smaller objects created a fresh set of smaller impact craters. Somewhat later a few regions were resurfaced by a fresh extrusion of icy material from the interior. The extensive faulting seems to have been produced by a global extension of the entire crust, which probably occurred during the last stages of the freezing of the ice in the interior of the satellite. After that, the surface of Titania froze into its present form. The few ray craters seen on the surface were formed by occasional impacts over the past 3 billion years.

MIRANDA

Miranda is the smallest of Uranus's five major moons. It has a diameter of only 480 km and orbits 129,000 km from the center of Uranus (Figure 14.13). It was discovered by Gerard Kuiper in 1948.

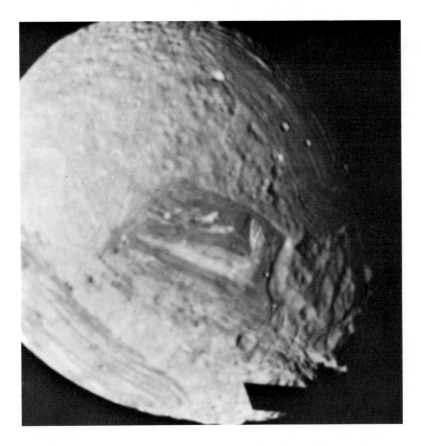

FIGURE 14.13 *Voyager 2* photomosaic of Miranda, taken at ranges varying from 30,000 to 40,000 km. The "trapezoid" feature is seen at the center, which is located approximately at the south pole of the moon. The feature known as the "banded ovoid" is at the lower left. The "ridged ovoid" is seen at the upper right. An ancient, heavily cratered terrain lies between these three features. In this view, Uranus is located toward the bottom, and the motion of the satellite is toward the left. (Photograph courtesy of NASA/JPL)

A major goal of the Voyager mission is a 1989 encounter with the planet Neptune and its large moon Triton. In order to reach Neptune it was necessary for the *Voyager 2* craft to pick up a gravitational assist during the Uranus flyby. The success of this maneuver depended on *Voyager 2* passing by Uranus at precisely the correct distance. This constraint naturally limited the choice of objects in the Uranian system that could be examined at close range. As luck would have it, the moon located nearest the point of Voyager's passage through the Uranian system was tiny Miranda.

It had been expected that Miranda would turn out to be just another uninteresting, heavily cratered ball of ice. Instead, this small moon turned out to have a most complex and interesting surface.

The surface of Miranda that was photographed by *Voyager 2* consists of two strikingly different types of terrain. One is a heavily cratered rolling terrain with relatively uniform albedo. This terrain is about as densely cratered as the lunar highlands and is therefore quite ancient, perhaps older than 4 billion years of age. In some places, this ancient cratered terrain appears to be crumpled in upon itself, perhaps due to intense crustal compression. Several of the larger impact craters have steep walls where large amounts of bright material have been exposed to the surface. In some places, this bright material extends to depths of at least 1 km beneath the average level of the surrounding rolling, cratered surface.

The other terrain type is completely different. It is characterized by roughly circular patterns of angular landforms surrounded by nearly parallel sets of alternatively bright and dark bands, scarps, and ridges (Figures 14.14 and 14.15). These features are far less densely cratered than the rolling cratered plains and must therefore be considerably younger. Three of these strange landforms were spotted in the Voyager photographs. The most prominent of these is located near the south pole and is trapezoidal in shape. The Voyager imaging team referred to it as the "trapezoid." The trapezoid contains an odd-looking angular feature in its interior that resembles a chevron. The outer boundary of the trapezoid as well as its internal patterns of ridges and bands displays numerous sharp corners. The outer band sharply truncates the ridges and grooves of the inner core. A second example of these strange features appears in the leading hemisphere of Miranda, just south of the equator. It is about 300 km wide and is termed the "banded ovoid." The outer margin of the banded ovoid has rounded corners and dark bands parallel to the boundary that curve smoothly around a more sharply rectangular inner core. The third example of these features occurs in the trailing hemisphere near the equator. It is called the "ridged ovoid." The topography of the inner core of the ridged ovoid consists of a complex set of intersecting ridges and troughs that are truncated by an outer belt. The outer belt is marked by roughly concentric linear ridges and troughs that blend fairly smoothly into the surrounding undulating cratered terrain.

A few days after the Voyager encounter, someone pointed out that the parallel sets of bright and dark ridges and bands that surround the central angular landforms had a striking resemblance to terrestrial racetracks, particu-

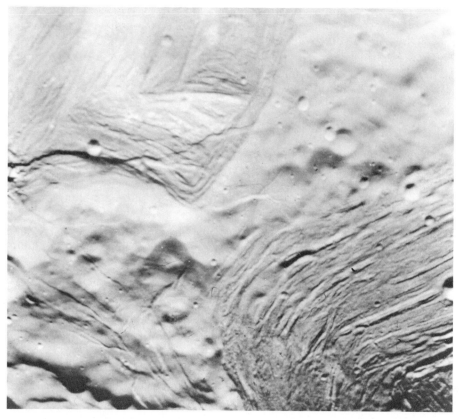

FIGURE 14.14 *Voyager 2* close-up view of trapezoid (left) and ridged ovoid (right) features on Miranda. The range is 42,000 km. An outer band of ridges and grooves wraps partly around an angular core of intersecting ridges and grooves in the interior of the trapezoid. The outer band sharply truncates the ridges and grooves on the inner core of the trapezoid. (Photograph courtesy of NASA/JPL)

larly those that were used to hold chariot races in ancient Rome. These landforms were therefore given the designation *circi maximi,* the Latin name for these ancient Roman stadia. Nothing like them has been seen anywhere else in the Solar System.

Miranda also has a series of enormous fault scarps that can often be traced across the entire width of the visible hemisphere. They slice through the cratered terrain, suggesting some sort of geological activity had transpired after the cessation of the meteoric bombardment. Some faults seem to be older than the circi maximi, but others are much younger. The most spectacular of the fault systems on Miranda is an enormous gorge that starts north of the equator (somewhere in the invisible hemisphere), extends along the margin of the trapezoid, and passes along the outer boundary of the banded ovoid until it finally terminates. Numerous patches of bright material are exposed along the length of this enormous fault system.

Voyager 2 passed close enough to Miranda to made a direct measurement of its mass. The density of Miranda is only 1.3 g/cm³, with an uncertainty of about 30 percent. Why should such a small, low-density object as Miranda have such a complex surface? In particular, what is the cause of the mysterious circi maximi? All of the circi maximi seem to have been formed by some sort of process that has built outward from a center by a series of geological faulting events. Intrusion and extrusion of fluid or plastic material may have played a role. Perhaps the most radical proposal put forth to date to explain the complex surface of Miranda is the hypothesis that the moon is a world that has turned itself inside out! According to this model, Miranda was originally a simple ice ball much like the Saturnian moon Mimas in structure and composition. Radioactive heating in the interior caused the denser silicate minerals to sink to the center, leaving an outer crust largely of water ice. Shortly thereafter, a catastrophic collision with a large asteroid-sized meteor shattered the moon into several angular fragments. A few years later, the chunks reassembled themselves to

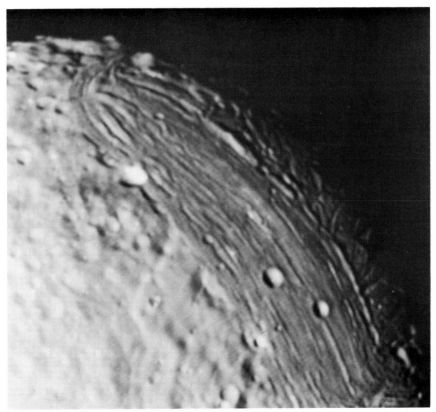

FIGURE 14.15 *Voyager 2* close-up view of ridged ovoid on Miranda. The range is 34,000 km. Possible volcanic flows can be seen near the terminator. These flows may have erupted and partly buried the grooves and ridges in the central core and outer bands of the ridged ovoid. (Photograph courtesy of NASA/JPL)

form the moon once again. Some chunks came together ice-side out, displaying the original impact craters. Others came together core side out, exposing the darker rocky material within the original core. These exposed core samples acted as nucleation centers for the formation of the circi maximi.

THE MYSTERIOUS HEAT

Why do so many of the moons of Uranus show signs of past geological activity? All of the Uranian satellites (with the possible exception of Miranda) are appreciably denser than the icy moons of Saturn, which indicates that they are relatively richer in rock content than their Saturnian counterparts. Energy released via the radioactive decay of heavy elements may indeed play a role in the heating of the interiors of these worlds. However, it is generally thought that icy worlds of such low masses and densities cannot produce sufficient radioactive heating to generate any significant amount of volcanism or other form of surface-altering activity. Alternatively, perhaps an Io–Europa or Enceladus–Dione type of resonant coupling between the orbital motions of the moons of Uranus is the cause of their internal heating. However, none of the orbits of the moons of Uranus are currently in resonance with each other, and theoretical calculations of the amount of internal heating currently produced by tidal forces fail by many orders of magnitude to come up with the heat energy that is required.

The paradox could be neatly resolved if the interiors of the moons of Uranus were rich in some sort of material that melts and flows at much lower temperatures than does ordinary water ice. The outer Solar System satellites are low-density objects that are assumed to have large amounts of water ice in their interiors. In the Jovian system the moons are roughly half rock, half water ice in composition. However, at greater distances from the Sun temperatures in the solar nebula may have been cool enough for appreciable amounts of methane or ammonia to have condensed along with the water ice. A mixture of water ice and ammonia has a melting point that is much lower than that of pure water ice, leading to the possibility of geological activity capable of being driven by extremely tiny amounts of heat. One problem with this theory is the absence of evidence for the presence of methane or ammonia on any of Uranus's moons. In addition, the densities of the moons of Uranus are a bit too large for their interiors to be modeled easily as a solar mixture of rock, water, ammonia, and methane. Finally, any truly successful theory of the origin of the geological activity must contend with Umbriel. This moon has been virtually inactive; its surface is now much the same as it was soon after it froze into solid form more than 4 billion years ago.

THE ANCIENT REAGGREGATION

The densely cratered surfaces of Umbriel and Oberon are clear evidence that the Uranian system (like that of every other planet studied to date) was subjected to an intense flux of large meteoroids very early in its history, probably earlier than 4 billion years ago. These objects were asteroid-sized bodies originally in heliocentric orbit — planetesimals that had managed to escape incorporation into large planets at the time they condensed. As in the case of the

Saturnian system of moons, calculations show that the meteoroid flux at this early time must have been sufficiently intense to be able to shatter any small moons initially located in this region of space into millions of fragments. Almost immediately thereafter, these fragments of rock and ice reassembled themselves to form new moons. The large gravitational field of Uranus provides a sort of focusing effect, so that the flux of large meteoroids is much larger in the inner parts of the system than it is in the outer. An object originally located at the current orbit of Umbriel was probably shattered only once, but anything located as close to Uranus as the current position of Miranda might have been destroyed as many as five times before it assumed its final configuration. The bizarre surface of Miranda may contain a record of the last of these episodes of satellite destruction and rebirth.

Like the icy moons of Saturn, the surfaces of the moons of Uranus show evidence for two distinct and separate episodes of meteoroid bombardment. The first of these episodes was caused by the influx of large, asteroid-size objects originating from outside the Uranus system. These large objects rained down upon the surfaces of the newly formed moons, excavating gigantic basins and craters, many of which were over 100 km in diameter. A record of the final years of this early bombardment episode is preserved on the ancient cratered plains of Umbriel and Oberon. Shortly after this meteoroidal bombardment had abated, Ariel and Titania became volcanically active, and most of the large craters initially present on their surfaces were obliterated. At a somewhat later time, all the moons were subjected to a new episode of meteoroid bombardment, this time from a population of smaller objects originating from within the Uranian system itself. These objects may have been ejecta from earlier cratering episodes, or else they could have been debris left over from the last episode of satellite destruction and reaccretion.

NEW MOONS—A MEMORIAL IN SPACE?

The *Voyager 2* television cameras made a careful search of the Uranian system for previously unseen moons. It found 10, all of them inside the orbit of Miranda (Figures 14.16 and 14.17). These moons are uniformly tiny, none being any larger than 170 km across. They are all exceedingly dark, having reflectivities no larger than 5 percent. This makes them no brighter than a piece of coal.

The largest of these 10 newly discovered satellites is 1985U1. It has a diameter of about 170 km and is almost perfectly spherical in shape. A few craters are visible on its surface, the largest being about 45 km in diameter. There is no observable albedo contrast. If 1985U1 is predominantly an icy object with only relatively small amounts of dark contaminants covering the outer surface, it is rather odd that no bright craters are seen where a meteoroid has penetrated the dark outer crust to expose the brighter icy material beneath. Perhaps some process erases or darkens the bright icy material almost as soon as it reaches the surface. Alternatively, 1985U1 may be completely dark throughout its entire interior. If so, it must be made of something entirely different from water ice.

As yet, the newly discovered moons of Uranus are known only by num-

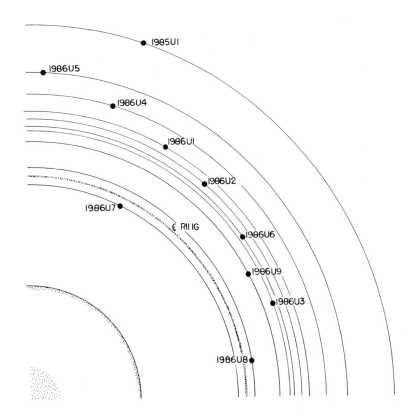

FIGURE 14.16 Orbits of the 10 newly discovered small moons of Uranus.

bers, pending the assignment of names. A week after the Voyager flyby, the *Challenger* disaster took place. It was proposed that a fitting memorial to the astronauts who perished in the explosion would be to name some of the new moons of Uranus after them. This would be a break with tradition, since the five large moons of Uranus bear the names of characters from the works of Shakespeare and Pope. The International Astronomical Union, which has the final say in such matters, has not yet made a decision.

THE RINGS OF URANUS—
THE DARK MYSTERY

It had been known for some time that Uranus would occult a particularly bright star (SAO 158687) on March 10, 1977. Since this event would provide astronomers with a unique opportunity to probe the upper atmosphere of Uranus, many observatories (including the *Kuiper Airborne Observatory,* a unique flying research platform carried aboard a converted Lockheed C-141 Starlifter) prepared extensive programs of observation in anticipation of the occultation.

What actually happened surprised everyone. Just before the scheduled occultation, the light from the star showed a series of unexpected dips. These were reproduced in exactly the same pattern in reverse order after the star emerged on the other side of the planetary disk, making it exceedingly unlikely that unseen moons could have been responsible. A more likely cause is the presence of a previously unknown ring system, much like that of Saturn but far less extensive.

In the years following the initial discovery event, several other stellar occultations have been observed, making it possible to map out the structure of the Uranian ring system in considerable detail. At least nine separate ring components have been identified. Each ring component has been assigned a number or Greek letter — 6, 5, 4, alpha, beta, eta, gamma, delta, and epsilon — in order of increasing distance from Uranus (Figure 14.18, Figure 14.19, and Table 14.1). The rings occupy a comparatively narrow band, 16,000 to 26,000 km from the Uranian cloud tops. The outermost ring (epsilon) lies 78,000 km inside the orbit of Miranda, Uranus's nearest large moon. The individual rings are exceedingly narrow (only 5 to 15 km wide) except for the outermost one (epsilon),

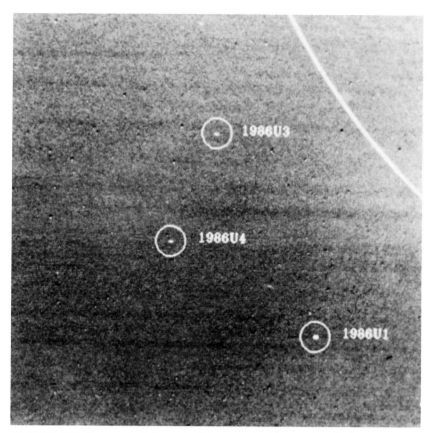

FIGURE 14.17 *Voyager 2* photograph of three newly discovered small moons of Uranus. They are 1986U3, 1986U4, and 1986U1. The bright band on the upper right is the epsilon ring. At the time that the photograph was taken, the spacecraft was 8 million km from the planet. (Photograph courtesy of NASA/JPL)

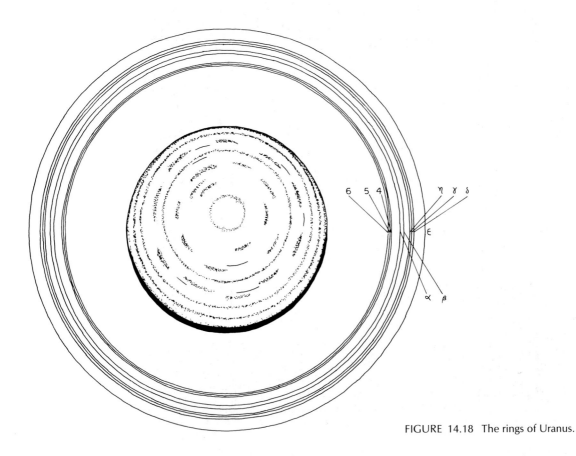

FIGURE 14.18 The rings of Uranus.

which has a variable width averaging 60 km across. The entire Uranian ring
system contains only as much material as is found inside Cassini's division of the
rings of Saturn, which in the pre-Voyager era was thought to be entirely free of
matter.

The ring plane is very nearly parallel to the Uranian equator, although
there are slight departures from strict parallelism between one ring and another.
All but the gamma, delta, and eta rings have small but measurable departures
from perfect circularity. The epsilon ring is the most extreme example; its eccentricity is about 0.008. In addition, the width of the epsilon ring varies from 22 km
at periapsis (point closest to Uranus) to 93 kilometers at apoapsis (point farthest
from Uranus). All nine rings vary in width and in optical depth with longitudinal
position around their circumferences. The orientations of the elliptical rings
change continuously with time in response to the gravitational forces produced
by Uranus's oblateness.

The Voyager mission confirmed the presence of the ring system and discovered a few additional components. A faint 10th ring, 50,000 km from the
planet's center, was found lying between the epsilon and delta rings. It was

FIGURE 14.19 Mosaic of *Voyager 2* photographs of the rings of Uranus. In this view, all nine rings can be seen. The bright band on the left is the epsilon ring, the outermost of the rings. Just to the right of it are the three narrow rings delta, gamma, and eta. Still farther right are the beta and alpha rings. At the far right are the three narrow innermost rings 4, 5, and 6. The newly discovered ring feature 1986U1R is barely visible midway between the epsilon and delta rings. (Photograph courtesy of NASA/JPL)

given the temporary name 1986U1R. A faint, broad band of material 2500 km across was found 1600 km inside the innermost (6) ring. It bears the designation 1986U2R.

As the *Voyager 2* spacecraft flew past the planet, its radio signals passed through the ring system on their way to Earth. In addition, the spacecraft's optical instruments observed the fading of conveniently located stars as they were occulted by the various components of the ring system. Analysis of the results of these experiments revealed the presence of a fine structure within the Uranian ring system. The outer edge of the epsilon ring is exceedingly sharp, going from fully packed to empty space in only 50 m. The inner edge is 500 m wide. The epsilon ring also has an extensive ringlet structure reminiscent of that of the rings of Saturn. A couple of dozen individual ringlets were found, the width of a typical ringlet being of the order of a few hundred meters. The gamma ring is exceedingly narrow, having a width of only 600 m. The delta ring had three components visible at one occultation but only one at the other. The alpha ring was found to have three broad components and one narrow one. The beta ring had two broad components and one narrow one. A number of other

TABLE 14.1 URANIAN RING DATA

RING COMPONENT	DISTANCE FROM CENTER OF URANUS (KM)	ECCENTRICITY	INCLINATION (°)	WIDTH (KM)
1986U2R	37,000 – 39,500	$<10^{-4}$	$<10^{-3}$	2500
6	41,850	0.0010	0.063	1 – 3
5	42,280	0.0019	0.052	2 – 3
4	42,580	0.0011	0.032	2 – 3
alpha	44,730	0.0008	0.014	7 – 12
beta	45,670	0.0004	0.005	7 – 12
eta	47,180	$<10^{-4}$	0.002	0 – 2
gamma	47,630	$<10^{-4}$	0.011	1 – 4
delta	48,310	$<10^{-4}$	0.004	3 – 9
1986U1R	50,040	$<10^{-4}$	$<10^{-3}$	1 – 2
epsilon	51,160	0.0079	0.001	22 – 93

SOURCE: Adapted from Smith et al. (1986).

possible rings or partial rings (arcs) have also been identified in the occultation data.

The rings of Uranus consist of myriads of small particles, each particle in an individual Keplerian orbit around the planet. Unlike the rings of Saturn, the Uranian rings do not scatter very much sunlight in the forward direction. This must mean that the rings of Uranus are relatively deficient in small, micron-sized dust particles. Analysis of the results of the Voyager radio-occultation experiment confirmed that the majority of the ring particles are larger than 1 m across. However, the Voyager photographs did show the presence of a broad but highly diffuse band of tiny dust particles lying between the orbit of 1986U1R and the outer edge of 1986U2R. In addition, photographs taken by *Voyager 2* when the Uranian rings were back-lighted by the Sun show the presence of several narrow, well-defined bands of low optical depth that are rich in dust. The brightness of these dust bands does not correlate very well with the locations of the known rings.

The relatively low density of dust in the ring system is somewhat puzzling. Several imaginative theories have been suggested. Uranus has an extended hydrogen envelope that is slowly leaking away from the planet. Viscous dragging forces exerted by this gas envelope on the ring particles will affect smaller particles more drastically than larger ones. Gas drag may have caused the particles smaller than 1 cm or so to spiral inward toward the upper atmosphere and be destroyed. Another possibility is that the particles in the rings are "sticky." They coalesce into larger lumps when they collide with each other rather than shatter into smaller fragments. A third possibility is that Uranus's magnetosphere is responsible. When ionizing radiation strikes an object there is a reasonable probability that the irradiated object will acquire a net electric charge. A charged particle moving in a magnetic field will experience a net force. Uranus's rapidly rotating, off-center magnetic field may have swept the smallest of these charged particles of dust completely out of the ring plane and pulled them inward to the upper atmosphere, where they were destroyed by air friction. A similar phenomenon may be responsible for the spokes in Saturn's B ring.

The rings of Uranus are exceedingly dark, which explains why they are so difficult to see from Earth in the telescope. *Voyager 2* discovered that the particles that make up the rings of Uranus have an extraordinarily low albedo, estimated to be only 4 to 5 percent. They contrast sharply with the much brighter particles in the ring systems of Jupiter and Saturn. The ring particles are much darker than any of the surfaces of Uranus's five major moons; the albedo of the ring particles quite closely matches that of the surfaces of the 10 newly discovered inner small moons of Uranus. The ring particles certainly cannot, therefore, be made entirely of pure water ice.

Something much darker must be present in abundance within the ring particles. It is as yet impossible to say whether the ring particles are made entirely of dark material or whether the dark material is simply an outer coating over an inner core of nearly pure water ice. Dark material does seem to be ubiquitous throughout almost the entire Uranian system. It is present on the ring particles, on the 10 small moons, and on the surfaces of Umbriel and Oberon. It is uncertain just what this dark material is, but its universality throughout the Uranian system suggests a common origin. One possibility is that this material is an exceedingly ancient low-temperature condensate left over from the initial formation of the Solar System nearly 4.7 billion years ago. The most-often quoted theory is, however, that the dark substance in the Uranian system is organic matter!

How did organic material ever get into the Uranian system? Certainly nothing alive could have ever been in this region of space. It is generally assumed that the Uranian magnetosphere and the associated radiation are responsible for the dark material. It is thought that the region of space in which the rings and moons of Uranus formed was sufficiently cold that substantial amounts of methane were incorporated into the water ice when these bodies condensed. The moons and rings sit in the middle of the magnetosphere and their surfaces are continually bombarded by high-energy electrons and protons. Experiments performed on Earth have demonstrated that methane molecules frozen inside water ice will combine with each other (*polymerize*) when the ice is irradiated with high-energy particles. Darkly colored hydrocarbons are thereby formed. In some cases, the methane molecules are completely stripped away by the radiation, leaving behind a deposit of pure carbon. The particles that make up the rings of Uranus may not be all that different in composition from chunks of coal. The surfaces of the moons of Uranus may be covered with a thin layer of frozen petroleum! One problem with this model is the lack of evidence for the presence of methane on any of the moons of Uranus. However, the radiation intensity in the past may have been so intense that all of the surface methane has by now been converted entirely into organic matter.

Why are Uranus's rings so narrow and thin? A similar problem is encountered in attempting to explain the presence of Saturn's extensive ringlet structure. As in the case of Saturn, most theorists have assumed that small shepherd moons lying within the ring system itself are responsible. The motions of these moons act to confine the ring particles to relatively narrow strips. The theoretical astronomers Dermott, Gold, and Sinclair have proposed that each separate ring actually has a small satellite lying within it that traps small particles in nearby orbits. These small moons may have originally formed farther away from Uranus. They were gradually pulled closer and closer to the planet by tidal forces and by fraction with the thin hydrogen envelope. As these moons neared Uranus, the tidal forces became strong enough to tear them apart, releasing many small chunks of material that spread out to encircle the entire planet. Meteorites striking the surfaces of these small moons may also have played a role in generating the ring debris. The ring system may be quite young, having

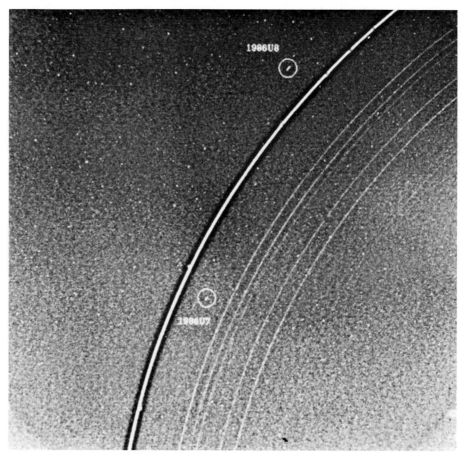

FIGURE 14.20 Two "shepherd" moons 1986U7 and 1986U8, which straddle the epsilon ring. (Photograph courtesy of NASA/JPL)

been formed by the fragmentation of these tiny moons in the relatively recent past. The ring structure may be steadily changing with time. It may be destined for a short lifetime, as friction with the thin Uranian hydrogen gas envelope gradually pulls the ring particles from their orbits and drags them down into the upper atmosphere of the planet where they are destroyed by the heat of air friction.

The shepherd hypothesis of ring formation was given fresh impetus when it was discovered that the ring particles and the inner moons have similar albedos, suggesting that they are made of the same material. The Voyager images were carefully searched for evidence of shepherd moons. Only two were found 1986U7 and 1986U8, which straddle the epsilon ring (Figure 14.20). As in the case of Saturn, the search for shepherds has largely turned up negative results. Most theorists continue to insist that they must be there, but are too small to be seen in the Voyager photographs. In point of fact, calculations indicate that the Voyager photographs would have missed any object in the ring system smaller than 10 km in radius, assuming it to have the same albedo as the two known shepherds. Perhaps their existence will be confirmed once more detailed photographs are available.

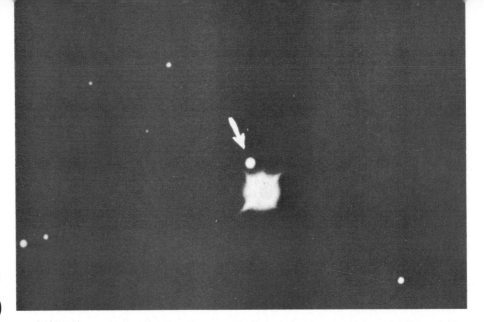

15

Neptune and Triton —
Worlds To Explore

The discovery of Neptune was a major triumph for modern mathematical astronomy. Small irregularities in the motion of Uranus led John Couch Adams (1819–1892) in England and Urbain Leverrier (1811–1877) in France independently to propose that there must be another planet outside the orbit of Uranus. The planet was discovered in 1846 by the German astronomer Johann Galle (1812–1910), who found it in the sky just about where it had been predicted to be.

Neptune occupies a nearly circular orbit averaging 30.1 AU from the Sun. The orbit makes an angle of 1.8° with the ecliptic, and the length of the year is 158.5 Earth years. The planet moves so slowly across the sky that it has not yet had time to complete even a single orbit since the time it was discovered.

SIZE AND ROTATION

Neptune appears in the telescope as a small greenish disk with no apparent surface markings. Because of its greater distance from the Sun, Neptune is even harder to study than Uranus. Since no spacecraft has yet flown past the planet, most details about its size and shape are subject to relatively large uncertainties.

Neptune and Triton, photographed with a 120-in. telescope. (Photograph courtesy of Lick Observatory)

314

The equatorial radius is approximately 25,000 km, the precise number depending on the measurement technique used. Neptune is therefore a virtual twin of Uranus in terms of size (Figure 15.1). The oblateness is hard to measure because of the extreme distance. However, it seems to be fairly small, on the order of 2 percent.

In the pre-Voyager era, there was a long history of controversy about the rotation rate of Uranus. The same has been true for the spin of Neptune. Moore and Menzel reported a period of 15.8 hours in the late 1920s, based on their observations of the Doppler shifts in the light reflected from different parts of the planet's disk. They concluded that the equator of Neptune is approximately parallel to its orbital plane (the angle is 28°) and that the rotation is in the direct sense.

As in the case of Uranus, the Moore–Menzel measurements of the rotation rate of Neptune were accepted virtually without question for many years. In the late 1970s, new measurements with more advanced equipment seemed to suggest a much slower rotation rate. Periods ranging from 18 to 22 hours have been reported by various workers. This slower rotation is probably more consistent with the low oblateness of Neptune than is the original 15.8-hour period obtained by Moore and Menzel.

Recently, a team led by Bradford Smith of the University of Arizona attached a charge-coupled device (originally intended for use on the orbiting Space Telescope) to a telescope in Chile. The high sensitivity of this device enabled them to see some long-lived discrete cloud features on Neptune. They could watch these features move across the planetary disk as Neptune rotated and concluded that the rotational period is 17 hours, 50 minutes. More definitive measurements must await the launch of the Space Telescope or the arrival of *Voyager 2* in 1989.

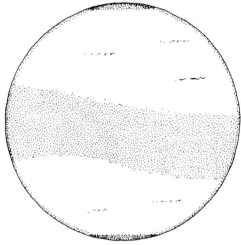

FIGURE 15.1 The relative sizes of Neptune and the Earth. Neptune is so distant that almost nothing is known about surface detail. However, some telescopic photographs suggest the existence of a cloud system in the equatorial regions of the planet.

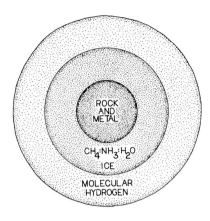

FIGURE 15.2 The interior of Neptune.

THE INTERIOR

Neptune has two known moons, so its mass can readily be determined. It is 1.2×10^{29} kg (17.23 Earth masses), somewhat larger than Uranus's mass. Since Neptune appears to be slightly smaller than Uranus, it must have an appreciably higher average density: 1.67 g/cm³. As on Uranus, most of the matter in the planet is likely to be low-density material such as hydrogen, methane, ammonia, and water.

The internal structure is probably similar to that assumed for Uranus: an outer hydrogen ocean, an icy mantle, and a rocky core (Figure 15.2). Rough calculations indicate that the outer 6500 km of the planet is mostly liquid hydrogen, although there is probably an uncertain amount of liquid helium as well. The total mass of this liquid layer is estimated to be only 6 percent of the total bulk of the planet. Underneath the liquid-hydrogen ocean is an 8000-km-thick shell of solid water, methane, and ammonia under enormous temperatures and pressures. This material is the dominant component of Neptune, making up approximately two-thirds of the total mass of the planet. Neptune probably has a rocky central core of approximately 4.3 Earth masses. The radius of this core is 8000 km and the average density is 14 g/cm³. The pressure and temperature at the center of the core are of the order of 2.2×10^7 atm and 7000°C, respectively.

The effective temperature of Neptune (as measured by studying the spectrum of the light reflected from the planet) is −218°C. The expected temperature of a body at Neptune's distance from the Sun is only −227°C, so Neptune, like Jupiter and Saturn, must have some sort of internal energy source. The origin of this excess heat is unknown.

THE ATMOSPHERE

Not much is known about Neptune's atmosphere, although it is probably quite similar to that of Uranus. The dominant component is hydrogen gas mixed with an undetermined amount of helium. The atmosphere has a considerable amount of methane (CH_4), which accounts for Neptune's green color. Ethane (C_2H_6) has also been detected, but not ammonia. However, ammonia is almost certainly present. It may eventually be found further down in the atmosphere in the form of frozen ice crystals in the clouds.

Mathematical models of the atmosphere predict a temperature of $-228°C$ at the level where the pressure is 0.1 atm. The temperature should steadily increase with depth (reaching $-203°C$ at a pressure of 1 atm and $-103°C$ at a pressure of 10 atm) until the atmosphere eventually merges with the liquid hydrogen layer below (Figure 15.3).

Details about the cloud structure and weather in the upper atmosphere of Neptune are uncertain because of the large distance involved. The upper cloud deck is probably so deep in the atmosphere that the planet must appear just as bland as Uranus when viewed from above. However, a dark equatorial band has been noticed in some telescopic observations, and there is evidence for some discrete cloud features. In addition, there seems to be a haze layer at a depth of 0.7 atm.

WITNESSES TO A CATASTROPHE?

Neptune has two known moons, Triton and Nereid. The orbits of these two moons are perhaps the most unusual of any satellites in the Solar System,

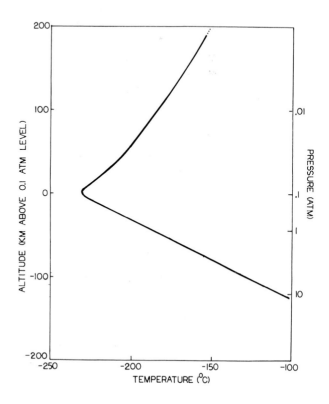

FIGURE 15.3 The vertical profile of Neptune's atmosphere.

indicating that something violent and spectacular must have taken place within the Neptune system in the distant past.

TRITON—AN OCEAN-COVERED WORLD

Triton is one of the largest moons in the Solar System. It was discovered by William Lassell in 1846, only a month after Neptune itself was first seen.

Triton's orbit is nearly circular, averaging about 350,000 km from Neptune (Figure 15.4). In some respects, Triton's orbit around Neptune is similar to the Moon's orbit around Earth. Like the Moon, Triton has an orbit that is approximately parallel to the orbital plane of its primary. Like the Earth, Neptune has an appreciable axial tilt. The plane of Triton's orbit therefore makes a rather steep angle (approximately 28°) with respect to the planet's equator. Triton's orbit does differ from that of the Moon in one important respect. The motion of Triton is retrograde; it revolves about Neptune in a direction opposite to the planet's rotation.

A satellite in a retrograde orbit is unstable. Since the tidal bulge raised on Neptune by Triton's gravitational forces is carried in the opposite direction by the planet's rotation, Neptune exerts a small but significant component of gravitational force that acts in a direction opposite to the moon's motion (Figure 15.5). Under the influence of this force, Triton will slowly spiral inward toward Neptune, eventually to be destroyed. Early calculations projected that Triton will collide with the planet sometime during the next 10 to 100 million years. However, more recent studies have indicated that the rate of orbital decay is actually much slower and that Triton could last as long at 10 billion years.

Although Triton is far too distant to be anything other than a small speck of light in the telescope, careful studies of the variations in the intensity of the light reflected from the moon have confirmed that Triton's rotation is synchronous with its rotation. Triton, like most other moons in the Solar System, perpetually turns the same face toward its primary.

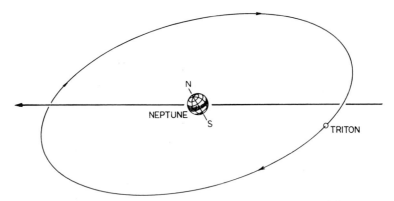

FIGURE 15.4 The orbit of Triton. In this view, Neptune, Triton, and the width of the orbit are shown in their correct relative sizes. The long arrow shows Neptune's orbit around the Sun. Triton's orbit is approximately coplanar with Neptune's orbit around the Sun.

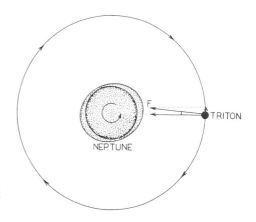

FIGURE 15.5 The decay of Triton's orbit. Triton travels around its orbit in a direction opposite to that of the rotation of Neptune. Consequently, the tidal bulge of Neptune will gradually retard the orbit of Triton, pulling it in closer and closer to the planet.

Because of the extreme distance, accurate measurements of the radius of Triton are extremely difficult to make. Values as small as 1600 km and as large as 2600 km have been obtained by various measurement techniques. The most recent and perhaps most reliable value is 1600 km, obtained by the method of speckle interferometry. This makes Triton approximately the size of Io and the Moon but definitely smaller than either Ganymede or Titan.

The mass of Triton can be crudely estimated by studying the tiny wobble that its motion induces in the orbit of Neptune. These techniques give a mass for Triton of about 1.9 times that of Earth's Moon, although the error is probably considerable. This mass presents astronomers with a serious theoretical problem. It requires that the density of the moon be quite high, suggesting that Triton is made largely of heavy metals and silicates rather than lighter methane, ammonia, or water-ice volatiles. This result is somewhat perplexing, as all other outer Solar System satellites studied to date are made mostly of low-density frozen volatile material. However, the errors inherent in the measurements of the mass and radius of so distant a world are considerable, and it is possible that the density of Triton has been grossly overestimated. More accurate measurements are necessary before it is possible to resolve the dilemma.

Because of its extreme distance from the Sun, the surface of Triton must be exceedingly cold. The temperature there cannot be much greater than −220°C. Even though this world is exceedingly frigid, there is, nevertheless, the distinct possibility that Triton possesses a significant atmosphere and perhaps even an ocean!

Since Triton presumably formed in the outer reaches of the solar nebula, it was anticipated that water ice should be an important component of its surface. The first crude spectra of the light reflected from Triton were taken in 1979 by Dale Cruikshank and Peter Silvaggio. Surprisingly, they failed to turn up any evidence for the presence of water ice. However, the reflectance spectra did show evidence for the presence of a different type of volatile material — methane.

The phase in which this methane exists is still uncertain. Triton's methane could be in the gaseous state as part of an atmosphere or it could be in the solid state as a frozen snow permanently stuck to the surface. The temperature on Triton is so cold that most of its methane is probably frozen solid. The intensity of the methane spectrum varies as the moon rotates, which suggests that the methane coverage is not uniform. The leading face of Triton may have more methane than the trailing one.

There is an additional feature in the infrared reflectance spectrum of Triton that may be due to the presence of nitrogen. It is so cold on Triton that any nitrogen actually present must exist primarily in the liquid state. This raises an intriguing possibility. Triton might be covered with vast oceans of liquid nitrogen!

The intensity of the nitrogen band varies as Triton rotates, suggesting that the nitrogen ocean is not so deep that it covers the entire surface. At the exceedingly cold temperatures that exist on Triton, nitrogen has a higher vapor pressure than does methane. Consequently, a significant nitrogen atmosphere may be present over the liquid. The surface pressure on Triton could be as high as 0.1 atm. Triton may be a frozen world dominated by liquid nitrogen oceans and frozen methane continents. Perhaps giant methane icebergs eternally drift across nitrogen oceans. We can only speculate until more data becomes available.

Triton appears in the telescope to be distinctly reddish in color. Frozen methane and frozen ammonia are both colorless, so something else must be present in abundance on the surface of Triton. It has been suggested that the reddish component consists of organic compounds, perhaps produced by the ultraviolet irradiation of mixtures of methane, ammonia, and water dissolved in liquid nitrogen. Like the Saturnian moon Titan, Triton may also be an organic chemistry laboratory, with the molecules essential to life being present in great abundance.

The decaying retrograde orbit of Triton suggests that the moon is not indigenous to Neptune but originated somewhere else in the Solar System and was captured by Neptune during an accidental close approach. The anomalously high density of Triton may mean that it is made up largely of metals and silicates rather than frozen volatiles. However, dense Earth-like metal–silicate worlds are typically not found in the outer regions of the Solar System. Triton may actually have formed in a region of space much closer to the Sun. Early in the history of the Solar System, a catastrophic encounter may have thrown Triton into a highly eccentric orbit that took it far from the Sun. During one of its outward swings, it happened to pass close to Neptune and was permanently captured in a retrograde orbit.

NEREID

Neptune's other moon, Nereid, was first detected by Gerard Kuiper in 1949. It is a good deal smaller and occupies a highly eccentric orbit (e = 0.7) that makes a rather steep angle with the equator of Neptune. This orbit has a higher eccentricity than that of any other moon in the Solar System, but Nereid's motion is in the direct sense.

Nereid is much smaller than Triton. Its diameter is estimated to lie somewhere between 150 and 525 km. Almost nothing is known about its physical properties. The eccentric orbit of Nereid may mean that it too was captured during an accidental close approach.

THE RINGS OF NEPTUNE?

The discovery of rings around Uranus has spurred a search for a similar ring system about Neptune. The occultation of the star BD 17° 4388 by Neptune in 1968 did not show any evidence for a ring system around the planet. Another stellar occultation in 1983 also failed to produce any characteristic intensity dips that could be ascribed to unseen rings. However, University of Arizona astronomer Harold Reitsema did note one momentary flicker in the combined brightness of Neptune and a faint star as they passed each other in 1981. This event could have been produced by a third moon at a distance of 50,000 km from Neptune. Or it could have been produced by nothing more significant than a momentary disturbance in the upper atmosphere of the Earth. In 1984 two separate observatories in Chile each reported one brief flicker during an occultation event. This might be evidence for a narrow ring 70,000 to 80,000 km from Neptune. However, no secondary occultation was seen, so it is possible that the flicker was caused by one or more tiny moons orbiting close to Neptune.

In view of all the negative reports, if the ring of Neptune actually exists, it must be so thin and diffuse that it is invisible 90 percent of the time that astronomers search for it. Perhaps the ring is discontinuous or clumpy in structure. The general consensus of most planetary scientists, however, is that the confirmation of rings around Neptune is only a matter of time.

Like Uranus, much remains to be learned about Neptune. The spacecraft *Voyager 2* is scheduled to fly past Neptune on August 24, 1989, coming within 7400 km of the north pole of the planet. Five hours after its closest approach to Neptune the probe will pass within 10,000 km of Triton. Assuming that the electronic equipment aboard the probe survives for that long, many of the questions about Neptune and its intriguing moon Triton should be answered at that time.

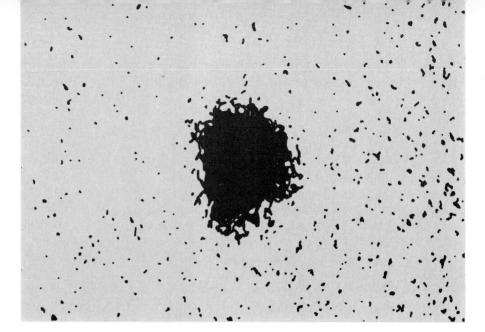

16

Pluto — the Edge of the Solar System

The outermost of the planets, Pluto, has the most eccentric orbit ($e = 0.248$), which averages 39.7 AU from the Sun. In fact, the orbit of Pluto is so eccentric that the planet actually comes closer to the Sun than does Neptune during part of its year, as it has now (Figure 16.1); it will reach perihelion sometime in 1989.

The orbit of Pluto is also unique in being more highly inclined with respect to the plane of the ecliptic (the angle is 17°) than any other planet. The length of Pluto's year is 250.3 Earth years. Pluto has not yet traversed even one-quarter of its circuit of the Sun since it was discovered, so its exact orbital parameters are still somewhat uncertain.

The discovery of Pluto in 1930 by the American astronomer Clyde Tombaugh of the Lowell Observatory in Arizona was the culmination of a long photographic search. Years earlier, the astronomer Percival Lowell (the same man responsible for publicizing the Martian canals) noted that there were tiny irregularities in the motions of both Uranus and Neptune that could be explained by the presence of another planet farther out. People searched the skies in vain for years until Tombaugh finally found a tiny dot of light on a photographic plate. The fact that it was a previously unknown planet was confirmed by noticing its slight motion on photographic plates taken on successive nights.

Photograph of Pluto taken by the U. S. Naval Observatory in 1978. The oblong shape of the image suggests the presence of a heretofore unknown satellite. (Photograph courtesy of U. S. Naval Observatory).

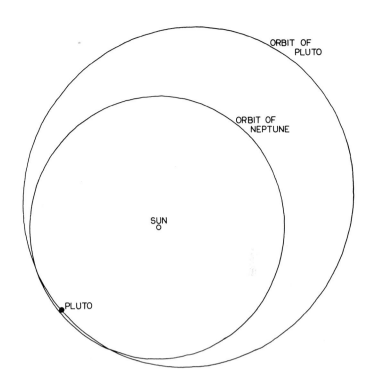

FIGURE 16.1 The elliptical orbit of Pluto. At present, Pluto is located inside the orbit of Neptune.

It was given the name Pluto, after the ancient Greek god of the underworld. This choice of name indirectly honored Percival Lowell, as the first two letters of the name Pluto are his initials.

THE SIZE–MASS CONTROVERSY

There has been a long-raging controversy about the size of Pluto. Pluto is so small and distant that it has an angular diameter of only about 0.25 second of arc, near the limit of resolution for most large telescopes. This gives a diameter of 6000 km, but this number could be in error by as much as 50 percent. A more accurate measurement of the size could be made if Pluto were to occult a star. This has never happened, but a near miss occurred in 1965. This gave an upper limit for Pluto's diameter of 6800 km. In 1979 a relatively new technique known as speckle interferometry (which has been successful in measuring the diameters of some larger nearby stars) provided the first direct measurement of Pluto's diameter. A value of 3000 to 3600 km was obtained, making Pluto an object smaller than Earth's Moon.

Pluto's mass has been an even greater problem. Lowell (in 1915) estimated that his proposed ninth planet should have a mass of about 6.6 Earth masses, based on his observations of the orbits of Uranus and Neptune. William Pickering revised the estimate of the mass of the as-yet-unseen planet down to about 2

Earth masses in 1919. Based on further observations of the motions of Uranus and Neptune, he brought the estimated mass down still further to 0.7 Earth masses in 1928.

Following the discovery of Pluto, several astronomers during the 1930s and 1940s all seemed to agree on a value of about 1 Earth mass for the planet, based on progressively more and more accurate measurements of the orbits of Uranus and/or Neptune. A value of 1 Earth mass for the mass of Pluto did, however, present astronomers with one critical problem. It would require that the density of Pluto be of the order of 55 g/cm^3, 5 times the density of lead. The planet would have to be made of collapsed matter!

The advent of digital computers during the 1960s made it possible to make far more precise analyses of orbital motions. Based on accurate computer calculations of the orbits of Uranus and Neptune, the mass of Pluto was once again revised sharply downward, this time to about 0.11 to 0.18 Earth mass. This gave a much more reasonable density of about 5 g/cm^3. This is probably still too high, since most outer planets (as well as their moons) have densities ranging between 1 and 2 g/cm^3.

PLUTO'S NEARBY MOON CHARON— THE CONTROVERSY RESOLVED

The resolution of the Pluto mass controversy came in an entirely unexpected manner. In 1978 James W. Christy, an astronomer working at the U. S. Naval Observatory, happened to notice that the image of Pluto appeared inexplicably elongated in some telescopic photographs. This meant either that Pluto is very much longer than it is wide or else that it has a nearby moon that cannot be separately resolved. The second alternative is more likely, as a planet-sized body so elongated should quickly tear itself apart as it rotated. This moon was named Charon, in honor of the mythical boatman who ferried souls of the departed across the river Styx to the underworld. The orbital period of Charon is about 6.4 Earth days, measured by watching the changing shape of Pluto's image on successive photographic plates.

Computer analyses of telescopic observations using highly sensitive charge-coupled devices as the light detection element have been able to separate the visual images of the two worlds. Charon occupies a nearly circular orbit only 15,000–20,000 km from the planet (Figure 16.2). The moon Charon is estimated to be only about one-fifth as bright as Pluto. If both planet and satellite are made of the same material, Charon should then be approximately 40 percent as large as Pluto. This would give Charon a radius of the order of 1000 km and a mass about one-tenth of that of Pluto.

The presence of the previously unseen moon enables the mass of Pluto to be estimated with some degree of accuracy for the first time. The result is 0.0015 to 0.0024 Earth masses (depending on the assumed separation), much less than any previous estimate. The density is therefore somewhere between 0.3 and 2.5 g/cm^3, depending on the value that is taken for the radius. Probably the most

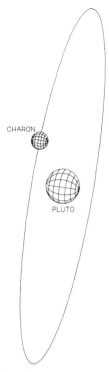

FIGURE 16.2 The orbit of Charon. In this diagram, Charon, Pluto, and the width of Charon's orbit are shown in their correct relative sizes. The plane of Charon's orbit around Pluto is approximately perpendicular to the plane of Pluto's orbit around the Sun.

reliable value for the radius is the one obtained by speckle-interferometry measurements. This gives Pluto a density of 0.4 to 1.0 g/cm^3.

Such a low density must mean that Pluto is made up largely of materials such as frozen methane, ammonia, or water rather than silicate rocks or heavy metals. The spectrum of solid methane has been found in the light reflected from Pluto. It is indeed possible that the planet is actually made up almost entirely of frozen methane. A methane "snowball" the size of Pluto should have a density of only 0.53 g/cm^3, consistent with the new density estimates. Over Pluto's brief history as a known planet, its perceived character has changed from that of a superdense object made of some as-yet-unknown substance to that of an icy snowball made of frozen slush.

THE CELESTIAL DANCE

Like most of the other moons in the Solar System, Charon is almost certainly tidally coupled to Pluto, perpetually showing the same face to its primary. However, the intensity of the light reflected from Pluto varies with a period of 6.3874 Earth days, the same as the orbital period of Charon. Pluto spins on its axis at exactly the same rate at which the moon Charon orbits around it. Pluto and its moon are both permanently locked into synchronous rotation; they perpetually exhibit the same faces toward each other. Billions of years from now, Earth and Moon will probably suffer the same fate.

The plane of Charon's orbit is estimated to be inclined at an angle of 106° with the ecliptic, although the error in the estimate is probably considerable. This makes Charon yet another example of a moon in a retrograde orbit. Early in 1985 workers at observatories in Texas, California, and Hawaii all reported that momentary dips were beginning to appear in the intensity of the combined image of Charon and Pluto in the telescope. This must mean that the two objects were passing in front of each other as seen from the Earth. Since Charon's orbit is approximately perpendicular to the plane of Pluto's orbit, such eclipses can be seen from Earth only twice every Plutonian year (Figure 16.3). It is anticipated that the current series of eclipses will last for several years. These eclipses are extremely fortuitous; they provide us with perhaps the only opportunity in this century for careful measurements of the orbits, sizes, and masses of these two intriguing worlds.

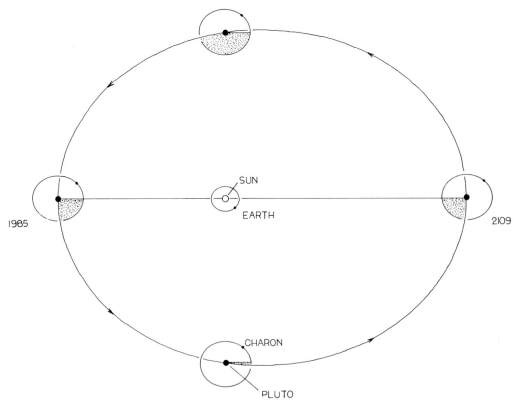

FIGURE 16.3 The orientation of Charon's orbit remains unchanged as Pluto travels around the Sun. Eclipses of Charon by Pluto can be seen from Earth only twice every Plutonian year (once every 125 Earth years).

There is considerable uncertainty in the orientation of Pluto's equator. Most astronomers suspect that in all probability it shares the same plane as the orbit of Charon, but it is difficult to be certain. Whatever the angle of inclination of Pluto's equator turns out to be, it is likely to be rather steep (greater than 50°), as there has been a long-term decrease in Pluto's mean brightness over the last few decades as well as an overall increase in the amplitude of the reflected light curve.

PLUTO'S ATMOSPHERE?

The effective temperature of the bright side of Pluto is $-230°C$, as determined by a spectral analysis of the light reflected from the planet. Some infrared reflectance spectra seem to indicate the presence of a thin methane atmosphere, although Pluto's low mass probably makes any sort of extensive atmosphere rather unlikely. The surface pressure was estimated to be about 0.0001 atm, with an upper limit of 0.05 atm. The 6.3874-Earth-day variation in the reflected light curve must in all probability mean that the atmosphere is not so thick that the surface of Pluto is completely obscured.

If this methane atmosphere is maintained by the sublimation of a surface methane frost, the average surface temperature must be at least $-215°C$, considerably warmer than the effective temperature. In order to maintain such a high temperature, some sort of heat source must be present, perhaps a greenhouse effect.

If Pluto *does* have a significant methane atmosphere, it presents astronomers with a serious problem. A gas as light as methane will rapidly escape from the atmosphere of so small a world. Pluto's atmosphere should have disappeared a long time ago, unless of course there is some means by which the methane lost to space is continually replenished. One possible source is, of course, the sublimation of the frozen methane on the surface. If this is actually how Pluto's atmosphere has been maintained over the billions of years of its existence, the planet itself should have completely disappeared millions of years ago!

Why then is it still there? One possibility is that there is an additional unseen heavier gas (perhaps nitrogen or argon) present in the Plutonian atmosphere that keeps any significant amount of methane from escaping into outer space. Another possibility is that Pluto has an appreciable atmosphere only when the planet is near perihelion. The orbit of Pluto is quite eccentric, and the surface temperature at perihelion could be significantly higher than it is at aphelion. The temperature at perihelion could be as high as $-200°C$. Since methane freezes at $-223°C$, the surface could get warm enough at the time of perihelion passage that some of the methane frozen on the surface vaporizes and forms a temporary atmosphere. When Pluto moves away from the Sun, this atmosphere freezes back onto the surface, forming a solid methane crust perhaps a few centimeters thick. The atmosphere alternatively appears and disappears once every Plutonian year. Since Pluto is currently nearing perihelion, we may simply be viewing the planet during the relatively brief time that its atmosphere is actually there.

THE ORIGIN OF PLUTO

There is an ingenious theory, due jointly to Raymond Lyttleton and Gerard P. Kuiper, that Pluto is not really a planet at all in the strict sense of the term, but is rather an escaped moon of Neptune. According to this model, Pluto and Triton originally occupied direct orbits around Neptune. By accident, they happened to pass quite close to each other. During the encounter, Triton was kicked into its present retrograde orbit and Pluto was ejected from Neptune altogether to travel in its own separate elliptical orbit around the Sun.

This theory does have some attractive features. The orbits of Pluto and Neptune do cross. In fact, the orbits appear to be in a 2 : 3 resonance, suggesting some sort of gravitational coupling between the two worlds. Pluto may actually be a lot like Triton in size and structure, suggestive of a similar origin. Furthermore, something highly catastrophic seems to have happened within the Neptune system in the distant past. However, there are significant problems. Even though the orbits of Pluto and Neptune do indeed cross, at no time do the two planets come any closer to each other than 16.7 AU. Pluto actually comes much closer to Uranus (10.6 AU) than it ever does to Neptune. Furthermore, the mass of Pluto is now known to be so much smaller (about 1/14) than that of Triton that even a near collision between the two worlds could not possibly have changed Triton's orbit into a retrograde one. In addition, there does not appear to be any significant amount of fine ring-type debris in orbit around Neptune, as was sure to have been created in vast amounts by so violent an encounter. Finally, the origin of Charon is unexplained.

Another possibility has been suggested by Dormand and Woolfson. They propose that Pluto was originally a "natural" satellite of Neptune. Triton was a dense, rocky asteroid-like body that originally formed in the inner Solar System. Triton was thrown into a highly elliptical solar orbit by a near collision with another asteroid. It accidentally entered the Neptune system during one of its orbits. During the encounter, Triton completely ejected Pluto from its orbit around Neptune. Triton lost enough energy during the encounter to become permanently bound to Neptune in a retrograde orbit. The moon Charon may have been torn from Pluto during the close encounter with Triton.

A 10TH PLANET?

The mass of Pluto still presents astronomers with a problem, as it is now known to be much too small to be the cause of the perturbations that are apparently present in the orbits of Uranus and Neptune. Unlike for Neptune, the discovery of Pluto was in all probability much more the result of a chance observation rather than a result of any prior knowledge of its position. Some astronomers have proposed that there is a 10th planet of about 2 to 5 Earth masses at a distance of 50 to 100 AU from the Sun that is responsible for the variations in the motions of Uranus and Neptune.

An alternative theory of Pluto's origin by U. S. Naval Observatory workers Robert Harrington and Thomas Van Flandern involves a near-collision between this 10th planet and Neptune. Pluto, Triton, and Nereid were all originally "natural" satellites of Neptune. During Neptune's close encounter with the 10th planet, the moons Triton and Nereid were deflected into their current odd orbits and Pluto was ejected from the Neptunian system altogether. During the violent ejection episode, tidal forces were strong enough to rip a chunk of matter from Pluto, forming the moon Charon. The catastrophic encounter with Neptune kicked the planet into an eccentric orbit, one which took it as far as 50 to 100 AU from the Sun.

17

Comets — the Primordial Snowballs

Among the most spectacular objects ever seen in the sky are the comets. They appear quite suddenly in the sky, put on a dazzling display for a few weeks or months, and then just as suddenly fade out and disappear. Not so long ago it was almost universally believed that the appearance of a comet in the sky was a supernatural event, most probably an omen of impending disaster. The apparition of a comet was therefore the object of superstitious dread. Today it is known that comets are objects of entirely natural origin that obey the same laws of celestial mechanics that the planets do.

For a time it was widely believed that a comet was actually some sort of upper-atmospheric disturbance. In 1577 the great astronomer Tycho Brahe conclusively demonstrated that comets were not connected with anything in the upper atmosphere. They must lie much farther out in space, even beyond the Moon.

Isaac Newton suggested in 1687 that comets were objects in heliocentric orbits that obeyed the same laws of celestial mechanics that the planets do. Edmund Halley (1656–1742) noted that a spectacular comet that had been seen in the year 1682 had a remarkable similarity to ones that had appeared in 1607 and in 1531. The same object seems to have returned to Earth's vicinity

Comet Bennett, a long-period comet, which came to perihelion in 1970. This comet has a very nearly parabolic orbit, which is inclined at an angle of nearly 90° with respect to the ecliptic plane.

330

once every 76 years. Halley predicted that this particular comet would come again in 1758. When it reappeared as scheduled, the natural origin of comets was confirmed. Subsequent examinations of ancient manuscripts found records of Halley's comet dating as far back as 240 B.C.

Most new comets are discovered by amateur astronomers, who make sky searching a hobby. When a new comet is reported it is initially designated by its date of discovery, followed by a letter indicating the order of discovery in that particular year (1910a, 1910b, and so forth). The new comet is also given the name of the person or persons who first reported it. Later, after the existence of the comet is confirmed and its orbit is well charted, it is designated by the year of its perihelion passage, followed by a Roman numeral giving the order of its arrival at perihelion for the comets appearing during that year.

SHORT- AND LONG-PERIOD COMETS

Comets are objects that travel around the Sun in elliptical orbits just as the planets do. However, the orbits of comets generally differ greatly from those of the planets. Cometary orbits fall under two broad general classifications: short period and long period.

Short-period comets are in highly elliptical orbits that lie entirely within the orbit of Pluto and have periods that are less than 200 Earth years. Most of these objects have been seen and recorded at least twice, and about 100 such comets are known. Halley's comet has a perihelion of 0.587 AU and an orbital eccentricity of 0.9672, so it comes closer to the Sun than does Venus and swings almost as far from the Sun as the orbit of Neptune (Figure 17.1a and b). Its closest approach to Earth was 0.04 AU in the year 837 A.D. The short-period comets generally have orbits that lie in or near the plane of the ecliptic. The mean angle of inclination is approximately 12°. However, about 1 in 20 have orbits that are retrograde; they travel around the Sun in a direction opposite to that of the planets. Halley's comet is an example of such a retrograde comet, with an orbit that makes an angle of 162.2° with the ecliptic. The distribution in the ascending nodes (the points where their orbits cross the ecliptic plane) of short-period cometary orbits is rather uniform, indicating that there is no preferred direction in the orientation of their orbits.

The *long-period comets* have extremely elongated orbits that stretch out far beyond Pluto's orbit. They have periods measured in thousands or even millions of Earth years. Their orbits are so elongated that they are difficult to distinguish from true parabolas (Figure 17.2). There are about 500 long-period comets known, but each has probably been seen by humanity only once. There are 86 long-period comets whose orbits have been sufficiently well charted to determine an aphelion and a period. The mean aphelion of these orbits is approximately 50,000 AU, nearly $\frac{1}{10}$ of the way to the nearest stars. The mean period is 4×10^6 Earth years. The long-period comets generally have orbits that make rather steep angles with the plane of the ecliptic. Their orbital inclinations seem to be essentially at random with respect to the ecliptic plane. About half are in retrograde orbits, half in direct orbits.

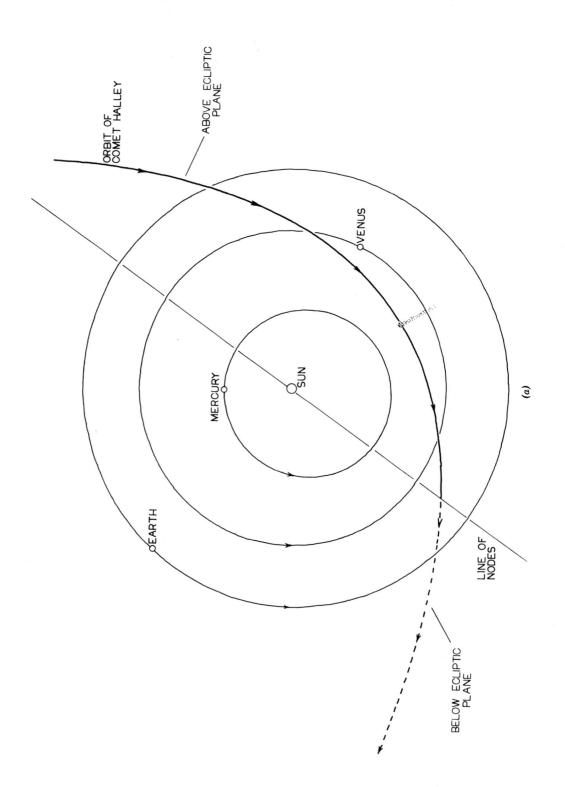

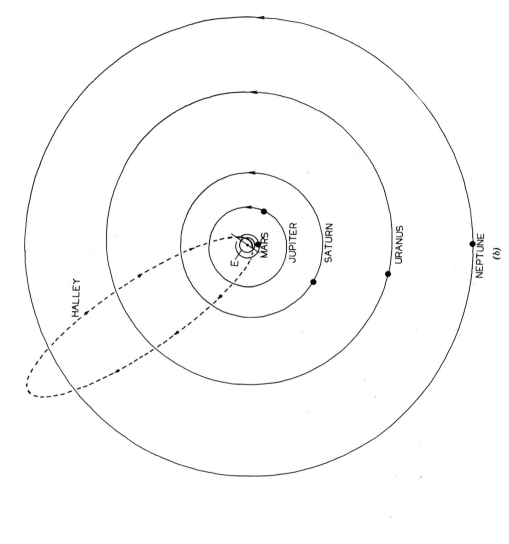

FIGURE 17.1 The orbit of comet Halley. (a) That part of the orbit that falls inside Earth's; (b) the full span of the orbit. The planets are shown in their correct positions at the time of comet Halley's passage through perihelion on February 9, 1986.

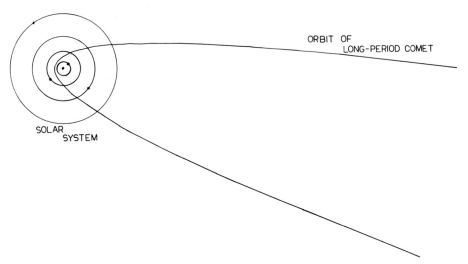

FIGURE 17.2 The orbit of a long-period comet. The perihelion of this comet orbit lies within Earth's orbit, but its aphelion is many thousands of astronomical units distant. Near the Solar System the shape of the orbit is approximately that of a parabola.

It is generally suspected that all short-period comets were originally long-period comets that happened to pass very close to one of the planets — in particular, Jupiter. When such a close encounter occurs, the orbit of the long-period comet is vastly perturbed. In some near misses the comet is accelerated to such an extent that it is completely ejected from the Solar System, never to return. In others the comet is slowed down and is placed in a less-elongated short-period orbit (Figure 17.3). Examples of both types of planetary near-misses have actually been observed.

About 60 percent of all short-period comets have aphelia that are located near the orbit of Jupiter. An even larger percentage have one of their orbital nodes located near the orbit of Jupiter, suggesting that a Jovian encounter was responsible for placing them in their present orbits. A long-period comet occupying an orbit approximately parallel to the ecliptic plane is much more likely to pass near a planet than is one in an orbit that makes a large angle with the ecliptic plane. This probably explains why most short-period comets are found in orbits that lie close to the plane of the ecliptic.

NUCLEI, COMAE, AND TAILS

Comets must have extremely small masses as compared to other significant Solar System bodies. No one has ever directly measured the mass of a comet, but there have been a few occasions when comets have been observed passing through Jupiter's system of moons. During such encounters, the paths of the comets are invariably perturbed, but the motions of the moons are unaffected.

The structure of a comet depends on how far away it is from the Sun.

Comets have three primary structural components: a nucleus, a coma, and a tail (Figure 17.4).

When a comet is far from the Sun (10 AU or more), its size must be quite small, perhaps only 2 to 10 km across. This object is called the *nucleus*. The nucleus is so small that it is difficult, if not impossible, to see from Earth even when using the largest and most powerful astronomical telescopes. There have been a few occasions when a comet has happened to pass in front of the Sun as seen from Earth. On each and every one of these occasions, the comet was not visible against the bright solar disk.

In November of 1980, Comet Encke approached within 0.3 AU of Earth. At that time the Arecibo radio telescope in Puerto Rico was able to detect the presence of the nucleus by radar. Analysis of the reflected radar pulse indicated

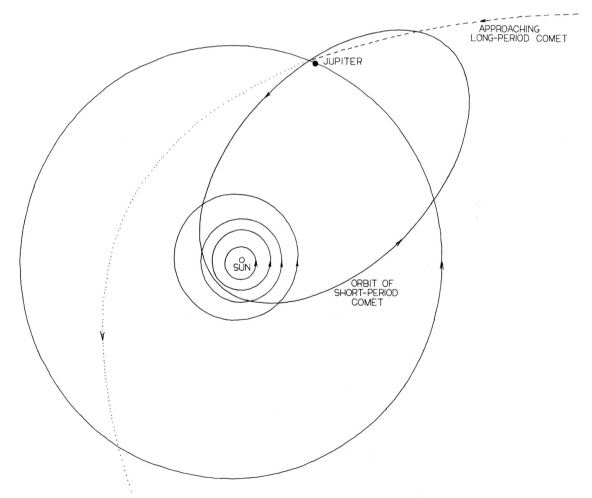

FIGURE 17.3 Origin of short-period comets. A long-period comet approaching Earth in a near-parabolic orbit is shown making a close encounter with Jupiter. This encounter places the comet into a much less elongated, short-period orbit.

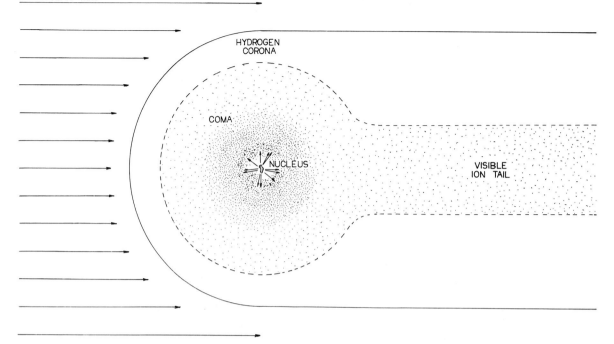

FIGURE 17.4 Schematic view of the nucleus, coma, and tail of a typical comet. (This diagram is not to scale.)

that the radius of the nucleus of Encke was 1 to 2 km. The nucleus of Halley's comet is estimated to be about 5 km across. There may be a few long-period comets that have nuclei that are as much as 100 km across. The Great Comet of 1729 may be an example of such a large-nucleus comet. At the time of its perihelion, this comet was so bright that it was easily visible to the naked eye, even though it never came any closer to the Sun than the outer reaches of the Asteroid Belt.

When a comet comes nearer to the Sun its appearance changes dramatically. At a distance of about 10 AU from the Sun, a bright, diffuse cloud begins to appear. This luminous cloud is called the *coma*. The coma expands and grows brighter as the comet comes closer and closer to the Sun. By the time that the comet reaches perihelion, the coma is often as large as the planet Jupiter.

The most spectacular part of a comet is its *tail*, which can be up to 1 AU long. The tail is a giant, fan-shaped appendage that originates from the bright cometary coma. It grows in length and brilliance as the comet nears the Sun. Comet tails always point away from the Sun, a fact which was first noted by the astronomer Petrus Apianus (1495–1552) in 1540.

After perihelion is passed, both the coma and the tail begin to decrease in size and eventually fade into invisibility as the comet recedes farther and farther from the Sun. A schematic view of the life cycle of a comet is shown in Figure 17.5.

THE DIRTY SNOWBALL

In 1950 the Harvard astronomer Fred Whipple introduced his "dirty snowball" theory of comets. He proposed that the nucleus is made up of ices of such volatile substances as water, carbon dioxide, hydrogen cyanide, formaldehyde, methane, and ammonia, mixed in with a lot of silicate dust and other rocky debris. This theory is now generally accepted. Cometary matter is thought to be the most pristine material still present in the Solar System, having remained essentially unaltered and undisturbed from the time the Sun and planets first formed nearly 4.7 billion years ago. The material within a comet is believed to be essentially of "solar composition." It has virtually the same relative elemental abundances as the outer atmosphere of the Sun, but lacks the excess hydrogen and helium. A hypothetical interior structure for a cometary nucleus is shown in Figure 17.6.

When far away from the Sun, a comet is cold enough so that all of its icy material is perpetually frozen in the solid state. However, as a comet nears the Sun, the rising solar heating vaporizes some of the near-surface ice and releases some of the trapped dust. This rapidly expanding cloud of gas and dust completely surrounds the nucleus and forms the coma. Ultraviolet light from the Sun breaks up some of the molecules in the coma to form such chemically exotic species as C, C_2, CN, CS, HCN, CH_3CN, C_3, H, NH, NH_2, O, OH, and CH radicals and CO^+, N_2^+, CO_2^+, CH^+, H_2O^+, and OH^+ ions, many of which are created in highly excited energy states. These excited species emit fluorescent light, causing the coma to shine. The parent molecules of these species are thought to be H_2O, NH_3, CH_4, and CO_2, all of which exist in the solid form

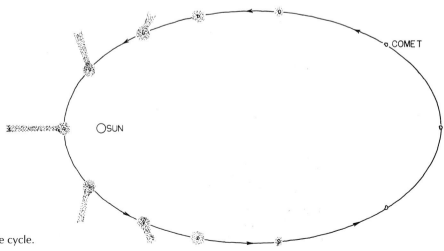

FIGURE 17.5 A comet's life cycle.

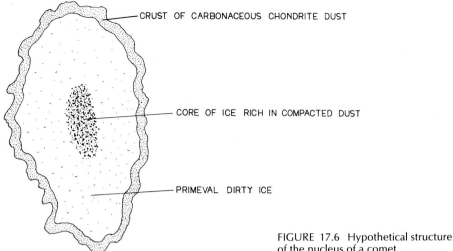

CRUST OF CARBONACEOUS CHONDRITE DUST

CORE OF ICE RICH IN COMPACTED DUST

PRIMEVAL DIRTY ICE

FIGURE 17.6 Hypothetical structure of the nucleus of a comet.

within the cometary nucleus itself. In addition, the spectra of a large number of metallic ions (for example, Na, Ca, Cr, Co, Mn, Fe, Ni, Cu, V, Si, and K) have been identified in the light emitted by the coma. These ions probably originally came from fine dust particles that were intermixed with the ice. Surrounding the bright coma is an extended halo (or "corona") of H and OH radicals, probably originating in the ultraviolet break-up of water molecules within the coma itself. On occasion such halos can exceed the Sun in size.

There is a strong possibility that the ultraviolet photochemistry responsible for the exotic chemical species found within the coma may have also produced a large number of even more complex organic molecules within the nucleus of the comet itself. A cometary nucleus may actually be a thick frozen organic soup, rich in such biologically interesting materials as hydrocarbons, amino acids, sugars, and nucleic acids. Fred Hoyle and Chandra Wickramasinghe have even suggested that the first living organisms actually appeared within the nucleus of a comet rather than on the Earth's surface itself. Life was brought to the Earth when this comet collided with our planet. These organisms found a favorable environment on Earth, where they prospered and flourished.

THE FASCINATING COMET TAIL

The tail is perhaps the most interesting part of a comet. We know comet tails are quite diffuse and almost completely transparent, since background stars can often be seen through them. A comet tail generally has two distinct components, termed type 1 and type 2. These two components appear to be completely different in structure and composition from each other and they are believed to be formed by entirely separate and distinct physical mechanisms (Figure 17.7).

The *type-1* tail component is a bright streamer that always points directly away from the Sun. Like cometary comae, type-1 tails shine by the emission of fluorescent light from species that are in highly excited states. However, emission spectra of type-1 tails are generally dominated by ions rather than by neutral atoms or molecules. Type-1 tails often have an intricate dancing pattern of narrow, threadlike streaks. This pattern is quite changeable and can vary from week to week, from day to day, and even from hour to hour. Sometimes "knots" appear, which accelerate down the tail away from the Sun, reaching speeds as high as 250 km/sec. On other occasions an ion tail streamer will suddenly appear to detach from the coma, only to be quickly replaced by a fresh one.

The other tail component (termed *type 2*) is a more dim and diffuse part that curves outward somewhat behind the first. Type-2 tails are less variable and generally exhibit less fine structure than do type-1 tails. Unlike type-1 tails, type-2 tails do not emit their own light. The spectrum of the light that comes from them closely resembles that of the light emitted by the Sun itself. Type-2 tails shine by reflected sunlight, just as the Moon and planets do. A detailed examination of the infrared light given off by type-2 tails indicates that they are largely composed of tiny particles of dust and are relatively poor in gas.

It was at one time assumed virtually without question that the pressure of solar radiation was responsible for pushing some of the material in the coma away from the Sun, forming the long tail. Radiation pressure is an exceedingly weak force, but, acting over a long time, it can have a measurable effect on small particles in free space. Theoretical calculations show that solar radiation pressure is fully able to account for the presence of the type-2 dusty tail. This

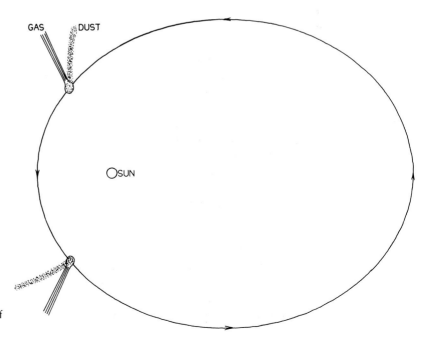

FIGURE 17.7 Two components of cometary tails.

radiation pressure gradually pushes the individual dust particles out of the coma and away from the Sun. Each dust particle then enters a separate Keplerian orbit around the Sun. Since these dust particles are all farther away from the Sun than is the comet itself, they all travel along their separate orbits at a slightly slower rate. The pattern of dust thus forms a fan-shaped appendage that always trails behind the head of the comet along its orbit.

However, radiation pressure is far too weak to account for the type-1 ion tail. In particular, it completely fails to account for the high velocities and rapid changes often seen in tails of this type. An entirely different mechanism must be responsible. During the 1950s, the Swedish scientist Ludwig Biermann proposed instead that the ion tail is produced by an invisible "wind" of charged particles emitted by the Sun. The existence of the solar wind was confirmed by the early lunar probes launched by the United States and Soviet Union during the late 1950s.

The solar wind consists primarily of electrons and protons that flow out of the solar corona at speeds of up to several hundred kilometers per second. Since electrons and protons carry electrical charges, there is a weak magnetic field (of a few milligauss in intensity) associated with the solar wind. This magnetic field interacts strongly with the ions present in the coma, dragging them away from the coma at high speeds. This produces a long, thin trail of ions streaming out of the coma in a direction pointing exactly away from the Sun.

IS KEPLER WRONG? THE SPINS OF COMETS

Several short-period comets appear to have significant departures from strictly Keplerian motion around the Sun, even after all possible gravitational interactions with the planets have been taken into account. During the 1800s, comet Encke persistently returned to perihelion 2.5 hours before precise orbital calculations predicted that it should. However, this discrepancy has steadily decreased over the years, so that this comet is now only a few minutes "early" each orbit. On the other hand, Comet Halley has persistently arrived at perihelion 4.1 Earth days "too late." Out of 20 short-period comets, all but 2 have small but significant deviations from pure Keplerian motion. Half arrive at perihelion too early, half too late.

These strange orbital deviations are probably not caused by the presence of some new and heretofore unknown exotic type of force. They seem instead to be caused by the presence of localized "hot spots" on the surface of the nucleus. When a comet approaches perihelion, the surface of the nucleus nearest the Sun becomes hot and some of the material is vaporized. This vaporized material is often emitted in the form of a high-velocity jet issuing from fissures or cracks in the outer surface. This directional jet causes a small but significant thrust to be exerted on the nucleus, slightly altering its orbit around the Sun.

The change that is produced in the orbit will depend on the precise direction of the high-speed vapor jet relative to the path of the comet. Because of solar heating, the rate of surface vaporization will always be greater on the daylight hemisphere than it is on the dark side. However, if the nucleus is rotating, there is an additional complication. The rotation of the nucleus causes the point of highest temperature to be tilted toward the "afternoon" side of the Sun-facing hemisphere. The vaporization rate will be at its highest when the orifice of the vapor jet is located there. Comets that are spinning in the pro-grade direction have their periods increased, whereas those spinning in the retrograde direction have their periods decreased.

These orbital changes can be used to derive crude inferences of the rotation rates of comets, even though the cometary nucleus itself is far too small to be seen. The rotation rates of as many as 50 comets have been deduced by this technique. The mean rotation period is about 15 hours. Rotation axes seem to be randomly oriented, with about half of the rotations prograde and half retrograde.

THE DEATHS OF COMETS

Each time a periodic comet passes close to the Sun a large amount of its surface material is vaporized away. A cloud of this gaseous and dusty debris travels parallel to the comet in its orbit. Perhaps the outer meter or so of a comet is stripped away every time it passes around the Sun, so a short-period comet should last only a few thousand years before it is completely destroyed.

Short-period comets do tend to become slightly fainter and less spectacular each time they return to perihelion. Comet Encke has been seen some 52 times since its discovery in 1786 and seems to be getting fainter by a factor of 2.5 per century. Comet Holmes faded into invisibility as it neared the Sun on its second return (as 1906III). It has not been seen since. Comet Brorsen made five appearances between the years 1846 and 1879 and then suddenly vanished altogether.

There have been a few occasions on which comets have disintegrated before the very eyes of observing astronomers. The great Sun-grazing comet of 1882 split into several fragments during its approach to perihelion (0.007 AU). Several different comets have subsequently been spotted along the same orbit. Biela's comet split into two between successive appearances in 1832 and 1845. Two pieces of this comet returned in 1852, but they both quickly faded into invisibility. Neither piece was ever seen again.

When comets disintegrate, their fragments accumulate in a disk of dust and debris that circles the Sun in a plane close to that of the ecliptic. This cloud of comet dust can actually be seen from Earth. On a dark, moonless night it appears as a diffuse band of light crossing the sky, passing through the 12 constellations of the zodiac. It is therefore known as *zodiacal light*. A particularly bright spot of zodiacal light can often be seen in the night sky, in a spot exactly 180° away from the Sun. This is called the *gegenschein* (German for "counter-glow"). This spot of light is produced by preferential reflection of light by cometary dust that is located in the antisolar direction at the time of observation.

THE OORT CLOUD

Where do comets come from? One theory proposes that comets are simply the remnants of the icy planetesimals that formed the outer planets Uranus and Neptune, just as asteroids are the remnants of the metallic and rocky planetesimals that formed the inner terrestrial planets. Close encounters with the accreting outer planets threw these loose chunks of ice into near-parabolic long-period orbits. There may be many millions of these icy objects still orbiting the Sun. They will be gradually reduced in number as they are either slowly vaporized by successive trips around the Sun or else collide with one or the other of the planets.

In 1950 the Dutch astronomer Jan Oort proposed that comets actually originate much farther out in the Solar System. He suggested that there is a permanent cloud of dirty snowballs that surrounds the entire Solar System at a distance of many thousands of astronomical units, way beyond the orbit of Pluto and nearly $\frac{1}{10}$ the distance to the nearest stars. This distant swarm of comets has come to be known as the *Oort cloud* (Figure 17.8). A large fraction of the long-period comets have aphelia of about 50,000 AU, indicating that the densest part of the Oort cloud lies at this distance from the Sun. So far, the Oort

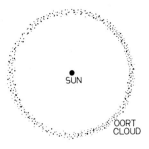

●STAR

●STAR

●STAR

FIGURE 17.8 the Oort comet cloud.

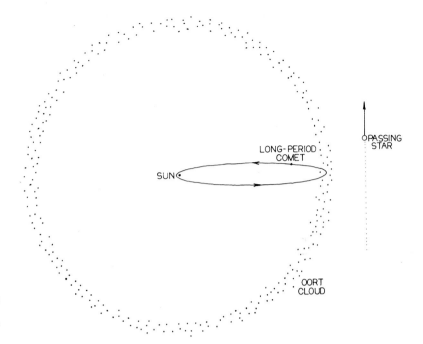

FIGURE 17.9 Origin of long-period comets. The motions of passing stars perturb the motions of the comets in the Oort cloud, sending some of them toward the inner regions of the Solar System.

cloud has only a hypothetical existence, as it is quite out of the question to detect such small objects as cometary nuclei at such great distances.

The Oort cloud was probably formed at the same time as the Solar System itself. These distant objects are the remnants of icy planetesimals that condensed from volatile materials originally located in the very coldest outermost reaches of the Solar System, far from the warmth of the Sun. The comets in the cloud circle the Sun in elliptical orbits with periods of many millions of years.

Most of the comets in the Oort cloud never come anywhere near the Sun. However, these icy snowballs are so distant from the Sun that nearby stars can exert significant gravitational forces on them, perturbing their orbits around the Sun. In a few cases, the perturbations produced by the random motions of these stars can be strong enough to deflect the comets into paths that send them inward toward the Solar System. These become the long-period comets (Figure 17.9).

One problem with the Oort theory is that there must be an enormously large number of comets in the cloud in order to account for the numbers of comets that are currently observed. Short-period comets generally last only a few thousand years before they are destroyed. In order to have maintained a steady supply of these transient objects for a time as long as 4.7 billion years, the total mass of all of the comets within the Oort cloud must be perhaps 10 to 100 times the combined mass of all of the planets in the Solar System!

VISITS TO A DISTANT COMET

Comet Halley came again to perihelion on February 9, 1986. Its appearance was one of the most eagerly anticipated and heavily publicized astronomical events in recent history. The number of books, magazine articles, television programs, newspaper stories, T-shirts, and other memorabilia dealing with the coming of Halley became a veritable flood. Unfortunately, this particular passage of Halley was one of the most unfavorable for viewing in recorded history. When the comet was near perihelion and hence at its most spectacular, it was on the opposite side of the Sun from Earth. The comet was visible only to those people fortunate enough to be out in the country, well away from city lights. Millions of people (including the author) were not able to see the comet at all.

Although the 1986 passage of Halley was rather disappointing to most people, it provided the focus for one of the most rewarding programs of international cooperation in space exploration yet to take place. When Halley returned to perihelion this time, an international flotilla of five spacecraft was waiting for it.

The most sophisticated of these craft were two identical probes launched by the Soviet Union. Named *Vega 1* and *Vega 2,* the probes had initially been dispatched toward Venus. After dropping off landing probes during the Venus flyby phase of the mission, both Vegas were deflected toward encounters with comet Halley. Both of the craft passed through the coma of Halley, missing the nucleus by only 8000 to 9000 km. The Vegas carried numerous instruments designed to study the magnetic fields and plasmas in the vicinity of the comet. Each Vega also carried equipment to measure the composition of the gas and the dust in the coma. Included aboard each craft was a dust counter and mass analyzer supplied by a group of American scientists. The craft were also equipped with television cameras that were to send images of the nucleus back to Earth. The spacecraft themselves were controlled by the Soviets, but the scientific program during the Halley encounter was directed by a committee representing nine different nations.

The Japanese Institute of Space and Astronautical Science (in Japan's first deep-space venture) sent two separate probes to the vicinity of Halley. The goal of these missions was a detailed study of the plasma environment and hydrogen corona of the comet. The first of these probes was named *Sakigake* (Pioneer), or simply MS-T5. It carried a set of instruments to study plasmas and magnetic fields in the vicinity of the comet. By design, *Sakigake* came no closer than 7×10^6 km to the nucleus. *Sakigake* was complementary to another quite similar probe named *Suisei* (Comet) or simply Planet A. *Suisei* passed within 151,000 km of the nucleus of Halley. The craft carried a camera for observing the ultraviolet light emitted by the coma and hydrogen corona of Halley as well as instruments to study the behavior of the solar wind in the vicinity of the comet.

The European Space Agency (ESA) also became involved in the Halley exploration program. Their probe was named *Giotto,* in honor of the Florentine painter Giotto di Bondone (1266 – 1337). Giotto witnessed the 1301 passage of Halley and was so impressed that he gave the comet a prominent place in his well-known Nativity scene *Adoration of the Magi,* with the comet playing the role of the Star of Bethlehem.

Giotto passed within 610 km of the nucleus of Halley, the closest approach of any of the five spacecraft. It carried instruments to measure the density and composition of the gas, dust, and plasma in the inner coma near the nucleus. In addition, *Giotto* was equipped with a television camera to relay pictures of the nucleus back to Earth during the close approach. Although the spacecraft was equipped with a set of shields to protect its delicate instruments from the impacts of dust particles, it was feared that *Giotto* might be destroyed during passage through the dusty inner regions of the coma. Although many of its instruments were either completely disabled or heavily damaged by dust impacts, the spacecraft itself did manage to survive the encounter.

Unlike most previous planetary missions, the results of the five Halley encounters provided few major surprises. Most of what we thought we knew about comets was confirmed. In particular, the Whipple dirty-snowball theory of nuclear structure was dramatically verified.

The Soviet Vegas and the European *Giotto* determined the chemical composition of the gases in the coma. As expected, the water molecule is the dominant species within the coma, making up 80 percent (by volume) of all gases present. The density of the water vapor in the region of the coma nearest the nucleus is about 300 billion times less than the density of air at sea level on Earth. This water vapor density is consistent with a gas production rate at the nucleus of approximately 10^{30} molecules/sec. This is equivalent to approximately 30,000 kg of water being driven off the nucleus every second. The velocity at which typical water vapor molecules departed from the vicinity of the nucleus was approximately 1 km/sec.

A large percentage of the remainder of the coma gases is carbon dioxide. The chemical identity of the other vapors is less certain, but they are relatively low-molecular-weight gases made up mostly of carbon, oxygen, hydrogen, and nitrogen atoms. The most probable candidates are ammonia, nitrogen, and methane.

The Halley probes encountered the first ions of cometary origin at a distance of 10^6 km from the nucleus. Ions such as CO^+, CH^+, C^+, H_2O^+, CO^+, H^+, C_2^+, OH^+, H_3O^+ were detected, as well as radicals such as OH, CH, C_2, C_3, CN, NH, and NH_2. These ions and radicals are undoubtedly the products of the photochemical breakup of water and carbon dioxide (and perhaps ammonia and methane as well) parent gases emitted from the nucleus of the comet.

Giotto and both the Vegas collided with numerous tiny particles of dust during their passage through the coma. The outer boundary of Halley's dust coma was encountered at a distance of 250,000 to 300,000 km from the nucleus. The dust density increased steadily as the probes came closer and closer to the nucleus. *Giotto* alone counted more than 12,000 dust impacts. *Giotto* probably swept up as much as 150 mg of comet dust during passage through the inner coma, enough to produce a small but measurable deceleration in the probe's velocity.

The instruments aboard these probes were able to count the number of dust impacts and were able to measure the masses of the dust particles. In that manner, the density of dust within the coma could be determined. The dust-to-gas ratio is somewhere between 1 : 10 and 1 : 4 (by mass), which makes Halley a "moderately dusty" comet. The average rate of dust production is of the order of 3000 kg/sec but is highly variable. The largest dust particle encountered had a mass of 40 mg. The smallest had a mass of 10^{-17} g, the limit of sensitivity of the instruments. The presence of such large numbers of low-mass dust particles in the coma was unexpected, as some telescopic measurements taken from Earth had seemed to suggest that there should be few, if any, dust particles less than 10^{-14} in mass.

The Halley probes were also able to determine the chemical composition of the dust grains that were encountered. The results provided somewhat of a surprise. Eighty percent of the grains of dust that were studied were composed almost exclusively of the lighter elements carbon, nitrogen, oxygen, hydrogen, and sulfur. The remaining 20 percent were also rich in these light elements, but they also had appreciable amounts of heavier elements such as sodium, magnesium, silicon, potassium, calcium, and iron. Comet dust seems to be largely organic in composition, with only relatively small amounts of metals or silicate minerals. Most dust particles have a "fluffy" structure, with a typical density of only 0.35 g/cm³. A typical particle of comet dust probably consists of a central "grain" of silicate minerals and/or metals, surrounded by a "cocoon" of organic matter.

The organic matter might resemble tar, carbon black, or coal dust in structure and composition. The presence of such large amounts of organic matter in comet dust was unexpected. This organic material may be some of the oldest matter in the Solar System. It may have been created by the ultraviolet irradiation of the primitive water, ammonia, carbon dioxide, and methane ices present in the Solar System before the solar nebula had fully condensed. The intense ultraviolet light emitted by the newly formed Sun caused the molecules of these volatile ices to react with each other, forming higher-molecular-weight, nonvolatile organic materials.

The most eagerly awaited results of the Halley missions were the television pictures of the nucleus. The photographs sent back to Earth by *Giotto* and by the Vegas were somewhat disappointing, as we have by now become so accustomed to crisply detailed views of distant worlds. The nucleus of Halley was almost completely hidden from view by dense clouds of inner-coma dust. Special processing of the television images was required to bring out even the coarsest detail.

However, enough could be seen so that the overall size and shape of the nucleus could be determined. The photographic images show the nucleus to be an irregular, potato-shaped object approximately 14 km long and 7.5 km wide. This size is somewhat larger than expected. The surface appears to be irregular and shows signs of the presence of circular features, which might be impact craters. The surface is extremely dark, with a reflectivity of only 2 to 4 percent. This albedo is similar to that of the rings and moons of Uranus as well as that of the dark material on the Saturnian moon Iapetus. The surface of the nucleus cannot, therefore, be pure ice in composition.

As in the case of many other planets and moons in the Solar System, much information about the internal structure and composition of the nucleus of comet Halley could be inferred if we knew its average density. The average density in turn requires knowledge of both the size and the mass of the object. Size information is available from the spacecraft photographs of the nucleus, but the mass can be determined only by observing the deflection of the trajectory of a spacecraft as it flies near the nucleus. Unfortunately, the friction exerted by the dense clouds of gas and dust in the inner coma affected the trajectories of the passing spacecraft far more than did the tiny gravitational field exerted by so small an object. Consequently, the density of the nucleus of Halley is still unknown.

Although the density of the nucleus could not directly be measured by the Halley spacecraft, the Vegas may have obtained a possible clue to the interior structure by the use of a different technique. These two spacecraft were able to analyze the infrared light emitted by the nucleus of Halley, thereby measuring its temperature. It was in the range of 30 to 120°C. This temperature is somewhat of a puzzle. The surface of the nucleus is much hotter than it should be if the outer layers of the nucleus are made up largely of water ice intermixed with only minor amounts of darker contaminants. The conversion of solid water ice to the vapor phase (a process known as *sublimation*) is an efficient cooling source. It carries away the heat absorbed by sunlight at such a high rate that the surface temperature of a frigid, icy body approaching the Sun from the outer Solar System should never rise above −70°C, even at perihelion. Another problem for theorists is provided by the measured rate of water loss from the nucleus of Halley. It is only 0.1 of that which would be expected from the sublimation of an icy object of that size at the relatively high temperatures that were measured.

The international team of Halley investigators has resolved the dilemma by proposing that the dark surface of the nucleus is a relatively thin layer of nonvolatile organic material mixed with lesser amounts of heavier metals and silicate minerals. It surrounds an interior that is much richer in volatile ices. The layer of surface material must have an extremely low thermal conductivity, so low that it keeps the volatile interior of the nucleus cool, even at perihelion. The external boundary layer receives the solar flux and slowly transfers heat downward into the interior. Vaporized ice rises slowly upward through cracks or pores in the crust until it reaches the surface and escapes into space. The thickness of the insulating crust is estimated to be approximately 1 cm.

Both the *Giotto* and Vega images show jets of gas and dust bursting outward from six or seven small areas on the surface of the nucleus. The long-suspected existence of cometary jets was dramatically confirmed. Two of the jets accounted for the bulk of the observed gas and dust emission. The jets appear to have fine structure, suggesting that they could be composed of clusters of many smaller jets or else that they come from multiple sources. The active areas from which the jets emerge occupy less than 10 percent of the total area of the nucleus. The rest of the surface appears to be completely inactive, producing little if any dust or gas emission. Jets were seen emerging only from the "day" side of the comet, the side facing the Sun. None were seen coming out of the "night" face of the comet.

The lack of any identifiable jets coming from the dark side of the nucleus indicates that areas of active dust and gas emission quickly become inactive as soon as the solar heating is removed. As the nucleus rotates, jet emission sources that are moving into daylight turn on, whereas those that are moving into darkness turn off. Since the small regions that emit most of the gas and dust are not distributed uniformly over the surface, the rate of gas and dust emission varies as the nucleus rotates. The Halley probes found that major variations in gas emission rate take place in times as short as 1 or 2 hours. *Vega 1* encountered the coma during a highly active period of gas and dust emission. The craft may have passed directly through one of the jets. *Vega 2* passed through the coma during a relatively quiescent period. Consequently, *Vega 2* got a better view of the nucleus than did *Vega 1*. The arrival of *Giotto* fortunately seems to have coincided with a period of relative coma inactivity. Had *Giotto* encountered the coma during an active period, the spacecraft might well have been destroyed by the heavy flux of impacting dust.

The variability in the rate at which dust and gas are driven off the nucleus can be used to estimate the rate at which the nucleus is rotating. Photographs taken during the 1910 passage of Halley showed that features within the inner coma vary with a 53-hour period, suggesting that the nucleus rotates once every 53 hours. The Japanese spacecraft *Suisei* found that the hydrogen corona surrounding Halley out to a distance of about 3.0×10^7 km "breathed" — that is, it periodically expanded and contracted. The "breathing period" was 53 hours — presumably the rotation period of the nucleus. A rotation period of 53 hours was confirmed by making a comparison of the images of the nucleus taken by *Vega 1* with those taken by *Vega 2*. The rotation is in the direct sense, with the axis being approximately perpendicular to the plane of the comet's orbit. As expected, the rotation axis is nearly perpendicular to the long axis of the nucleus.

The comet probes also studied the interaction between the comet and the solar wind. Unlike the Earth, the nucleus of Halley does not have an intrinsic magnetic field. Consequently, the comet does not have a magnetosphere in the same sense that our planet does. In particular, it lacks the radiation belts associated with magnetic planets such as Jupiter and the Earth.

Even though Halley lacks an intrinsic magnetic field, its presence nonetheless has a profound and significant effect on the solar wind in its immediate vicinity. Since the solar wind consists of a current of charged particles, a magnetic field is associated with the flow. The flux lines of this interplanetary magnetic field lie perpendicular to the direction of the solar wind. Because the ions in the coma of Halley also have an electric charge, they cannot readily travel across the magnetic field lines that are carried by the solar wind. Consequently, these ions tend to be caught up in the flow of the solar wind and are carried downstream along with it. This process is shown schematically in Figure 17.10.

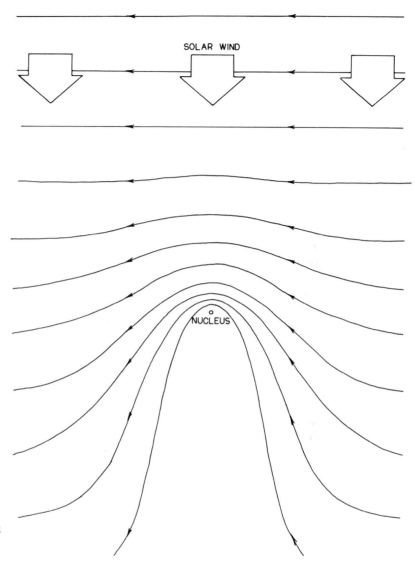

SOLAR WIND

NUCLEUS

FIGURE 17.10 Magnetic fields in the vicinity of a comet. The solar wind carries with it a weak magnetic field. When the wind encounters a comet, the magnetic field lines bunch up in front of the comet and drape around the sides in the antisolar direction. Ionized particles from the coma flow along these magnetic field lines, forming the tail.

The sweeping up of coma ions by the solar wind creates a bow shock or bow wave, much like the bow wave produced by a boat passing through water. The Halley probes found that the bow wave was approximately 400,000 km from the nucleus in the sunward direction. Because coma ions are continually being produced within the coma, the solar wind is forced to carry more and more material with it as it gets closer and closer to the nucleus. This "loading" of the solar wind causes the wind speed to slow down as it penetrates farther and farther toward the nucleus. This deceleration in solar wind velocity in turn causes the magnetic field lines carried by the solar wind to crowd closer and closer together as one gets nearer to the nucleus. The solar wind deceleration

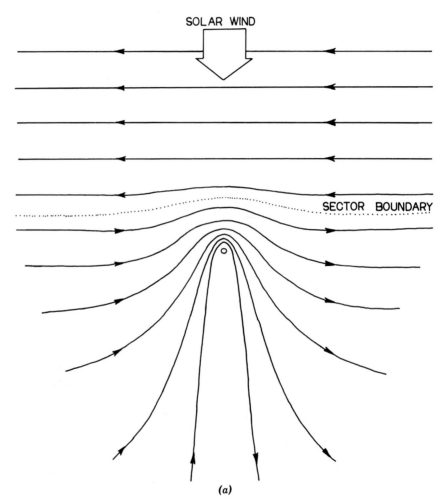

SOLAR WIND

SECTOR   BOUNDARY

(a)

FIGURE 17.11 Hypothetical origin of tail-detachment events. A sector boundary is a
region of the solar wind where the associated magnetic field changes sign. *(a)* When the
sector boundary passes through the comet's coma, the magnetic field is momentarily
weakened, and ions cease flowing into the tail. *(b)* The tail is re-formed once the sector
boundary has passed.

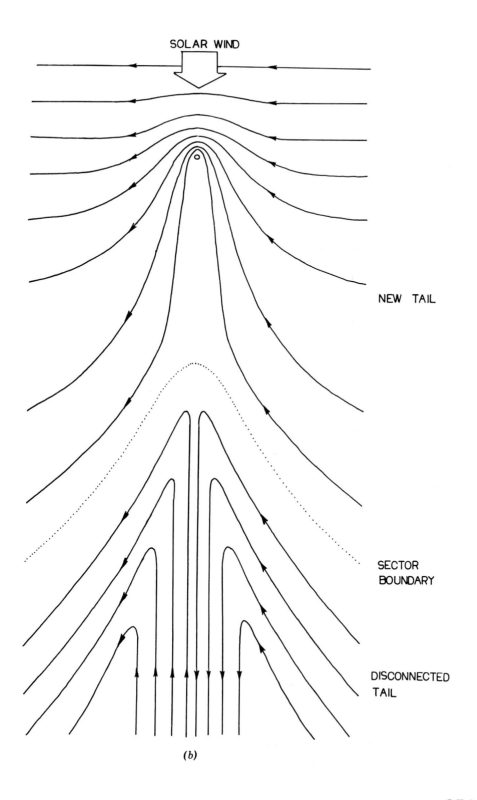

SOLAR WIND

NEW TAIL

SECTOR
BOUNDARY

DISCONNECTED
TAIL

(b)

Comets — the Primordial Snowballs **351**

continues until the inward pressure exerted by the solar wind is exactly balanced by the outward pressure exerted by the ions of the coma. This boundary is known as the *contact surface*. Since the solar wind cannot penetrate the contact surface, a cavity with no magnetic field is created in the vicinity of the nucleus. The width of the cavity is approximately 8500 km, as measured by *Giotto*.

To the sides of the comet, the number of cometary ions drops off rapidly, allowing the solar wind particles to pass on by relatively undeflected. This causes the magnetic field lines produced by the solar wind to drape themselves around the body of the comet. This draping of the field lines channels the flow of coma ions into an ion tail.

The Japanese probe *Sakigake* may have determined the cause of the mysterious comet "tail-detachment" events, which are sometimes visible from Earth. In these events, a segment of a comet ion tail will suddenly appear to separate from the main body of the tail and fall away. A new ion tail soon regrows. These events seem to be caused by the passage of the comet across a *sector boundary* in the solar wind. A sector boundary is a region in interplanetary space where the magnetic field associated with the solar wind abruptly changes polarity. When a sector boundary arrives at the vicinity of the comet, the coma momentarily encounters a zero magnetic field and the ions are for a short time no longer channeled into an ion tail. The ion tail grows back once the sector boundary has passed through the coma and the magnetic field is present once again. This is diagrammed in Figures 17.11*a* and *b*.

A SAMPLE OF COMET ICE

The United States did not attempt to send a spacecraft to comet Halley. A Halley rendezvous mission conceived during the 1970s was canceled due to budgetary constraints. However, the United States did manage to achieve the distinction of the first encounter of a spacecraft with a comet. The target was the comet Giacobini–Zinner. This object is a short-period comet (6.6 Earth years) with a perihelion of 1.03 AU and aphelion of 6 AU. The spacecraft that made the encounter was the *ISEE-3*, a joint NASA–ESA venture originally launched in August 1978. The spacecraft had originally been placed in a "halo orbit," which stationed it permanently in space at the point between the Earth and the Sun where the gravitational pulls of the two bodies on the craft exactly counterbalanced each other. *ISEE-3* had been engaged in a long-term study of solar phenomena and had long ago fulfilled all of its mission objectives. Robert Farquhar of NASA/Goddard Space Flight Center happened to notice that this spacecraft was in a perfect position to reach the orbit of comet Giacobini–Zinner. In June of 1982 the remaining fuel on board the spacecraft was used to nudge *ISEE-3* out of its halo orbit and place it on a path that intercepted that of the comet. The craft was then renamed *ICE* (for International Comet Explorer) in recognition of its new mission. *ICE* encountered Giacobini–Zinner on September 11, 1985, passing through the tail. Unfortunately, *ICE* carried no television cameras, but its other instruments were able to perform valuable analyses of the properties of cometary comas and tails.

NASA fired the thrusters of *ICE* one more time in February 1986. This maneuver placed the craft on a trajectory that will send it back toward the Moon. In the year 2012 it will encounter the lunar gravitational field and will be kicked into Earth orbit. The *ICE* craft will then be in position for retrieval, hopefully containing a load of dust from Giacobini – Zinner.

Even though NASA passed up its opportunity to travel to comet Halley, it is now considering several options for extremely advanced missions to comets. One of these is termed Comet Rendezvous/Asteroid Flyby (CRAF). According to the initial plan, the CRAF spacecraft is to rendezvous with a suitably chosen comet well before its perihelion and remain with the comet throughout most of its active phase. Included as a secondary mission objective is a flyby of an asteroid while en route to the comet. The craft may possibly be equipped with a comet-nucleus penetrator system that will fire a meter-long projectile into the comet's nucleus in order to measure the atomic composition of the comet as well as its thermal and mechanical properties. Launch of CRAF will probably not take place before 1992.

In late 1985, NASA began long-range planning for a comet-sample return mission during the 1990s, one which could collect material from at least four different comets. The project is named Multicomet Sample Return and will make extensive use of the American space station, which hopefully will be operational early in the 1990s. According to the plan, a flotilla of three spacecraft will be placed into near-Earth solar orbit sometime during 1992. The largest of the three craft will be devoted primarily to long-term study of solar phenomena and will have only a secondary comet exploration capability. However, the other two are to be dedicated coma probes, each equipped with special dust collectors. The two coma probes will be directed toward different comets. During each of the encounters, the coma probes will gather up samples of comet dust and gas. Following the encounters, the coma probes will return to Earth orbit and will be retrieved by a remotely controlled vehicle and delivered to the new space station. The coma samples will then be studied by a team of astronauts. The coma probes themselves are to be refurbished aboard the space station and sent off again into space to gather samples from different target comets. Samples from as many as four different comets could be collected during the years 1995 – 2005. If for some reason the operational status of the space station is delayed, the coma probes could be "parked" indefinitely in Earth orbit until they can be retrieved.

The European Space Agency has also begun planning for its own comet sample return mission. The initial concept, termed *Caesar* (for Comet Atmosphere and Earth Sample Return), is to use a modified *Giotto* spacecraft equipped with panels containing collectors that will gather up dust and gas during passage through the coma of a comet. The craft would then return to the vicinity of the Earth. During Earth flyby, a capsule containing the collected gas and dust will be placed in Earth orbit for later retrieval by the space shuttle orbiter. The concept awaits approval for a new mission start.

18

Death from the Skies

Comets were at one time thought to be harbingers of impending doom and disaster and were therefore objects of superstitious dread. In the seventeenth century, it was discovered that these spectacular bodies obeyed the same laws of physics that the planets did. Since their motion across the sky could be mathematically predicted with high precision, much of the fear and dread that accompanied the appearance of a comet was removed.

However, in the twentieth century, we have found that the ancients might not have been too far off the track in their fear of comets. Evidence has recently been accumulated to indicate that the passage of comets has had profound and catastrophic effects on life on Earth and could pose great dangers for the future.

THE TUNGUSKA EVENT

On June 30, 1908, a remote area of Siberia near the basin of the Tunguska River was devastated by a fireball intense enough to fell trees for distances up to 40 km from the impact point. People at a distance of 60 km away felt intense heat and heard a tremendous noise. Seismic disturbances from the impact were recorded all over the world and atmospheric pressure waves were detected by stations as far away as the British Isles. Visible phenomena were observed as far

View of north polar region of the Saturn moon Rhea taken by the probe *Voyager 1*. The surface is heavily scarred by impact craters. Every world in the Solar System has been bombarded by an intense flux of meteorites in the past, suggesting that meteorites have played an important role in the history of the Earth and the other planets. (Photograph courtesy of NASA/JPL)

354

as 700 km from the event and loud explosions were audible at distances of up to 1000 km. Apparently no one was seriously injured; although the burnt carcasses of reindeer were later found scattered throughout the impact area. For a few months after the event, the nighttime skies over Europe and western Siberia were abnormally bright, people reportedly being able to read newspapers unaided outside at night.

In 1927 an organized expedition led by the Soviet scientist L. A. Kulik entered the area. They found trees blown down over a 30- to 40-km radius. All of the felled trees were lying in a radial pattern that pointed away from a central impact point. To their surprise, the expedition found no large impact crater at the center. Trees were still standing there, although they had been stripped of their bark and branches. The scientists dug deep holes in the ground in search of the remains of an impacting object; none was found.

The Tunguska event has puzzled scientists ever since and has given rise to many imaginative theories. The most controversial of these was proposed in 1946 by the Soviet writer Alexander Kazantsev. He noted the apparent similarity between the Tunguska event and the then-recent atomic bomb destructions of Hiroshima and Nagasaki and proposed that the explosion had been produced by a malfunctioning nuclear-powered spaceship from some distant world attempting to make an emergency landing. The energy released by the explosion was estimated to have been equivalent to that of a 0.2- to 20-megaton nuclear weapon.

Other theories have also been proposed. Some have suggested that the Tunguska object could have been a natural piece of fissionable material of subcritical mass that was induced to produce a nuclear explosion by the intense heat of entry into the atmosphere. Others have proposed that the object could have been a chunk of antimatter that reacted violently with the air as it entered the atmosphere. However, antimatter is not believed to exist in the gross state, at least not in our particular part of the universe. Some have even proposed that the Tunguska event could have been caused by the collision of a miniature black hole. However, such a small black hole should have drilled a hole right through the Earth and emerged a few minutes later on the other side, somewhere in the North Atlantic. An explosion of equal magnitude should have occurred at the exit point.

The possibility of some sort of radiation-releasing process being responsible for the Tunguska event was suggested by some initial measurements of the amount of carbon 14 in tree rings. These seemed to indicate that there was a sudden jump in worldwide radioactivity in the years 1908-09. However, more careful later measurements could not confirm such an increase. Furthermore, Soviet scientists were not able to find any more radioactivity present in the Tunguska region than anywhere else, indicating that a nuclear explosion is an unlikely explanation for the event.

Today, most scientists believe that the Tunguska event was caused by the impact of a spent comet that was too fragile to survive the passage through the atmosphere. It disintegrated in midair at an altitude of several kilometers. The explosion devastated the surrounding countryside but left no tell-tale crater in the ground. Analysis of eyewitness accounts of the fireball's approach indicated that the object must have originally been in a retrograde solar orbit, one with a considerable inclination with the ecliptic. Such an orbit is consistent with a comet but is unlikely for an asteroid.

In 1947 there was another such event, again in Siberia. An area 5 km across was destroyed. This time, an errant asteroid was probably responsible, as several tons of iron fragments were recovered. It would have caused considerable death and destruction had it struck a populated area.

THE LATE CRETACEOUS EXTINCTION

It is estimated that a body of radius 1 km or larger will strike the Earth only about once every 250,000 years. Such an impact would be equivalent to the energy released by the explosion of 10,000 ten-megaton hydrogen bombs. Fortunately, no such event has occurred in recorded human history. However, such impacts must have occurred many times in the past, with profound effects on life.

Something utterly horrible seems to have happened on Earth about 65 million years ago, at the end of the Cretaceous period. The fossil record shows a sudden and abrupt extinction at this time of many species that had lived on Earth for millions of years. The large clams, the diverse ammonites, the belemnites, the coccolithids, several types of algae and microscopic plants, marine reptiles, flying reptiles, and both orders of dinosaurs were entirely wiped out. About half the genera existing on Earth at that time were suddenly extinguished in one relatively brief instant. No terrestrial vertebrate larger than 25 kg in mass was able to survive. What happened?

Many geologists and biologists blame the late Cretaceous extinction on a sudden, worldwide climatic change brought about by the breakup of Pangaea. This disruption began about 180 million years ago, and one by one the continents began to drift apart. The Atlantic Ocean opened up for the first time at the end of the Cretaceous, and a rapid onrush of frigid water into the Atlantic could have produced a sudden onset of colder weather that spelled disaster for those creatures unable to adapt. Other scientists have theorized that there was a series of massive volcanic eruptions at this time. These volcanoes threw enormous amounts of dust and ash into the air, producing a sudden cooling of the entire planet that killed off large numbers of species of living creatures.

Other scientists have looked for extraterrestrial causes of the late Cretaceous catastrophe. A sudden decrease in the solar luminosity could have produced a worldwide cold snap lasting hundreds or thousands of years. The explosion of a nearby supernova could have been the cause, releasing a flood of radiation that momentarily stripped away the Earth's protective ozone blanket and allowed life-threatening ultraviolet radiation to reach the ground.

Perhaps the most intriguing theory of all is that of Walter Alvarez, Luis Alvarez, Frank Asaro, and Helen Michel, all of the University of California at Berkeley. Walter Alvarez had found that a particular thin sedimentary deposit near Gubbio in the Italian Apennines has about 30 times more iridium metal than any other. This particular deposit occupies a niche that separates marine limestone of the late Cretaceous from that of the early Paleocene. Iridium excesses have subsequently been found in late Cretaceous deposits from such widely separated places as Denmark, Spain, and New Zealand. They are also found in deep-sea core samples taken from both the Atlantic and Pacific Oceans. Could the iridium excess be a clue to the cause of the death of the dinosaurs?

Iridium is quite rare in Earth rocks. It is a noble metal, one which does not readily combine with other elements to form compounds. Most of Earth's supply of iridium must have sunk to the molten core many years ago. Iridium is, however, much more abundant in meteorites. The Berkeley team suggests that the impact of a small asteroid or a cometary nucleus was the cause of the late Cretaceous extinction. The impact of an object as large as a few kilometers in diameter could have kicked enough dust into the upper atmosphere to have created a continuous night on Earth that lasted for several years. Gradually, the dust settled back to Earth, but not until photosynthesis had been suppressed for a long enough time to cause the collapse of several important food chains.

The absence of any obvious remains of a significant impact crater dating from this period is a problem for the theory. However, it is possible that the impact took place in the ocean or in a continental region where subsequent geological activity has totally erased the crater. By now over half of the late Cretaceous oceanic crust has disappeared into subduction zones.

NEMESIS, THE MESSENGER OF DEATH

In 1984 University of Chicago researchers David Raup and John Sepkowski reported on their extensive studies of the fossil records of extinct species. Far more species of living creatures have become extinct than are living today. They found that the rate at which species of living creatures disappeared was exceedingly slow throughout most of geological history. However, there were relatively brief episodes during which the extinction rate was much higher than usual. For some odd reason, the episodes of massive extinction appeared to occur in a periodic fashion, about once every 26 million years.

What could be the cause of these periodic episodes of massive extinction? There is no known terrestrial phenomenon with a period as long as 26 million years. Could the cause then be extraterrestrial? Perhaps the periodic extinction was caused by the impacts of large meteorites. Walter Alvarez and R. A. Muller found that there was an odd periodicity in the ages of large terrestrial meteorite impact craters. There ages all tend to group around specific values, ones which closely match the dates of the mass extinctions.

But why should there be a periodicity in the rate at which large meteorites strike the Earth? In 1984 two groups of scientists independently proposed that the cause of the periodic rain of meteorites is an invisible star that circles the Sun once every 26 million years (Figure 18.1). The orbit of the Sun's invisible com-

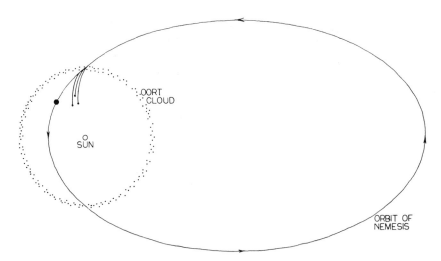

FIGURE 18.1 The orbit of Nemesis, a hypothetical stellar companion of the Sun. It passes through the Oort cloud once every 26 million years, sending a lethal flood of comets toward the inner Solar System.

panion is pictured as being highly eccentric. The object travels as far as 170,000 AU from the Sun at aphelion, a range of almost 1/10 of the distance to the nearest stars. Since the hypothetical companion of the Sun has never been seen, it was suggested that it is a "black dwarf" subluminous star, one with a mass somewhere between 0.002 and 0.07 solar masses. Over half of all the stars in the sky are members of double or triple systems; perhaps our Sun is as well.

The invisible companion spends most of its time quite distant from the Sun and is consequently quite harmless. However, once every 26 million years, it turns and heads toward the Sun. During its approach to the Sun, the companion passes through the deep inner region of the Oort comet cloud, supposedly located in space some 20,000 AU from the Sun. During passage, it scatters nearby comets in all directions. Some of these comets are deflected into highly elliptical orbits which send them into the inner reaches of the Solar System where the Earth and the other planets reside. So, once every 26 million years, Earth and the other planets are subjected to a shower of lethal comets.

The invisible companion was named *Nemesis*, after the Greek goddess who relentlessly punished the excessively proud and arrogant. Since Nemesis' mass is so small, hydrogen fusion must be so weak that starlight from the object is exceedingly feeble. Such a faint object might never be detectable in ground-based optical telescopes. Nevertheless, Nemesis might radiate a significant amount of light in the infrared. In 1983 an Infrared Astronomy Satellite *(IRAS)* was sent into Earth orbit to search for extraterrestrial sources of infrared light. Enough interest was generated by the Nemesis hypothesis to induce a few astronomers to search for evidence of such an object in the data sent back to Earth by the satellite. Nothing was found.

At present, the Nemesis hypothesis stands on very shaky ground, if for no other reason than that no direct evidence for its existence has been found. The hypothesis is in trouble for other reasons too. Nemesis swings so far from the Sun that its orbit would be severely perturbed by passing stars and by the general gravitational field of the Galaxy. Such perturbations would either strip Nemesis completely away from the Sun or else cause its period to vary appreciably from one perihelion passage to the next.

PART FIVE

A QUESTION OF ORIGIN

The Pleiades star cluster in the constellation Taurus. These bright stars are quite young, having only recently condensed from a vast cloud of interstellar gas and dust. Some wisps of dust left over from this primordial cloud still surround these stars. (Photograph courtesy of Hale Observatories)

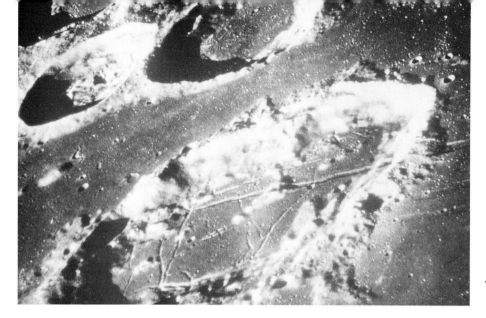

19

Meteors and Meteorites — Visitors from Outer Space

On any clear night dozens of bright, luminous streaks can be seen rapidly crossing the sky. These streaks are caused by small particles entering Earth's upper atmosphere at high velocities. They are called *shooting stars* or *meteors*. These particles enter the upper atmosphere at speeds of 35 to 100 km/sec and are rapidly heated to incandescence by air friction at altitudes of 50 to 100 km. The vast majority of all meteors are entirely consumed by the heat long before they can reach the ground. The few meteors that survive the trip through the atmosphere and strike the surface are known as *meteorites*. It is estimated that a total mass of about 5×10^6 kg of extraterrestrial material falls upon the Earth every day.

When an object enters the atmosphere at high velocity, its front surface is heated and the surrounding air becomes ionized. This creates a bright fireball than can be many meters in diameter. On some occasions, a fireball can be as bright as the Sun. Most fireballs occur at altitudes of 120 km and last only a few seconds. A grazing approach can produce a fireball lasting up to a minute. Fireballs are often accompanied by sonic booms.

The heat of atmospheric friction melts the outside surface of the meteor

Goclenius, a crater 55 km in diameter located at the western edge of Mare Fecunditatis on the Moon. This photograph was taken by the crew of *Apollo 8* in 1968. All of the planets and moons in the Solar System were bombarded by an intense flux of meteorites, over four billion years ago, at the beginning of the Solar System. (Photograph courtesy of NASA) and the melted material is swept away, leaving a trail of dust particles behind the meteor as it streaks through the air. This selective removal of the outermost layer of material is known as *ablation*. The ablation of surface material effectively carries heat away from the meteor, so the interior actually remains rather cool. However, an irregular shape or an internal weakness can cause the meteor to split up into fragments under the influence of the strong air friction forces.

As the meteor penetrates deeper into the atmosphere, the object decelerates and the atmospheric friction abates. By the time that an altitude of 10 to 30 km is reached, the object is moving so slowly that the fireball has vanished. Assuming that there is anything left of the object after its fiery trip through the upper atmosphere, it falls the rest of the way to the ground darkly.

METEOR SHOWERS—COLLISIONS WITH COMETARY DEBRIS

On any given night, an observer can usually see from 5 to 10 meteors every hour, provided there is no bright Moon in the sky and no clouds to block the view. However, there are a few nights during which the meteoric activity is considerably higher, sometimes as great as 30 to 50 per hour. Such events are called *meteor showers*.

Meteor showers occur at the same time every year, and the meteors in a given shower all appear to be bombarding the Earth from a single point in the sky. They are hence given the name of the constellation of stars from which they appear to be originating. For example, the meteors comprising the Cepheid shower (which lasts from November 7 to 11) all appear to terrestrial observers to be streaming out of a point located in the constellation Cepheus. The most spectacular of the meteor showers are the Perseids (August 2–23), the Geminids (December 10–16), and the Quadrantids (January 2–4). On very rare occasions, veritable blizzards of meteors have been seen. In 1833 and again in 1966, the Leonid shower (November 15–19) produced as many as 100,000 meteors per hour!

These meteors do not, of course, originate in the stars but only appear to come from there because they are all traveling in very nearly parallel paths when they strike the Earth. They must be objects in heliocentric orbit whose paths cross that of the Earth once every year. About 100 years ago, it was noticed that the meteors which comprise the Leonid shower all travel in orbits that are quite similar to that of comet 1866I. Since then, about a dozen meteor showers have been identified with the orbits of short-period comets.

As a short-period comet makes successive trips around the Sun, more and more of its volatile material is stripped away. The disintegration products become spread out both ahead and behind the dying comet in its orbit. If the Earth happens to cross the orbit of the disintegrating comet, some of these particles will enter the atmosphere and produce a meteor shower (Figure 19.1). Few of these particles ever reach the ground, as most cometary debris is probably too brittle and fragile to survive the fiery trip through the atmosphere.

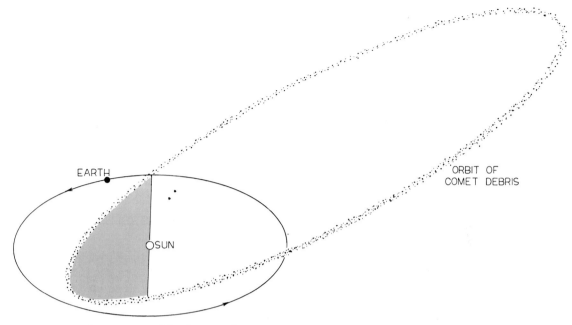

FIGURE 19.1 The cometary origin of meteor showers.

METEORITES—THE COSMIC FOSSILS

Meteorites are those objects that are sufficiently robust to survive their entry into the atmosphere and reach the surface. In order to reach the ground un-scathed, meteorites must be rather substantial bodies, probably quite different in composition from the objects responsible for meteor showers. About 500 meteorites strike the Earth every year, but only about 10 of these are ever recovered. Recovered meteorites are usually named after the city nearest the spot where they fell. It is customary to distinguish between those meteorites that are recovered soon after their observed impact (the *falls*) and those discovered only after an indefinite period on the ground (the *finds*).

Where do meteorites come from? For many years, virtually all scientists believed that it was impossible for stones to fall from the sky. Eyewitness reports of meteorite impacts were dismissed as hallucinations or hoaxes. However, in 1803 a meteorite impact took place in plain sight of a member of the French Academy of Science. From then on, it was no longer possible to deny their reality. However, most scientists assumed that meteorites must be rocks that had been hurled over large distances by the violent eruptions of active volca-noes. Today, we know that meteorites do not originate anywhere on Earth. They are visitors from outer space.

On a few rare occasions, eyewitness accounts of a meteorite's fall have been sufficiently detailed to make it possible to calculate the orbit in which the object was traveling before it hit the Earth. There are even a few cases in which networks of meteor cameras were able to track the fiery path of a meteorite as it streaked across the sky. Detailed analyses of these photographs made it possible to chart the original orbit of the meteorite. In all cases, the original orbits of these meteorites were similar to those of the Apollo asteroids, whose paths cross that of Earth (Figure 19.2).

STONES AND IRONS—THE METEORITE CLASSIFICATION SCHEME

Aside from Moon rocks, meteorites are the only samples of extraterrestrial material that we have. Meteorites are classified according to their composition. There are three broad categories, termed *stony, iron,* and *stony iron.*

Most meteorites fall into the stony category. About 1400 stony meteorites are known. Most of them are falls rather than finds, perhaps because they look so much like ordinary terrestrial rocks that they tend to be overlooked. They can be further subdivided into two categories, the *chondrites* and the *achondrites.*

Most stony meteorites are chondrites. They acquire their name because of the presence of small rounded grains in their interiors. These grains, known as chondrules, are typically 1 to 5 mm across and are rich in olivine and pyroxene minerals. The typical chondrite contains (by weight) 45 percent olivine, 25 percent pyroxene, 10 percent plagioclase, 5 percent troilite (iron sulfide, FeS), and from 10 to 20 percent iron–nickel metals. The iron–nickel metal found in chondrites usually appears in the form of lumps of kamacite and taenite alloys, but there is considerable variation in metal content from one chondrite to the next. Plagioclase, olivine, and pyroxene minerals are quite common in terrestrial

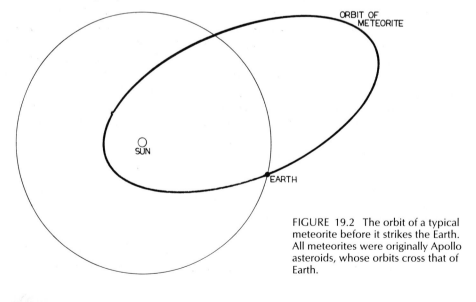

FIGURE 19.2 The orbit of a typical meteorite before it strikes the Earth. All meteorites were originally Apollo asteroids, whose orbits cross that of Earth.

rocks (particularly in oceanic crust), but kamacite and taenite are both rather rare on Earth and troilite is completely nonexistent.

There are a few chondrites that have particularly high concentrations of aluminum, magnesium, and calcium. In addition, they are richer in oxygen and have almost no nickel–iron metal. These meteorites have a dark color and are called *carbonaceous chondrites*. About 30 such meteorites are known. A typical carbonaceous chondrite meteorite has a density of 2.6 g/cm^3. Many of them have a matrix of serpentine, a type of water-bearing silicate mineral intermixed with such familiar compounds as magnetite, epsomite, and other low-temperature minerals. The water content can be as high as 20 percent. Many carbonaceous chondrites are rich in carbon, organic compounds, and other volatile materials.

Carbonaceous chondrites are especially interesting to planetary scientists because their overall elemental composition very closely matches the composition of the "condensible" part of the Sun. This is the material that one would get if one took a sample of the solar atmosphere, allowed it to cool and condense into a solid, and then discarded the material (primarily hydrogen and helium) that remained gaseous. Such a mixture is often termed *chondritic*. Compared to the Sun, however, carbonaceous chondrites tend to be short of nitrogen, carbon, and oxygen. These three elements are highly volatile or else form volatile compounds such as carbon dioxide.

The rarest type of stony meteorite is the *enstatite chondrite*. These have so little oxygen that most of the iron that is present occurs either in the metallic form or as a sulfide in the mineral troilite. Their name comes from the most abundant mineral in their interiors — enstatite, a magnesium silicate with the formula Mg$_2$Si$_2$O$_6$. There are lesser amounts of kamacite, troilite, and plagioclase minerals.

Chondrites seem to have been formed by a rapid cooling from an initial high-temperature state, but there is no evidence that they were ever molten. They must have condensed at high temperatures from the vapors of the primitive solar nebula and then rapidly cooled. Since carbonaceous chondrites have an abundance of water, carbon, and other volatile materials, it is thought that they must have formed at lower temperatures than did the ordinary chondrites. Following their initial formation, the chondrites underwent relatively little additional heating. Many chondrites have chondrules with sharply defined, dark borders, but others have chondrules that are so indistinct that they are all but invisible. The progressive fading of chondrules is taken as evidence of prolonged heating of the meteorite to successively higher temperatures.

Stony meteorites other than chondrites are termed achondrites, which simply means "without chondrules." Most resemble terrestrial basaltic rocks. This must mean that they have cooled from an initially molten state. The most numerous members of this class are termed *eucrites*. Eucrites are composed primarily of the plagioclase feldspar anorthite (CaAl$_2$Sl$_2$O$_8$) and the pyroxene pigeonite (MGFeSi$_2$O$_6$). Eucrites are usually brecciated; that is, they consist of individual fragments of rock and dust that have been welded together by heat and high pressure. Eucrites tend to be poorer in sodium, potassium, and other volatile elements than are terrestrial basalts. They are also less highly oxidized. Occasionally, grains of metal are found in their interiors.

The second major class of meteorites is the iron variety. Most iron meteorites are finds, probably because their unusual appearance and greater resistance to weathering and erosion makes them easy to identify when found. About 700 are known. Iron meteorites are dominated by the two iron–nickel alloys—kamacite and taenite. Most irons contain between 5 and 6.5 percent (by weight) nickel, but a few have as much as 13 percent nickel. These variations in nickel content can lead to the formation of sharply different crystalline structures. Gallium, germanium, and iridium are present in small but variable amounts. Many irons have inclusions of stony silicate minerals.

The last major class of meteorites is the stony-iron class. They are the least prevalent of the three types; only 80 are known. Stony iron meteorites are approximately half nickel–iron metal and half silicate minerals. They typically have a density of about 3.5 g/cm³.

The various meteorite classes are summarized in Table 19.1.

THE ORIGIN OF METEORITES

How old are meteorites? The length of time that a meteorite has existed as a solid body can be estimated by using the same radioactive clock that was used to establish the ages of terrestrial rocks and Moon dust. Most meteorites turn out to be incredibly old, having formed 4.5 to 4.7 billion years ago. All types—the chondrites, the irons, and the stony irons—have similar ages. No meteorite has ever been found with an age any greater than 4.7 billion years. The ages of the oldest meteorites seem to correlate quite closely with estimates for the ages of the Earth, the Sun, and the Moon. This suggests that all of the matter in the Solar System condensed at approximately the same time—4.7 billion years ago.

Interplanetary space is flooded with cosmic rays. These ultra-high-energy atomic nuclei impinge upon the Solar System from all directions of interstellar space. A meter or so of matter will stop all but the most energetic of these particles. When a cosmic ray strikes a material body, it collides with some of the atoms of the crystalline matrix. These high-energy collisions produce unusual isotopes of such elements as helium, neon, and aluminum. These isotopes remain trapped in the interior of the body. By measuring the amounts of these unusual isotopes in the interior of a meteorite, the length of time that the object has been exposed to cosmic radiation can be estimated.

It turns out that most meteorites have "cosmic-ray ages" on the order of millions of years, but not billions. One meteorite was found to have a cosmic-ray age of less than 25,000 years. Most of the meteorites that strike the Earth today have been wandering around in outer space for only a relatively short time. They have spent the vast majority of their lives somewhere else.

Where could they have been for all those years? They must have been locked up inside much larger bodies, shielded from bombardment by cosmic rays. But what kind of larger objects could have produced such a wide variety of materials? One clue can be obtained by checking the meteorite's interior for

TABLE 19.1. METEORITE CLASSIFICATIONS

STONY—At least 75%, by weight, silicate minerals. The rest, metal alloys.

   Chondrites: Aggregates of silicate minerals and metals; contain rounded grains of silicates known as chondrules; rich in plagioclase, pyroxene, and olivine minerals; some have glass; all condensed from vapor phase.

     Ordinary: 2%–20% total iron, at least 75% silicate minerals.

       Three different groups:

         H: 27% total iron; 12%–20% nickel–iron metal

         L: 23% total iron; 5%–10% nickel–iron metal

         LL: 20% total iron; 2% nickel–iron metal

       Oxygen content increases in sequence H>L>LL

       Five different textural and mineralogical classifications within each group:

         Type 3: Sharply defined chondrules; opaque, fine-grained matrix; clear glass present; olivine is of variable composition; pyroxene is mainly disordered; no plagioclase; very rapid cooling, and no subsequent heating above 700° C.

         Type 4: Distinct chondrules; matrix opaque but less fine-grained; turbid glass present; olivine is of uniform composition; pyroxene is mainly ordered; no plagioclase; slower cooling and/or reheating.

         Type 5: Indistinct chondrules; granular matrix; no glass; olivine is of uniform composition; pyroxene is ordered; plagioclase in turbid crystals; very slow cooling or else reheating to 850° C.

         Type 6: Sparse, indistinct chondrules; granular, coarse-grained matrix; no glass; olivine is of uniform composition; pyroxene is ordered; plagioclase in clear, well-defined crystals; extremely slow cooling or else reheating to 950° C.

         Type 7: No chondrules; coarse-grained texture; no glass; olivine is of uniform composition; pyroxene is ordered; plagioclase in clear, well-defined crystals; prolonged heating to 1200° C.

     Carbonaceous: Almost no nickel–iron metal; high relative amounts of magnesium, aluminum, and calcium; very little heating above 500°C.

       Four different groups:

         CM: Dark matrix of serpentine plus other low-temperature minerals; 10% water; many complex organic compounds.

         CV: Relatively little carbon; irregular whitish inclusions of high-temperature calcium, aluminum, and titanium-rich minerals set in olivine-rich matrix; 1% water; large chondrules often containing sulfides.

         CO: Relatively little carbon; great abundance of small chondrules; some have mineral assemblages like the white inclusions in CV; 1% water.

         CI: No chondrules; olivine very rare; many low-temperature, water-bearing minerals; 8% to 22% water; carbonates and sulfates; chemical composition close to that of "condensible" part of Sun.

TABLE 19.1 (*continued*)

> *Enstatite:* Almost no oxygen; rich in metal; all iron occurs as metal or as sulfide; low magnesium: silicon ratio; almost no olivine; formed at 1600° C.

Achondrites: Igneous rocks cooled from a melt; no chondrules; most are highly fractured.

> *Eucrites:* Resemble lunar lava flows and terrestrial basalts; equal proportions of plagioclase and calcium-bearing pyroxene; small amounts of nickel–iron metal and troilite; lower sodium and potassium content than terrestrial rocks.

> *Hypersthenes:* Cumulates of different minerals; rich in "hypersthene", a type of pyroxene with little calcium; 12% iron by weight.

> *Howardites:* Formed from eucrites and hypersthene minerals, with a minor chondritic component.

> *Ureilites:* Cumulates of mainly olivine crystals, with minor amounts of calcium-bearing pyroxene; carbon is present, some of it as diamond.

> *Enstatite:* Similar in composition to enstatite chondrites, but lacking chondrules.

> *SNC* Shergottite, Nakhlite, and Chassignite; cumulate of pyroxene–olivine minerals; young crystallization age.

IRON: Almost entirely metallic; cumulates of nickel–iron metals, primarily in the form of kamacite and taenite alloys.

Twelve different chemical groups, based on abundances of nickel, gallium, germanium, and iridium.

STONY-IRON: Approximately 50% nickel–iron metal, 50% silicate minerals.

> **Pallasite:** Clusters of olivine crystals set in metal matrix

> **Mesosiderite:** Mixtures of materials from several possibly unrelated sources; angular chunks of rocks composed of plagioclase and calcium-bearing pyroxene; rounded masses of olivine; metal sometimes present as round slugs, sometimes as veins.

SOURCE: Based on Hutchison (1983).

minerals that can form only at high pressures. If a meteorite contains minerals that can crystallize only at pressures of 10,000 atm or more, then we can conclude that that particular object must at one time have been deep in the interior of a large, planet-sized mass such as Earth, Mars, or the Moon. On the other hand, if the meteorite has only low-pressure minerals, it is reasonable to presume that the object came either from near the surface of a large planet or else from deep in the interior of a relatively small object no larger than a few hundred kilometers in diameter.

No meteorite has ever been found with minerals that crystallized at pressures greater than 12,000 atm, acting over a long period of time. In particular, the high-pressure crystalline modifications typical of terrestrial rocks that came from the upper mantle are entirely absent in meteorites. So all meteorites must have come from the interiors of small bodies no larger than a few hundred kilometers wide, definitely smaller than the Moon. These dimensions are typical of the larger asteroids still found in the belt between Mars and Jupiter.

METEORITES AND ASTEROIDS

Meteorites must be the debris left over from collisions between asteroids that took place in the relatively recent past. These collisions fragmented the asteroids into numerous chunks of rock, each of which entered its own separate orbit around the Sun. Some of these orbits were highly eccentric and crossed that of Earth. After an indeterminate period in space, these chunks of rock collided with Earth, ultimately to end up in a museum or research laboratory.

Several scientists have noted that there is a sharp correlation between the optical spectra of meteorites and those of certain asteroids. The ordinary chondrites have reflectance spectra similar to those of the S-type asteroids. The S asteroids tend to congregate in the innermost parts of the Asteroid Belt. The carbonaceous-chondrite meteorites have reflectance spectra similar to those of the C-type asteroids, which tend to be found in the outer parts of the main Asteroid Belt. Iron meteorites have been compared to the M-type asteroids, which are found throughout the Asteroid Belt. It is thought that M-type asteroids are the remnant metal-rich cores of ancient planetoids that were stripped of their silicate mantles by catastrophic collisions in the far distant past. Stony-iron meteorites seem to have been formed by the flow of molten metal into a solid rocky silicate mass. They were perhaps once part of planetoids in which metallic core material was forced into the pores of silicate minerals by heat and high pressure.

The eucrites appear to be in a class by themselves. These objects must have been molten at the time of their formation. The molten material that formed these meteorites may have originally been deep in the interior of a particularly large asteroid. Under the influence of heat and high pressure, the lava was forced upward through cracks in the interior. It flowed out over the surface, where it cooled and solidified. Later meteor impacts knocked chunks of this solidified lava off the surface of the asteroid, sending some of them on a collision course with the Earth. The asteroid 4 Vesta seems to be covered with a large amount of solidified lava; it is possible (but not proven) that most if not all eucrites originally came to Earth from this single planetoid!

The SNC achondrites are perhaps the most unusual meteorites of all. Only nine are known. SNC stands for "Shergottite–Nakhlite–Chassignite," taken from the names of three towns near the places where falls of meteorites of this type took place. SNC meteorites typically have crystallization ages of only 1.3 billion years, much younger than the ages of other meteorite types. These SNC meteorites must have solidified quite recently. The relative abundances of the elements in these meteorites are quite different from those in most other types of meteorites, which probably means that they have an entirely different origin. A careful measurement of the relative abundances of isotopes of xenon and nitrogen in gases trapped in the interior of one of these meteorites has led several workers to a startling conclusion. These isotope abundance ratios closely match the ratios currently found in the gases of the Martian atmosphere. This meteorite did not come from the Asteroid Belt — it came from Mars! Perhaps it was violently ejected from Mars during a recent volcanic episode.

DID A SUPERNOVA TRIGGER THE FORMATION OF THE SOLAR SYSTEM?

Many meteorites (particularly the chondrites) show little evidence of thermal processing subsequent to their initial formation 4.7 billion years ago. They must have rapidly cooled off soon after they condensed and remained cool forever after. However, numerous other meteorites do show clear evidence of intense heating soon after their formation. The eucrites, the irons, and the stony-iron meteorites all seem to have cooled from an initially molten state. This raises a troubling paradox: meteorites come from parent bodies that were never any larger than a few hundred kilometers across. Nevertheless, they were melted by some sort of intense heat source.

What kind of energy source could have melted such small bodies? The energy released by the decay of long-lived radioactive isotopes (such as ^{235}U, ^{238}U, ^{40}K, and ^{87}Rb) is responsible for the heating inside a large planet-sized object such as the Earth. The Earth is so large that internal heat leaks outward to space at an exceedingly low rate. Quite soon after it formed, there was enough radioactive heating to melt the entire planet. The interior of the Earth has remained entirely molten ever since. However, smaller bodies of asteroid size radiate away their internal heat much more efficiently, and the temperature increase produced by such slow radioactive decay is very slight.

How then could such small objects ever melt? Although the origin of the interior heating of small asteroids is still largely unknown, a possible clue may have been found in a large carbonaceous-chondrite meteorite that crashed near the small Mexican town of Pueblito de Allende in 1969. An examination of the Allende meteorite turned up an anomalously large amount of ^{26}Mg, a stable isotope of magnesium. Similar magnesium isotope anomalies have turned up in other meteorites. Why should meteorites have an excess of ^{26}Mg? Magnesium 26 happens to be the product of the beta decay of ^{26}Al, a highly radioactive isotope of aluminum. Perhaps large amounts of radioactive aluminum were incorporated into the meteorite at the time that it condensed. The unusual aspect of all this is that ^{26}Al has a half-life of only 740,000 years. The radioactive aluminum present inside the Allende meteorite must have been synthesized at most only a couple of million years before the meteorite itself condensed!

How did such a short-lived radioactive isotope of aluminum ever happen to get inside a large body in the first place? The only way this could have happened is for there to have been a sudden infusion of material into the Solar System immediately prior to or even during its formation. The most likely source of the radioactive aluminum is a supernova that happened to explode quite close to the collapsing cloud of gas and dust that was in the process of forming the Solar System. Perhaps the supernova itself triggered the collapse of the cloud by sending out an expanding compressional shock wave. Many elements were created by the supernova explosion, some of them highly radioactive. These newly synthesized elements were injected into the region of the forming Solar System by the force of the explosion. These new elements intermixed with the material already there. Within a few thousand years, the gas and dust condensed into the solid matter that was to form the planets and the asteroids.

The energy released by the decay of these short-lived isotopes could have melted any object larger than a few kilometers across in a time as short as 100,000 years. The rapid heating would have destroyed any permanent record of the initial ^{26}Al excess, since the ^{26}Mg decay product would have been completely intermixed with the abundant amounts of nonradiogenic ^{26}Mg already present. Only in the very smallest objects, which had never melted, would any record of an initial ^{26}Al excess be preserved.

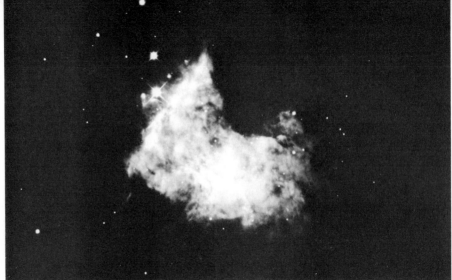

20

The Origin of the Solar System

The question of the origin of the Solar System has intrigued scientists for hundreds of years, almost since the time that the scientific view of astronomy itself first became widely accepted. However, prior to the mid-twentieth century, there was little reliable information available about the Earth, to say nothing of the other planets. In the absence of such knowledge, theories of planetary origin could proliferate without restraint. In the past 15 years, however, the opening of the Solar System to exploration by unmanned spacecraft has provided much detailed information about the Earth and the other planets. The astounding discoveries made by these spacecraft have added new and more rigid constraints that any truly successful theory of planetary origin must satisfy. These are summarized in Table 20.1.

THE NEBULAR HYPOTHESIS

Perhaps the first attempt to explain the origin of the Solar System in any systematic way was the so-called *nebular hypothesis,* put forward in various versions in the seventeenth and eighteenth centuries. In 1644 the French philosopher/scientist René Descartes (1596–1650) proposed that the Sun and the planets had originally formed from a vast nebula of gas. He pictured the planets as being

Telescopic view of the central region of the Great Nebula in Orion. This is a region of space in which vast clouds of gas and dust are condensing to form new stars and perhaps new planets, just as the Sun and the Earth did nearly five billion years ago. (Photograph courtesy of Hale Observatories).

TABLE 20.1 CONSTRAINTS ON THEORIES OF PLANETARY ORIGIN

1. The planets and the Sun are all about the same age.
2. The planets are all in roughly circular orbits that lie in very nearly the same plane.
3. The planets all travel around the Sun in the same direction.
4. Most of the planets have equators nearly parallel to their orbital planes and most rotate in the direct sense.
5. The Sun has most of the mass in the Solar System, but the planets have most of the angular momentum.
6. The inner planets are dense and rocky, whereas the outer planets are gaseous or icy.
7. The spacing of the planetary orbits is regular.
8. Three planets (Jupiter, Saturn, and Uranus) have regular systems of satellites that are small-scale replicas of the Solar System itself.
9. The planets and moons all experienced a period of heavy meteoric bombardment early in their histories.
10. Three planets (Jupiter, Saturn, and Uranus) have ring systems.
11. The presence of the Asteroid Belt must be explained.
12. The existence of comets must be explained.

products of the condensation of "vortices" that developed in the primitive gaseous nebula as it contracted. In 1755 the German philosopher Immanuel Kant (1724 – 1804) adopted most aspects of this picture, but proposed in addition that the gaseous nebula had become very hot during the condensation process. In 1796 the French scientist Pierre Laplace (1749 – 1827) added further refinements to the model, suggesting that the primitive nebula had initially been rotating before contraction began. The nebula then began to shrink under the influence of its own gravity and took on the shape of a flat disk as it got smaller and smaller. As the contraction process progressed further, the disk spun faster and faster. Eventually the disk was spinning so fast that it become mechanically unstable. At that point, the rotating disk began to shed a succession of rings at its edges. The gas in these rings eventually condensed to form the planets. This process is shown in Figure 20.1.

THE NEBULAR HYPOTHESIS DISCREDITED

These nebular theories gave a reasonable explanation of the circular planetary orbits, their coplanarity, and their regular spacing. Nebular theories held sway for many years, but their obvious successes led many to overlook some very serious shortcomings. It is a curious fact that the Sun has 99.9 percent of the mass in the Solar System but has only 2 percent of the angular momentum. Nebular theories predict that the process of contraction should have taken place in such a manner that the mass at the center (that is, the Sun) should have

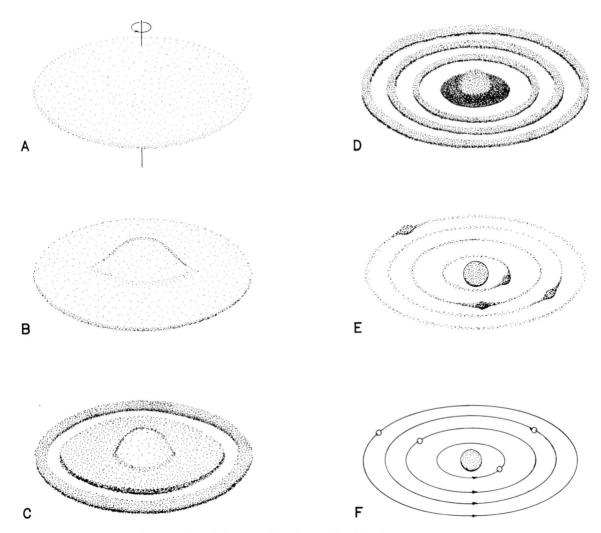

FIGURE 20.1 Diagram depicting the nebular-origin hypothesis of the Solar System.

ended up spinning much faster than is presently observed. In fact, the Sun should actually possess most of the angular momentum in the Solar System. Another problem was encountered while considering the details of the condensation process itself. In the mid-nineteenth century, James Clerk Maxwell pointed out that it is physically impossible for rings of gas ejected from a collapsing nebula to condense to form a solid body.

The closer people looked into the various versions of the nebular hypothesis, the more difficulties seemed to be encountered. This forced people to turn to *catastrophic* theories of planetary origin. The original catastrophic theory had actually been proposed as early as 1745 by the French scientist Georges Louis Leclerc, Comte de Buffon (1707–1788). He suggested that some sort of large body (perhaps a comet) had collided with the Sun, producing a massive eruption of gas from the surface. This gas entered orbit around the Sun, cooled, and condensed to form planets.

Buffon's proposal did not attract much attention when it was first presented, but in 1880 Bickerton was attracted to the general idea. He suggested that the Sun had undergone a near-collision with a passing star. A "ribbon" of gaseous material had been drawn out from both stars by the strong gravitational tidal forces produced during the encounter. After the encounter, this ribbon of star-stuff broke up into smaller globules, each of which cooled and formed a planet. This process is shown in Figure 20.2.

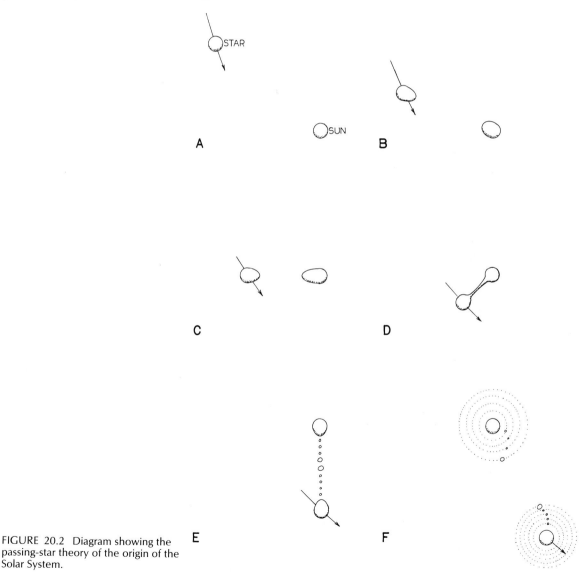

FIGURE 20.2 Diagram showing the passing-star theory of the origin of the Solar System.

Catastrophic theories were very much in vogue during the early years of the twentieth century, but they soon encountered serious problems on their own. Angular momentum was still a major stumbling block. Detailed mathematical analyses demonstrated that no type of encounter between the Sun and a passing star that was close enough to pull out material was able to give the planets the required large angular momentum. In addition, the English astronomer Sir Harold Jeffreys proved mathematically that protoplanets drawn off from the Sun in this manner could not possibly end up spinning faster than the Sun itself, in sharp contrast to what is actually observed. The details of the condensation from the vapor were still a problem. Lyman Spitzer demonstrated theoretically that a column of gas drawn out from the Sun should either fall right back or else disperse into space. It could never coalesce to form a solid body. Last, but perhaps far from least, stars are so far apart that close encounters between them are extremely rare. If stellar near-misses are required for planetary formation, Solar Systems such as ours should be extremely rare in the universe.

Problems encountered with both nebular and catastrophic theories led to a brief period during which double- and even triple-star models were fashionable. In 1935 the American astronomer Henry Norris Russell suggested that the Sun had originally been part of a binary system. The Sun's companion was disrupted by a near-collision with another star. The debris from the collision that did not immediately disperse into interstellar space coalesced to form the planets. Raymond Lyttleton proposed in 1936 that the Sun had originally been part of a triple system of stars. Five billion years ago, the other two stars in the Solar System collided with each other and were catastrophically disrupted. The debris left over from this ancient collision eventually formed the planets. In 1945 the famous English astronomer Sir Fred Hoyle suggested that the Sun originally possessed a massive companion that underwent a supernova explosion. The debris from this explosion contained a large concentration of heavier elements, some of which condensed to form planets.

The primary difficulty with all disruptive or catastrophic theories is that material ejected from a star should either fall back or else disperse into outer space rather than condense to form solid objects. Supernova theories have the additional problem of explaining the current absence in the Solar System of any neutron star or other such superdense object that is believed to be the end result of such a violent explosion.

THE NEBULAR HYPOTHESIS ASCENDANT

Since the end of the World War II, most theorists have followed the lead of such workers as Carl von Weizsacker, Gerard P. Kuiper, and Hannes Alfven in returning to revisions of the nebular hypothesis as providing the best explanation of the origin of the planets. A great deal of progress has been made in clearing up some of the more vexing difficulties with the earlier nebular theories, but much more work still needs to be done.

In the modern versions of the nebular hypothesis, the Solar System is imagined to have originated in a vast interstellar cloud that existed in space nearly 5 billion years ago. This cloud was composed primarily of hydrogen and helium, but it also contained gases such as ammonia, methane, and water vapor. Interspersed within the vapors of the interstellar cloud were dust grains made up of metals, silicates, and other heavy elements. All of the elements in the cloud that were heavier in hydrogen and helium had originally been synthesized by hydrogen fusion deep in the interiors of massive stars. These elements had been dispersed to space when these stars exploded and died.

The motions of nearby stars and the explosions of supernovas produced turbulent vortices within the massive interstellar cloud of dust and gas. As a result of the turbulence, the giant cloud broke up into smaller clumps, each of the order of 500 to 5000 solar masses in size (Figure 20.3). Each of these clumps of gas and dust was sufficiently dense to begin contracting under its own gravitational attraction. As the contraction gained momentum, each cloud developed its own set of turbulent vortices and broke up into hundreds of even smaller fragments. Each of these fragments had a net rotation, and each was eventually to form a star (Figure 20.4).

One of these vortices was destined to form the Sun and the planets. This

FIGURE 20.3 Rotating vortices in interstellar cloud created by nearby supernovas and passing stars.

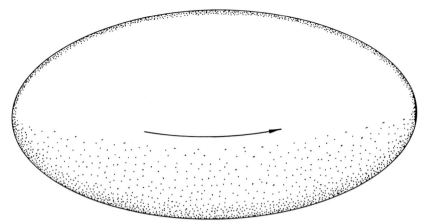

FIGURE 20.4 A rotating cloud
of protostellar gas.

turbulent vortex of swirling gas and dust must have been spinning rather rapidly. As it continued to contract under the force of its own mutual gravitation, the angular momentum of the shrinking cloud forced it to assume the shape of a flattened disk (Figure 20.5). Most of the material in the rotating disk eventually migrated toward the very center, where the Sun was destined to appear. Only a relatively small amount of matter remained distributed throughout the rim. Because of the large amount of dust, the disk must have been entirely opaque. Its temperature must have been exceedingly cold, probably not much greater than a few degrees above absolute zero.

Because of the viscous drag and the release of gravitational energy within the rotating cloud, its interior gradually became warmer as the collapse continued. Most regions of the cloud soon got so hot that the dust grains completely vaporized. As the collapse continued further, the density of matter at the very center became higher and higher and the temperature there became hotter and hotter. A *proto-Sun* composed of dense, hot gasses was beginning to form at the center of the disk.

Shortly thereafter, the very outermost regions of the gaseous disk began to cool off. As the temperature in the outer disk declined, some of the high-temperature refractory elements such as aluminum, silicon, and titanium began to condense out of the vapor, forming a cloud of fine dust at the rim of the rotating

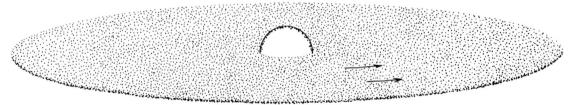

FIGURE 20.5 Protosolar disk of gas and dust. The massive object at the center ultimately formed the Sun. The thin disk surrounding the protosolar mass formed the moons and planets.

protoplanetary disk. As the temperature cooled still further, it got so cold in the outer parts of the disk that some of the more volatile vapors such as ammonia, methane, and water vapor were able to condense to form liquid droplets or even solid crystalline granules.

As the cooling process continued, a wave of condensation advanced closer and closer to the central proto-Sun. However, the region of space closest to the proto-Sun was much hotter than those regions nearer the outer rim of the protoplanetary disk. The temperature gradient existing across the width of the disk resulted in the condensation of different materials at different distances from the proto-Sun. The hot regions of the disk closest to the proto-Sun became rich in high-temperature refractory materials such as aluminum oxides, metallic iron, and silicate minerals. Highly volatile materials such as water, ammonia, and methane remained largely in the vapor state in this part of the nebular disk. The high-temperature materials that condensed in this region of space were destined to form the inner terrestrial planets Mercury, Venus, Earth, and Mars. The cooler regions of the disk more distant from the proto-Sun were of sufficiently low temperature so that large amounts of more volatile materials such as carbonates, water, and organics were able to condense. These lower-temperature materials eventually formed the asteroids, the cores of the giant planets Jupiter and Saturn, and their moons. The very outermost frigid regions of the disk were so cold that they acquired large concentrations of frozen ammonia and methane. The outer planets Uranus, Neptune, and Pluto, as well as the comets, are largely made up of this frozen material to the present day. The sharp differences found today in the compositions and internal structures of planets located at different distances from the Sun are a consequence of the steep temperature gradient that existed across the width of the protoplanetary disk in the ancient past.

Nearly 5 billion years ago, the Solar System was a vast disk of gas and dust that surrounded a massive, hot ball of vapor. Throughout this period, the proto-Sun had absorbed more and more matter as it continued its contraction to smaller and smaller dimensions and had gotten progressively hotter and hotter. Outside the environs of the proto-Sun, the newly condensed metallic particles and the globules of water, methane, and ammonia in the protoplanetary nebular disk must have undergone frequent collisions with each other. Many of these grains stuck to each other and formed larger and larger particles. Some of these particles may have been as large as 1 cm across. Because of viscous drag with the ample amounts of gas still remaining in the disk, the large particles slowly drifted toward the equator of the rotating protoplanetary disk, forming an ever-increasing concentration of rock and ice within the plane of the rotating nebula.

As the grains of material within the disk continued to collide with each other, progressively larger and larger bodies were formed. Some of them were perhaps as large as several kilometers across. These asteroid-sized objects all orbited the proto-Sun in nearly parallel paths. The early Solar System of nearly 5 billion years ago must have been a gigantic thin sheet of asteroid-sized bodies, somewhat similar to the rings of Saturn but on a much more gigantic scale (Figure 20.6). Within the inner parts of the disk, the rocks were largely made up

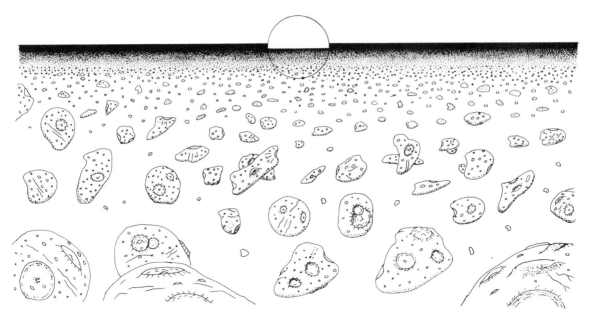

FIGURE 20.6 A sheet of asteroid-sized bodies orbited the Sun at the time of the initial formation of the Solar System.

of refractory metals and silicates. However, rocks with progressively larger amounts of volatile materials were found at larger and larger distances from the proto-Sun.

The asteroid-sized objects within the protoplanetary disk suffered frequent collisions with each other. In many cases, these collisions simply fragmented the larger objects into thousands of smaller pieces. However, other collisions resulted in the accumulation of many thousands of asteroid-sized masses into several loosely bound clusters. The continual collisions between the various rocks in the cluster led to a gradual relaxation in their relative motion, eventually producing a massive collapse of the individual particles down into a single large object (Figure 20.7). These massive bodies eventually became planets.

The rapidly growing protoplanets occupied virtually circular orbits, all lying roughly in the plane of the nebular disk. The planets all revolved in the same direction around the Sun, a reflection of the initial rotation of the protosolar nebula. The process of planetary formation was probably essentially completed by approximately 4.7 billion years ago.

The hydrogen and helium vapor in the original protoplanetary cloud that had managed to escape being absorbed into the massive proto-Sun remained in the vapor phase throughout the region of space now occupied by the newly formed planets. The outer planets happen to have accumulated much larger amounts of ice and rock than the inner planets did. The outer planets were in fact so massive that they were able to pull in significant amounts of primordial hydrogen and helium from nearby space. Massive amounts of primordial solar gases collapsed from space to cover the cores of these giant worlds with deep oceans of liquid hydrogen. These objects became the giant outer planets Jupiter, Saturn, Uranus, and Neptune. The inner planets were much smaller, apparently not massive enough to attract and hold any significant amount of hydrogen and helium. They became the rocky, hydrogen-poor worlds Mercury, Venus, Earth, and Mars.

The planets were probably nearly fully formed by the time that the central proto-Sun became hot and dense enough for thermonuclear fusion to begin. At the time that the Sun first began to burn hydrogen fuel, it was probably twice as massive as it is now. When the fusion energy conversion process turned on, the Sun became convectively unstable and began to emit matter at a prodigious rate from its surface. This hyperactive early stage of the Sun's youth is known as the *T-Tauri phase*. The primordial solar wind may have been so intense that the Sun lost half its initial mass in a time period as short as 1 million years.

At this time, there was still a lot of leftover gas and dust that had not been absorbed by one or the other of the planets, enough to make the young Solar System entirely opaque from the outside. The intense solar wind blew most of this residual gas and dust completely out of the Solar System, so that the planets and the Sun became visible from the outside for the first time (Figure 20.8).

Studies of other stars that are currently going through the T-Tauri stage

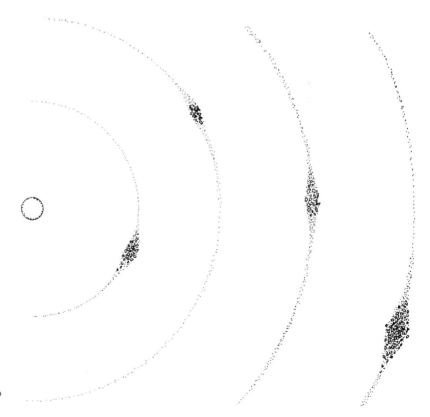

FIGURE 20.7 Collapse of individual asteroid-sized masses to form protoplanets.

indicate that the newly born Sun must have been spinning at a much faster rate than it is now. It must have also possessed an intense intrinsic magnetic field, one many orders of magnitude stronger than the relatively weak global field present today. The intense solar wind emitted by the young Sun interacted strongly with the solar magnetic field, braking the rotation of the Sun. Eventually, the T-Tauri violent phase of the Sun's adolescence came to an end, and the Sun settled down to a more or less stable existence as the steadily burning yellow star that we see today.

The newly formed terrestrial planets must have been far more uniform in their internal composition than they are now. Perhaps even before they had fully formed, the release of gravitational potential energy, the heat generated by the decay of radioactive elements, as well as the heat produced by the steady impacts of large asteroid-sized objects, caused the temperatures of the surfaces and interiors of these new planets to rise rapidly. The heating was sufficiently intense to melt virtually the entire masses of these young planets. Any atmospheres that had been initially acquired from the solar nebula were driven away to space by the intense heating. During this early hot, molten phase, the dense metallic material sank to the center of each planet to form a core, leaving the lighter silicate minerals at the surface to form an outer crust. In most cases, this

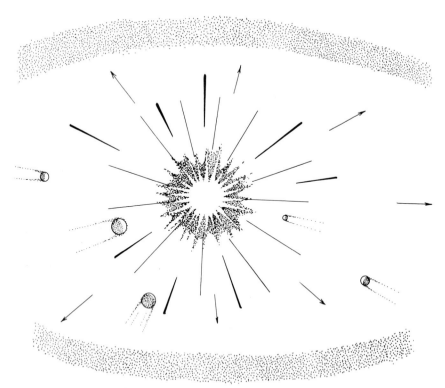

FIGURE 20.8 T-Tauri phase of Sun, in which an intense solar wind cleared the Solar System of most of its unbound gas and dust.

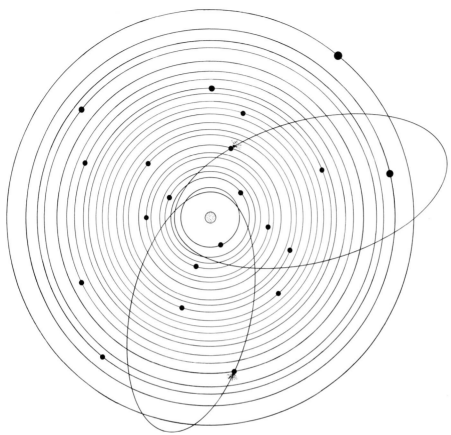

FIGURE 20.9 Schematic diagram of a sheet of asteroids orbiting the Sun in nearly circular orbits. Those asteroids occupying highly elliptical orbits have a much higher probability of undergoing collisions.

internal differentiation was probably completed by about 4.2 billion years ago. Evidence for an early episode of intense heating has been found on all planets so far studied by our spacecraft.

Eventually the inner terrestrial planets and the moons of the outer giants cooled sufficiently for solid surfaces to form. At this time, the newly formed planets were bombarded by frequent collisions with smaller bodies that had not been initially swept up into protoplanetary clusters. In this way, the Solar System was gradually cleared of smaller asteroid-sized masses, particularly those that occupied highly eccentric orbits or orbits with large inclinations with the ecliptic plane (Figure 20.9). The surfaces of most of the solid worlds in the Solar System still bear the scars of these ancient impacts. The most intense period of meteoric bombardment seems to have ended approximately 4 billion years ago, by which time most of the large loose chunks of matter had been swept up by a collision with one or the other of the planets. Since then, the rate of meteoroid bombardment has been quite low, and only an occasional small meteoroid strikes the terrestrial planets in the present day.

Appendix A

LIST OF LUNAR AND INTERPLANETARY SPACE PROBES

NAME	LAUNCH DATE AND LOCATION	BOOSTER	REMARKS
Pioneer 0	8-17-58 Cape Canaveral (USA)	Thor-Able 1 (127)	Attempt to place probe in lunar orbit. Ruptured fuel line caused booster to explode 77 sec after liftoff.
Pioneer 1	10-11-58 Cape Canaveral (USA)	Thor-Able 1 (130)	Attempt to place probe in lunar orbit. Guidance error resulted in too high a trajectory and a velocity too low for escape. Reached maximum altitude of 130,000 km and fell back to Earth. Measured extent of radiation belts.
Pioneer 2	11-8-58 Cape Canaveral (USA)	Thor-Able 1 (129)	Attempt to place probe in lunar orbit. Third stage failed to fire. Reached altitude of only 1500 km and fell back to Earth.
Pioneer 3	12-6-58 Cape Canaveral (USA)	Juno II (RTV-10)	Lunar flyby attempt. First stage of booster shut down prematurely. Reached altitude of 164,800 km and fell back to Earth. Discovered presence of two radiation belts.
Luna 1	1-2-59 Tyuratam (USSR)	SS-6 Sapwood/ SL-3/ A-1(L)	Lunar flyby mission. Flew past Moon 1-3-59, missing by 7460 km. Continued on into solar orbit. First spacecraft to escape from Earth.
Pioneer 4	3-3-59 Cape Canaveral (USA)	Juno II	Lunar flyby mission. Flew past Moon 3-4-59, missing by 59,520 km. Continued on into solar orbit. First American spacecraft to escape from Earth.
Luna 2	9-12-59 Tyuratam (USSR)	SS-6 Sapwood/ SL-3/ A-1(L)	Lunar impact mission. Crashed on Moon 9-13-59 at 29.1° N, 1.0° W, near craters Aristillus, Archimedes, and Autolycus at western edge of Mare Serenitatis (Sea of Serenity). Discovered absence of lunar magnetic field or radiation belts. First spacecraft to reach surface of Moon.
Luna 3	10-4-59 Tyuratam (USSR)	SS-6 Sapwood/ SL-3/ A-1(L)	Lunar far-side photographic mission. Probe placed in highly eccentric Earth orbit that went behind Moon on first pass. Flew behind Moon 10-6-59, missing by 7000 km. Photos taken of back side of Moon at distance of 60,000 to 70,000 km and transmitted to Earth 10-18-59. Photos extremely noisy. Many of the 400 lunar features reported were spurious. First look at lunar far side.

NAME	LAUNCH DATE AND LOCATION	BOOSTER	REMARKS
Pioneer P-3	11-26-59 Cape Canaveral (USA)	Atlas-Able IVB (20D)	Attempt to place probe in lunar orbit. Payload shroud separated 45 sec after liftoff. Upper stages torn from vehicle and fell into Atlantic.
Pioneer 5	3-2-60 Cape Canaveral (USA)	Thor-Able	Deep interplanetary spaceprobe. Placed in heliocentric orbit to study micrometeors and magnetic fields. Tests of transmission of scientific data over interplanetary distances.
Pioneer P-30	9-25-60 Cape Canaveral (USA)	Atlas-Able VA (80D)	Attempt to place probe in lunar orbit. Second stage failed to fire. Fell into Atlantic.
unnamed	10-10-60 Tyuratam (USSR)	Sapwood?	May have been attempted Mars probe. Did not reach Earth orbit.
unnamed	10-14-60 Tyuratam (USSR)	Sapwood?	May have been attempted Mars probe. Did not reach Earth orbit.
unnamed	10-24-60 Tyuratam (USSR)	Sapwood?	May have been attempted Mars probe. Huge explosion on launchpad during countdown. Many scores of people reported killed.
Pioneer P-31	12-15-60 Cape Canaveral (USA)	Atlas-Able VB (91D)	Attempt to place probe in lunar orbit. Premature ignition of second stage caused explosion of booster rocket 68 sec after liftoff.
"Sputnik 7"	2-4-61 Tyuratam (USSR)	SS-6 Sapwood/ SL-6/ A-2e	May have been failed Venus probe. Did not leave Earth parking orbit. Fell 2-26-61.
Venera 1	2-12-61 Tyuratam (USSR)	SS-6 Sapwood/ SL-6/ A-2e	Attempted Venus probe. Contact lost 2-27-61. Estimated to have passed within 100,000 km of Venus. Continued on into heliocentric orbit.
Ranger 1 *(P-32)*	8-23-61 Cape Canaveral (USA)	Atlas-Agena B (111D)	Attempt to launch test lunar spacecraft into highly elliptical Earth orbit in test for later Ranger lunar landers. Agena upper stage failed to restart in Earth parking orbit. Fell back into atmosphere after 111 orbits.
Ranger 2 *(P-33)*	11-18-61 Cape Canaveral (USA)	Atlas-Agena B (117D)	Attempt to launch test lunar spacecraft into highly elliptical Earth orbit in test for later Ranger lunar landers. Agena upper stage failed to restart in Earth parking orbit. Fell back into atmosphere after 6 orbits.

NAME	LAUNCH DATE AND LOCATION	BOOSTER	REMARKS
Ranger 3	1-26-62 Cape Canaveral (USA)	Atlas-Agena B	Attempt to take close-range lunar TV photographs and to land instrument package on lunar surface. Guidance system component failure 1 min after liftoff. Error in trajectory exceeded ability of midcourse correction system to compensate. Missed Moon by 35,000 km 1-28-62. No surface photos returned. Soft-landing package not released. Continued on into solar orbit.
Ranger 4	4-23-62 Cape Canaveral (USA)	Atlas-Agena B	Attempt to take close-range lunar TV photographs and to land instrument package on the lunar surface. Computer sequencing failure after leaving Earth parking orbit. Midcourse correction not possible. Impacted on far side of Moon 4-26-62 (12.9°S, 129.1°W) without releasing soft-landing package or returning any photos. First American spacecraft to reach surface of Moon.
Mariner 1	7-22-62 Cape Canaveral (USA)	Atlas-Agena B (145D)	Attempted Venus probe. Programming error in booster guidance system computer caused deviation from planned course. Booster destroyed by ground control after 5 min of flight.
unnamed	8-25-62 Tyuratam (USSR)	Sapwood?	May have been attempted Venus probe. Broke apart in Earth parking orbit.
Mariner 2	8-27-62 Cape Canaveral (USA)	Atlas-Agena B	Venus probe. Flew past Venus 12-14-62, missing surface by 34,000 km. First successful probe to another planet. Verified high surface temperatures on Venus. Continued on into solar orbit.
unnamed	9-1-62 Tyuratam (USSR)	Sapwood?	May have been attempted Venus probe. Did not leave Earth parking orbit.
unnamed	9-12-62 Tyuratam (USSR)	Sapwood?	May have been attempted Venus probe. Did not leave Earth parking orbit.
Ranger 5	10-18-62 Cape Canaveral (USA)	Atlas-Agena B	Attempt to take close-range lunar TV photos and to land instrument package on lunar surface. Power failure 8 hours after launch prevented midcourse correction. Missed Moon by 725 km 10-21-62 and went into solar orbit.
Mars 1	11-1-62 Tyuratam (USSR)	SS-6/ Sapwood/ SL-6/ A-2e	Attempted Mars probe. Contact lost 3-21-63 at distance of 1.06×10^8 km. Flew past Mars 6-19-63 and continued on into solar orbit.

NAME	LAUNCH DATE AND LOCATION	BOOSTER	REMARKS
unnamed	3-3-63 Tyuratam (USSR)	Sapwood?	May have been lunar probe attempt. Did not reach Earth orbit.
Luna 4	4-2-63 Tyuratam (USSR)	SS-6/ Sapwood/ SL-6/ A-2E	Lunar soft-landing attempt. Guidance error caused craft to miss Moon. Passed within 8000 km of lunar surface 4-5-63 and went into highly elliptical Earth orbit. No useful lunar data.
"Cosmos 21"	11-11-63 Tyuratam (USSR)	Sapwood?	May have been attempted Venus probe. Did not leave Earth parking orbit.
Ranger 6	1-30-64 Cape Canaveral (USA)	Atlas-Agena B (199D)	Attempt to take close-range TV photos of lunar surface. Impacted Moon 2-2-64 at 9.35°N, 21.4°E, in Mare Tranquillitatis (Sea of Tranquility), between craters Ross and Arago. TV cameras failed during final approach to Moon. No pictures returned.
unnamed	2-26-64 Tyuratam (USSR)	Sapwood?	May have been attempted Venus probe. Did not reach Earth orbit.
unnamed	3-4-64 Tyuratam (USSR)	Sapwood?	Believe failed Venus probe.
"Cosmos 27"	3-27-64 Tyuratam (USSR)	Sapwood?	May have been attempted Venus probe. Did not leave Earth parking orbit. Fell 3-28-64.
Zond 1	4-2-64 Tyuratam (USSR)	SS-6/ Sapwood/ SL-6/ A-2e	May have been attempted Venus probe. Radio contact lost shortly after launch.
unnamed	4-9-64 Tyuratam (USSR)	Sapwood?	May have been attempted lunar probe. Did not reach Earth orbit.
Ranger 7	7-28-64 Cape Canaveral (USA)	Atlas-Agena B (250D)	Close-range TV photos of lunar surface. Impact 7-31-64 at 10.7°S, 20.7°W near Fra Mauro crater at edge of Mare Nubium (Sea of Clouds). Returned 4316 photos of lunar surface during final approach. First completely successful U.S. Moon probe.
Mariner 3	11-5-64 Cape Canaveral (USA)	Atlas-Agena D	Mars flyby attempt. Payload shroud became snagged on spacecraft shortly after launch. Solar panels did not deploy. All contact lost.

NAME	LAUNCH DATE AND LOCATION	BOOSTER	REMARKS
Mariner 4	11-28-64 Cape Canaveral (USA)	Atlas-Agena D	Mars flyby probe. Passed within 9830 km of Martian surface 7-14-65. Returned 21 TV pictures of Mars. Discovered Martian craters, made radio-occultation measurements of atmosphere. Discovered absence of Martian magnetic field. First successful Mars probe.
Zond 2	11-30-64 Tyuratam (USSR)	SS-6/ Sapwood/ SL-6/ A-2e	Believed to have been attempted Mars probe. Radio contact lost 5-5-65.
Ranger 8	2-17-65 Cape Canaveral (USA)	Atlas-Agena D (196D)	Close-range TV photos of lunar surface. Impact 2-20-65 at 2.59°N, 24.77°E in Mare Tranquillitatis (Sea of Tranquility). Returned 7137 TV pictures of lunar surface during final approach.
Surveyor SD-1	3-2-65 Cape Canaveral (USA)	Atlas-Centaur AC-5 (156D)	Attempt to launch dummy lunar landing spacecraft into highly elliptical Earth orbit toward simulated Moon. Atlas booster engines shut down 2 sec after liftoff. Rocket fell back to pad and exploded.
"*Cosmos 60*"	3-12-65 Tyuratam (USSR)	Sapwood?	May have been attempted lunar probe. Did not leave Earth parking orbit. Fell 3-17-65.
Ranger 9	3-21-65 Cape Canaveral (USA)	Atlas-Agena D	Close-range TV photos of lunar surface. Impact on Moon 3-24-65 at 12.9°S, 2.4°W, inside crater Alphonsus at edge of Mare Nubium (Sea of Clouds). Returned 6150 live TV pictures of Moon during final approach.
Luna 5	5-9-65 Tyuratam (USSR)	SS-6 Sapwood/ SL-6/ A-2e	Failed lunar soft-landing attempt. Crashed on Moon 5-12-65 at 31°S, 8°E, near the crater Aliacensis in lunar highlands southeast of Mare Nubium (Sea of Clouds).
Luna 6	6-8-65 Tyuratam (USSR)	SS-6 Sapwood/ SL-6/ A-2e	Failed lunar soft-landing attempt. Midcourse correction maneuver performed improperly. Missed Moon by 160,000 km and went into solar orbit.
Zond 3	7-18-65 Tyuratam (USSR)	SS-6 Sapwood/ SL-6/ A-2e	Experiment in long-distance data transmission. Flew past Moon 7-20-65 and continued on into solar orbit. Photographs of far side taken during lunar flyby. Transmitted 25 photos to Earth 9 days later. Better quality than *Luna 3* photos. Found that lunar far side is different from near side.
Surveyor SD-2	8-11-65 Cape Canaveral (USA)	Atlas-Centaur (AC-6)	Launch of Surveyor test spacecraft into deep elliptical Earth orbit aimed toward simulated Moon. Eventually wandered into heliocentric orbit.

NAME	LAUNCH DATE AND LOCATION	BOOSTER	REMARKS
Luna 7	10-4-65 Tyuratam (USSR)	SS-6 Sapwood/ SL-6/ A-2e	Failed lunar soft-landing attempt. Crashed 10-7-65 at 9°N, 40°W in Oceanus Procellarum (Ocean of Storms) near crater Kepler.
Venera 2	11-12-65 Tyuratam (USSR)	SS-6 Sapwood/ SL-6/ A-2e	Attempted Venus flyby probe. Passed within 24,000 km of surface of Venus 2-27-66. Communication failure just before flyby. No useful Venus data returned.
Venera 3	11-16-65 Tyuratam (USSR)	SS-6 Sapwood/ SL-6/ A-2e	Attempted Venus atmosphere entry probe. Entered Venus atmosphere 3-1-66. Communication link failed just before atmospheric entry. No useful Venus data. First man-made object to reach surface of another planet.
"Cosmos 96"	11-23-65 Tyuratam (USSR)	Sapwood?	May have been attempted Venus probe. Did not leave Earth parking orbit. Fell 12-9-65.
Luna 8	12-3-65 Tyuratam (USSR)	SS-6 Sapwood/ SL-6/ A-2e	Failed lunar soft-landing attempt. Crashed 12-6-65 at 9.1°N, 63.3°W in Oceanus Procellarum (Ocean of Storms) near crater Galilaei.
Pioneer 6	12-16-65 Cape Canaveral (USA)	Delta DSV-3E (35)	Probe of solar phenomena. Entered solar orbit between Earth and Venus to study magnetic fields, cosmic rays, solar wind, radio wave propagation.
Luna 9	1-31-66 Tyuratam (USSR)	SS-6 Sapwood/ SL-6/ A-2e	First successful lunar soft landing. Landed 2-3-66 at 7.1°N, 64.3°W near crater Cavalerus at edge of Oceanus Procellarum (Ocean of Storms). Transmitted TV photos from surface, made measurement of background radioactivity.
"Cosmos 111"	3-1-66 Tyuratam (USSR)	Sapwood?	May have been lunar probe attempt. Did not leave Earth parking orbit. Fell 3-3-66.
Luna 10	3-31-66 Tyuratam (USSR)	SS-6 Sapwood/ SL-6/ A-2e	First artificial satellite of Moon. Entered lunar orbit 4-3-66. Measurements of meteors, gamma rays, magnetic fields, plasmas, infrared light. May have failed in attempt to return photos.
Surveyor SD-3	4-8-66 Cape Canaveral (USA)	Atlas-Centaur (AC-8) (184D)	Attempt to launch dummy Surveyor lunar-landing spacecraft toward simulated lunar target. Centaur second stage shut down prematurely after restart in parking orbit. Fell from orbit 5-5-66.

NAME	LAUNCH DATE AND LOCATION	BOOSTER	REMARKS
Surveyor 1	5-30-66 Cape Canaveral (USA)	Atlas-Centaur (AC-10)	First successful U.S. lunar soft landing. Landed 6-1-66 at 2.49°S, 43.32°W in Oceanus Procellarum (Ocean of Storms), north of crater Flamsteed. Returned 11,150 surface TV pictures.
Explorer 33	7-1-66 Cape Canaveral (USA)	Delta DSV-3E1 (39)	Attempt to place satellite in lunar orbit to gain data on terrestrial and solar magnetic fields and radiation. Velocity imparted by second stage too high for lunar orbit. Entered highly elliptical Earth orbit after flying past Moon.
Lunar Orbiter 1	8-10-66 Cape Canaveral (USA)	Atlas-Agena D	First U.S. lunar satellite. Entered lunar orbit 8-14-66. Photos of potential manned lunar landing sites, studies of lunar gravitational field. First photo of Earth as seen from Moon. Directed to lunar impact 10-29-66 on far side (6.7°N, 162°E).
Pioneer 7	8-17-66 Cape Canaveral (USA)	Delta DSV-3E1 (40)	Solar phenomena probe. Entered solar orbit close to Earth's to study cosmic rays, plasmas, radio propagation.
Luna 11	8-24-66 Tyuratam (USSR)	SS-6 Sapwood/ SL-6/ A-2e	Lunar satellite. Measurements of energetic electrons and solar wind. Apparently failed to return any photos.
Surveyor 2	9-20-66 Cape Canaveral (USA)	Atlas-Centaur (AC-7)	Failed lunar soft-landing attempt. Targeted toward Sinus Medii (Central Bay), near middle of Moon's visible face. Failure of vernier engines at time of midcourse maneuver. Craft began tumbling; all contact lost. Probably crashed on Moon at 5°N, 25°W, near crater Copernicus.
Luna 12	10-22-66 Tyuratam (USSR)	SS-6 Sapwood/ SL-6/ A-2e	Lunar satellite. Entered lunar orbit 10-26-66. Several lunar photos returned, measurements of lunar radiation environment, mapping of lunar gravitational field.
Surveyor SD-4	10-26-66 Cape Canaveral (USA)	Atlas-Centaur (AC-9)	Successful test of restart of Centaur upper stage in Earth parking orbit. Launched dummy Surveyor lunar landing spacecraft into eccentric Earth orbit aimed toward simulated lunar target. Reached apogee of 368,430 km.
Lunar Orbiter 2	11-6-66 Cape Canaveral (USA)	Atlas-Agena D	Lunar satellite. Entered lunar orbit 11-10-66. Returned 422 photos of potential lunar landing sites. Spectacular view of crater Copernicus. Directed to lunar impact 10-11-66 near Smyth's Sea just beyond visible edge of Moon (4°S, 93°E).

NAME	LAUNCH DATE AND LOCATION	BOOSTER	REMARKS
Luna 13	12-21-66 Tyuratam (USSR)	SS-6 Sapwood/ SL-6/ A-2e	Lunar soft-landing mission. Landed 12-24-66 at 18.9°N, 62.0°W, near crater Seleucus in Oceanus Procellarum (Ocean of Storms). Surface photos returned, gamma-ray measurements of soil density, mechanical penetrometer tests of soil hardness.
Lunar Orbiter 3	2-4-67 Cape Canaveral (USA)	Atlas-Agena D	Lunar satellite. Entered lunar orbit 8-14-66. Returned 307 photos of potential lunar landing sites. Measurements of orbital trajectory to study lunar gravitational field. Directed to lunar impact 10-9-67 near crater Einstein just beyond visible edge of Moon (14.6°N, 91.7°W).
"Cosmos 146"	3-10-67 Tyuratam (USSR)	Proton/ SL-12/ D-1-E	May have been attempted lunar probe. Did not leave Earth parking orbit. Fell 3-11-67.
"Cosmos 154"	4-8-67 Tyuratam (USSR)	Proton/ SL-12/ D-1-E	May have been attempted lunar probe. Did not leave Earth parking orbit. Fell 4-10-67.
Surveyor 3	4-16-67 Cape Canaveral (USA)	Atlas-Centaur (AC-12)	Lunar soft landing. Landed 4-19-67 at 2.98°S, 23.4°W, near Riphaeus Mountains in Oceanus Procellarum (Ocean of Storms), 612 km east of *Surveyor 1* site. Returned 6315 surface TV pictures. Performed trench-digging experiments.
Lunar Orbiter 4	5-4-67 Cape Canaveral (USA)	Atlas-Agena D	Lunar satellite. Entered near-polar lunar orbit 5-8-67. Returned 326 photos of potential lunar landing sites. Measurements of orbital trajectory to determine lunar gravitational field. First clear view of Mare Orientale. Directed to lunar impact 10-6-67.
Venera 4	6-12-67 Tyuratam (USSR)	SS-6 Sapwood/ SL-6/ A-2e	Venus atmosphere entry probe. Entered Venus atmosphere 10-18-67. Radio transmitter failed at altitude of 27 km. Confirmed high atmospheric temperatures.
Mariner 5	6-14-67 Cape Canaveral (USA)	Atlas-Agena D	Venus flyby probe. Passed within 3950 km of Venus 10-19-67. Measurement of radio occultation, magentic fields, charged particles, solar wind. Ultraviolet photometry of Venusian atmosphere.
Surveyor 4	7-13-67 Cape Canaveral (USA)	Atlas-Centaur (AC-11)	Failed lunar soft-landing. Contact lost 1.4 sec before touchdown 7-16-67 at 0.25°N, 1.20°W in Sinus Medii (Central Bay).

LIST OF LUNAR AND INTERPLANETARY SPACE PROBES (continued)

NAME	LAUNCH DATE AND LOCATION	BOOSTER	REMARKS
Explorer 35 *(IMP-5)*	7-19-67 Cape Canaveral (USA)	Delta DSV-3E1	Interplanetary Monitoring Platform. Entered lunar orbit 7-22-67. Magnetic field, radiation, plasma, and meteor measurements. Radio-telemetry signals used as radar beacon to bounce signals off Moon to determine radio reflectivity. Confirmed absence of lunar magnetic field.
Lunar Orbiter 5	8-1-67 Cape Canaveral (USA)	Atlas-Agena D	Lunar satellite. Entered near-polar lunar orbit 8-5-67. Photos of potential lunar landing sites, measurements of orbital trajectory to study lunar gravitational field. Analysis of data led to discovery of lunar mascons. Directed to lunar impact 1-31-68 in highlands southeast of Mare Crisium (Sea of Crises) (0°N, 70°W).
Surveyor 5	9-8-67 Cape Canaveral (USA)	Atlas-Centaur (AC-13)	Lunar soft landing. Successful landing 9-10-67 at 1.48°N, 23.23°E in Mare Tranquillitatis (Sea of Tranquility). Measurements of chemical composition and magnetic content of lunar soil. Vernier engines fired to shift position slightly. Returned 18,006 surface TV pictures.
Surveyor 6	11-7-67 Cape Canaveral (USA)	Atlas-Centaur (AC-14)	Lunar soft landing. Landing craft touched down 11-9-67 at 0.47°N, 1.48°W in Sinus Medii (Central Bay), 8 km from site of *Surveyor 4*. Measurements of chemical content of lunar soil. Returned 30,065 surface TV pictures. Vernier engines fired to lift craft from surface for short instant.
unnamed	11-22-67 Tyuratam (USSR)	Proton/ SL-12/ D-1-E	May have been attempt to launch recoverable spacecraft on circumlunar mission. Did not reach Earth orbit.
Pioneer 8	12-13-67 Cape Canaveral (USA)	Delta DSV-3E1 (55)	Solar phenomenon probe. Placed in heliocentric orbit near Earth's orbit to study solar phenomena and to map terrestrial magnetosphere.
Surveyor 7	1-7-68 Cape Canaveral (USA)	Atlas-Centaur (AC-15)	Lunar soft landing. Touched down 1-9-68 at 41.04°S, 11.43°W in highlands near crater Tycho. Study of chemical compostion and magnetic content of soil, trench-digging experiments, laser communication tests, stereo TV photography. Returned 21,274 surface TV photos.
Zond 4	3-2-68 Tyuratam (USSR)	Proton/ SL-12/ D-1-E	May have been attempt to launch recoverable spacecraft into highly elliptical Earth orbit in test for future manned circumlunar mission. Aimed at point in space directly opposite Moon. Contradictory reports on flight. Exact fate uncertain.

NAME	LAUNCH DATE AND LOCATION	BOOSTER	REMARKS
Luna 14	4-7-68 Tyuratam (USSR)	SS-6 Sapwood/ SL-6/ A-2e	Lunar satellite. Entered lunar orbit 4-10-68. Radiation and charged-particle measurements, radio transmission tests, gravitational field measurements. May have failed in attempt to return photos.
unnamed	4-20-68 Tyuratam (USSR)	Proton/ SL-12/ D-1-E	May have been attempt to launch recoverable spacecraft into highly elliptical Earth orbit in test for future manned circumlunar mission. Did not reach Earth orbit.
Zond 5	9-15-68 Tyuratam (USSR)	Proton/ SL-12/ D-1-E	Unmanned flight around back side of Moon and return. May have been test for future manned circumlunar mission. Passed within 1950 km of lunar far side 9-18-68. Capsule landed in Indian Ocean 9-21-68 and was recovered. First recovery of spacecraft returning from vicinity of Moon.
Pioneer 9	11-8-68 Cape Canaveral (USA)	Delta DSV-3E1 (60)	Solar phenomenon probe. Placed in heliocentric orbit inside Earth's to study solar phenomena, cosmic rays, electric fields.
Zond 6	11-10-68 Tyuratam (USSR)	Proton/ SL-12/ D-1-E	Unmanned flight around back side of Moon and return. May have been test for future manned circumlunar flight. Passed within 2420 km of lunar far side 11-14-68. Landed in USSR 11-17-68 and recovered. Carried biological samples, performed meteor-impact and cosmic-ray studies. Returned lunar far side photos.
Apollo 8	12-21-68 Cape Canaveral (USA)	Saturn 5 (SA-503)	Manned flight to lunar orbit and return. Crew of Frank Borman, Jim Lovell, Ed Anders. Entered lunar orbit 12-24-68. Detailed visual and photographic reconnaissance of Moon and test of Apollo spacecraft. Returned to Earth and landed in Pacific Ocean 12-27-68. First manned flight to vicinity of Moon.
Venera 5	1-5-69 Tyuratam (USSR)	SS-6 Sapwood/ SL-6/ A-1e	Venus soft-landing attempt. Entered Venus atmosphere 5-16-69 on planet's dark side. Radio signals from lander ceased at altitude of 24 to 26 km.
unnamed	1-8-69 Tyuratam (USSR)	Proton/ SL-12/ D-1-E	May have been attempt to launch recoverable spacecraft on circumlunar flight in test for future manned mission. Booster exploded at altitude of 40 km.
Venera 6	1-10-69 Tyuratam (USSR)	SS-6 Sapwood/ SL-6/ A-1e	Venus soft-landing attempt. Entered Venus atmosphere 5-17-69 on planet's dark side. Radio signals from lander ceased at altitude of 10 to 12 km.

NAME	LAUNCH DATE AND LOCATION	BOOSTER	REMARKS
Mariner 6	2-24-69 Cape Canaveral (USA)	Atlas-Centaur (AC-20)	Mars flyby probe. Passed within 3400 km of Martian surface 7-30-69, passing over equator. Returned 76 photos of surface; radio-occultation measurements of atmosphere.
Mariner 7	3-27-69 Cape Canaveral (USA)	Atlas-Centaur (AC-17)	Mars flyby probe. Flew within 3200 km of Martian surface 8-5-69, passing over south pole. Returned 126 photos of surface.
unnamed	4-69 Tyuratam (USSR)	Proton/ SL-12/ D-1-E	May have been attempt to launch lunar soft-lander capable of returning soil sample to Earth. Booster rocket exploded on launch pad.
Apollo 10	5-18-69 Cape Canaveral (USA)	Saturn 5 (SA-505)	Rendezvous and docking trials with lunar excursion module (LEM) in moon orbit. Crew of Tom Stafford, John Young, Eugene Cernan. LEM approached to within 15 km of lunar surface in extensive trials for later lunar landing. LEM discarded and directed to lunar impact. Crew returned to Earth in command module. Landed in Pacific Ocean 5-26-69. Many lunar and Earth photos returned.
unnamed	6-14-69 Tyuratam (USSR)	Proton?	May have been attempted lunar probe. Did not reach Earth orbit due to second-stage failure.
Luna 15	7-13-69 Tyuratam (USSR)	Proton/ SL-12/ D-1-E	May have been attempt to land unmanned station on Moon and return soil sample to Earth. Probably intended to upstage forthcoming *Apollo 11* moon landing. Entered lunar orbit 7-17-69. Crashed 7-21-69 at 17°N, 60°E in Mare Crisium (Sea of Crises). Apparent failure.
Apollo 11	7-16-69 Cape Canaveral (USA)	Saturn 5 (SA-506)	First manned landing on Moon. Crew of Neil Armstrong, Edwin Aldrin, Michael Collins. Entered lunar orbit 7-19-69. Landed 7-20-69 at 1.12°N, 23.82°E in Mare Tranquillitatis (Sea of Tranquility). Crew deployed solar wind collector, seismometer, and laser range reflector. Surface photographs taken; 35 kg of soil samples collected. Total of 21 hours on Moon. LEM ascent stage discarded and directed to lunar impact after orbital rendezvous. Command module landed in Pacific Ocean 7-24-69. Surface experiments turned off 8-27-69.
Zond 7	8-8-69 Tyuratam (USSR)	Proton/ SL-12/ D-1-E	Unmanned flight around back side of Moon and return. Passed behind lunar back side 8-11-69. Landed 8-14-69 in USSR and recovered. Photos of Earth and Moon returned.

NAME	LAUNCH DATE AND LOCATION	BOOSTER	REMARKS
Pioneer E	8-27-69 Cape Canaveral (USA)	Delta DSV-3L (73)	Attempt to launch probe into heliocentric orbit to study solar phenomena. Destroyed 8 min after launch due to hydraulic malfunction in launch vehicle.
"Cosmos 300"	9-23-69 Tyuratam (USSR)	Proton/ SL-12/ D-1-E	May have been attempt to launch unmanned probe capable of landing on Moon and returning soil sample to Earth. Did not leave Earth parking orbit.
"Cosmos 305"	10-22-69 Tyuratam (USSR)	Proton/ SL-12/ D-1-E	May have been attempted lunar probe. Disintegrated while in low Earth parking orbit.
Apollo 12	11-14-69 Cape Canaveral (USA)	Saturn 5 (SA-507)	Manned landing on Moon. Crew of Charles Conrad, Alan Bean, Richard Gordon. Entered lunar orbit 11-18-69. Landed 11-19-69 at 2.99°S, 23.40°W in Oceanus Procellarum (Ocean of Storms) at the site of *Surveyor 3* lander. Crew deployed nuclear-powered ALSEP package containing several experiments. Surface photos taken, soil samples gathered. Parts of *Surveyor 3* lander collected. Crew traversed a total distance of 1.3 km. Total of 31 hours on Moon. LEM directed to lunar impact after rendezvous, striking surface at a distance of 76 km from landing site. Command module landed in Pacific Ocean 11-24-69. ALSEP package turned off 9-30-77 due to budgetary crunch.
unnamed	2-19-70 Tyuratam (USSR)	Proton/ SL-12/ D-1-E	May have been attempted lunar probe. Did not reach Earth parking orbit.
Apollo 13	4-11-70 Cape Canaveral (USA)	Saturn 5 (SA-508)	Failed Moon landing attempt. Target was Fra Mauro region of Mare Nubium (Sea of Clouds). Crew of Jim Lovell, Fred Haise, John Swigert. Explosion in service module 4-14-70 forced cancelation of landing attempt. Looped around back side of Moon and returned to Earth. LEM remained attached for use as "lifeboat," and discarded just before Earth atmosphere entry. Command module landed safely in Pacific Ocean 4-17-70.
Venera 7	8-17-70 Tyuratam (USSR)	SS-6 Sapwood/ SL-6/ A-2e	Venus soft-landing mission. Lander entered atmosphere 12-15-70 and landed at 0°N, 10°W on night side of planet. Transmitted on surface for 23 min before failing. Measurements of temperature. First data from surface of another planet.

NAME	LAUNCH DATE AND LOCATION	BOOSTER	REMARKS
"Cosmos 359"	8-22-70 Tyuratam (USSR)	Sapwood?	May have been attempted Venus probe. Failed to leave Earth parking orbit.
Luna 16	9-12-70 Tyuratam (USSR)	Proton/ SL-12/ D-1-E	Lunar surface sample return mission. Entered lunar orbit 9-17-70. Landed 9-20-70 at 0.7°S, 56.3°E in Mare Fecunditatis (Sea of Fertility). Collected and returned 120 g of soil sample to Earth in capsule. Return capsule landed in USSR 9-24-70 and was recovered. First remotely controlled sample returned from Moon.
Zond 8	10-20-70 Tyuratam (USSR)	Proton/ SL-12/ D-1-E	Unmanned flight around back side of Moon and return. Passed within 1120 km of Moon 10-24-70. Landed in Indian Ocean 10-17-70 and recovered.
Luna 17	11-10-70 Tyuratam (USSR)	Proton/ SL-12/ D-1-E	Automatic lunar rover mission. Landed 11-17-70 at 38.3°N, 35.0°W in Mare Imbrium (Sea of Rains). Deployed Lunokhod 1 automated rover. Surface maneuvers, traversing a total distance of 10 km. Returned 20,000 photos. Laser reflector tests, chemical analyses of lunar soil.
Apollo 14	1-31-71 Cape Canaveral (USA)	Saturn 5 (SA-509)	Manned landing on Moon. Crew of Alan Shepard, Edgar Mitchell, Stuart Roosa. Entered lunar orbit 2-4-71. Landed 2-5-71 at 3.67°S, 17.47°W in Fra Mauro region of Mare Nubium (Sea of Clouds). Deployed nuclear-powered ALSEP package containing numerous experiments. Surface photos taken, 44 kg of soil samples collected. Traversed a total distance of 3.4 km. Spent 33 hours on Moon. LEM discarded and directed to lunar impact at a point 63 km from landing site. Crew landed in Pacific Ocean 2-9-71. ALSEP package turned off 9-30-77 due to budgetary crunch.
Mariner 8	5-8-71 Cape Canaveral (USA)	Atlas- Centaur AC-24 (5405C)	Attempted Mars orbiter. Autopilot fault in gyros of Centaur second stage caused craft to go off course and crash into ocean 1450 km from launch site.
"Cosmos 419"	5-10-71 Tyuratam (USSR)	Proton/ SL-12/ D-1-E	May have been attempted Mars probe. Did not leave Earth parking orbit. Fell 5-12-71.
Mars 2	5-19-71 Tyuratam (USSR)	Proton/ SL-12/ D-1-E	Combined Mars orbiter and lander. Lander separated from main craft during planetary approach phase 11-27-71. Radio contact with lander failed shortly after release. Lander probably crashed on Mars 500 km southwest of Hellas (44°S, 313°W). First spacecraft to reach surface of Mars. Main body of craft continued on to Mars orbit.

NAME	LAUNCH DATE AND LOCATION	BOOSTER	REMARKS
Mars 3	5-28-71 Tyuratam (USSR)	Proton/ SL-12/ D-1-E	Combined Mars orbiter and lander. Lander separated from orbiter during planetary approach phase and apparently landed successfully in heavily cratered southern hemisphere (45°S, 158°W) of Mars 12-2-71. Suddenly ceased transmitting 20 sec after touchdown. No useful surface data, but some useful atmospheric data. Rest of craft continued on to Mars orbit.
Apollo 15	7-26-71 Cape Canaveral (USA)	Saturn 5 (SA-510)	Manned landing on Moon. Crew of David Scott, James Irwin, Alfred Worden. Entered lunar orbit 7-29-71. Landed 7-30-71 at 26.1°N, 3.65°E, near Hadley Rille in Apennine Mountains at western edge of Mare Serenitatis (Sea of Serenity). Crew deployed nuclear-powered ALSEP package containing numerous experiments. Rover vehicle used to traverse a total distance of 28 km across surface, as much as 8 km from lander. Surface photos taken, 77 kg of soil samples collected. Total of 67 hours on Moon. After rendezvous, LEM was directed to lunar impact 8-2-71 (26.4°N, 0.3°E). Crew landed in Pacific Ocean 8-7-71. ALSEP package turned off 9-30-77 due to budgetary crunch.
Mariner 9	8-17-71 Cape Canaveral (USA)	Atlas-Centaur (AC-25)	Mars orbiter. Entered Mars orbit 11-13-71. First spacecraft to enter orbit around another planet. Returned 7329 photos of Martian surface and moons. Discovered volcanoes, canyons, dried-up rivers. Attitude control gas ran out 10-27-72, ending useful life.
Luna 18	9-2-71 Tyuratam (USSR)	Proton/ SL-12/ D-1-E	May have been attempt to land automated station on Moon and return lunar soil sample to Earth. Entered lunar orbit 9-7-71. Lander crashed at 3.57°N, 56.5°E in a mountainous area near edge of Mare Fecunditatis (Sea of Fertility). Apparent failure.
Luna 19	9-28-71 Tyuratam (USSR)	Proton/ SL-12/ D-1-E	Lunar gravitational field and surface mapper. Entered lunar orbit 10-3-71. Radiation measurements, radar altimetry, TV photography. Apparently no landing attempt involved.
Luna 20	2-14-72 Tyuratam (USSR)	Proton/ SL-12/ D-1-E	Lunar soil sample return mission. Landed at 3.4°N, 56.6°E, in a highland region near crater Apollonius, between Mare Crisium (Sea of Crises) and Mare Fecunditatis (Sea of Fertility). Several surface TV pictures returned. Capsule carrying 50 g of lunar soil fired off from Moon and recovered 2-25-72 in USSR.

NAME	LAUNCH DATE AND LOCATION	BOOSTER	REMARKS
Pioneer 10	3-2-72 Cape Canaveral (USA)	Atlas- Centaur (AC-27)	Probe of outer Solar System. Flew past Jupiter 12-3-73, coming within 130,000 km of surface. First spacecraft to reach vicinity of Jupiter. Returned 300 photos of Jupiter and Galilean moons. Extensive study of Jovian magnetosphere. During encounter with Jupiter, craft acquired enough additional velocity to escape from Solar System into interstellar space. First object to enter trajectory that will leave Solar System. Carried plaque with message for extraterrestrial intelligences.
"Cosmos 482"	3-21-72 Tyuratam (USSR)	Sapwood?	May have been attempted Venus probe. Did not leave Earth parking orbit.
Venera 8	3-26-72 Tyuratam (USSR)	SS-6/ Sapwood/ SL-6/ A-2e	Venus lander. Entered Venus atmosphere 7-22-72 at 10°S, 330°E, 600 km from terminator on day side of planet. Operated on surface for 50 min before failing. Took light-level readings, made temperature and pressure measurements, did studies of composition of atmosphere and soil.
Apollo 16	4-16-72 Cape Canaveral	Saturn 5 (SA-511)	Manned landing on Moon. Crew of John Young, Charles Duke, Thomas Mattingly. Entered lunar orbit 4-19-72. Landed 4-20-72 in Descartes region of lunar highlands (9.0°S, 15.52°E). Deployed ALSEP package containing numerous experiments. Lunar rover carried crew over a total distance of 26 km. Surface photographs taken, 98 kg of soil samples collected. Total of 71 hours on Moon. Incorrectly set switch prevented programmed crash of LEM onto Moon. Small subsatellite left behind in lunar orbit to study lunar magnetic field. Command module landed in Pacific Ocean 4-27-72. ALSEP package turned off 9-30-77 due to budgetary crunch.
Apollo 17	12-7-72 Cape Canaveral (USA)	Saturn 5 (SA-512)	Manned landing on Moon. Crew of Eugene Cernan, Harrison Schmitt, Ronald Evans. Entered lunar orbit 12-10-72. Landed 12-11-72 at 20.17°N, 30.77°E in Taurus Mountains near the crater Littrow at edge of Mare Serenitatis (Sea of Serenity). Crew deployed ALSEP package containing numerous experiments. Lunar rover used for surface maneuvering, traversing a total distance of 26 km. TV pictures returned, surface photos taken, 113 kg of soil samples collected. Discovered orange-colored glass beads on sur-

NAME	LAUNCH DATE AND LOCATION	BOOSTER	REMARKS
			face. Returned oldest sample of Moon yet found (4.5 billion years). Total of 75 hours on Moon. LEM directed to lunar impact 12-14-72 at point 9 km south of landing site. Crew landed 12-19-72 in Pacific. Last Moon landing in Apollo series. ALSEP package turned off 9-30-77 due to budgetary crunch.
Luna 21	1-8-73 Tyuratam (USSR)	Proton/ SL-12/ D-1-E	Lunar surface rover mission. Entered lunar orbit 1-12-73. Landed 1-15-73 at 25.9°N, 30.5°E, inside Le Monnier crater at eastern edge of Mare Serenitatis (Sea of Serenity). Deployed *Lunokhod 2* automated lunar rover. Rover traversed total distance of 37 km. Returned 80,000 TV pictures. Magnetic field, ultraviolet light, cosmic ray, soil chemistry, laser reflector measurements.
Pioneer 11	5-5-73 Cape Canaveral (USA)	Atlas-Centaur (AC-30)	Probe of planets of outer Solar System. Flew past Jupiter 12-3-74, coming within 42,800 km of surface. Photos of Jupiter and Galilean moons returned. Extensive studies of Jovian magnetosphere. Redirected to Saturn after Jovian encounter. Flew past Saturn 9-1-79, coming within 21,400 km of surface. First spacecraft to reach vicinity of Saturn. Passed through the ring plane just outside the visible edge. Photos of Saturn and rings returned. Extensive study of Saturn magnetosphere. Discovered additional ring components as well as a previously unknown moon (Janus or Epimetheus). After Saturn encounter, entered trajectory that will ultimately lead it out of Solar System. Carried plaque with message to extraterrestrial intelligences.
unnamed	5- -73 Tyuratam (USSR)	Proton?	May have been attempted lunar probe. Crashed into Pacific after second-stage malfunction.
Explorer 49 (*RAE-2*)	6-10-73 Cape Canaveral (USA)	Delta 1913 (95)	Radio Astronomy Explorer. Entered lunar orbit 6-15-73 to record deep-space radio signals. Operated until 4-30-77.
Mars 4	7-21-73 Tyuratam (USSR)	Proton/ SL-12/ D-1-E	Intended as orbital relay station for *Mars* 6 and 7 landers. Braking rockets did not fire and craft failed to enter Mars orbit. Missed Mars by 2200 km 2-10-73 and continued on into solar orbit. TV pictures returned during flyby.
Mars 5	7-25-73 Tyuratam (USSR)	Proton/ SL-12/ D-1-E	Intended as orbital relay station for *Mars* 6 and 7 landers. Entered Mars orbit 2-12-74. TV pictures of surface returned.

NAME	LAUNCH DATE AND LOCATION	BOOSTER	REMARKS
Mars 6	8-5-73 Tyuratam (USSR)	Proton/ SL-12/ D-1-E	Combined Mars flyby "bus" spacecraft and landing probe. Landing capsule released from bus 3-2-74 during approach phase, at distance of 3400 km from planet. Capsule entered atmosphere, but ceased transmitting just before reaching surface. Landed at 25°W, 24°S, northeast of Argyre Planitia. Flyby bus craft continued on into solar orbit.
Mars 7	8-9-73 Tyuratam (USSR)	Proton/ SL-12/ D-1-E	Combined Mars flyby bus spacecraft and landing probe. Landing capsule released from bus 3-9-74. Separation maneuver performed improperly, and lander missed planet by 1200 km. Both lander and bus continued on into solar orbit.
Mariner 10	11-3-73 Cape Canaveral (USA)	Atlas-Centaur (AC-34)	Probe of planets of inner Solar System. Flew within 5770 km of surface of Venus 2-5-74. Photographs of Venus upper atmosphere returned. Discovered intense winds in upper atmosphere. Deflected to Mercury after Venus encounter. Three flybys of Mercury (3-29-74, 9-21-74, 3-16-75), closest approach being 330 km. First photos of surface of Mercury. Discovered magnetic field of Mercury. Attitude control gas ran out after third Mercury encounter, ending useful life.
unnamed	1-11-74 Cape Canaveral (USA)	Titan III E Centaur (TC-1)	Test of Titan-Centaur launch vehicle capability. Carried dynamic model of Viking Mars orbiter/lander. Centaur second stage failed to fire. Destroyed by ground control after 12 min of flight.
Luna 22	5-29-74 Tyuratam (USSR)	Proton/ SL-12/ D-1-E	Lunar orbiter. Entered lunar orbit 6-2-74. Studies of lunar gravitational field, meteors, solar radiation, gamma rays from lunar surface, plasmas. Radio altimetry and laser reflector experiments. Extensive orbital maneuvering, approaching to within 30 km of lunar surface.
Luna 23	10-28-74 Tyuratam (USSR)	Proton/ SL-12/ D-1-E	Attempted lunar landing and soil sample return mission. Entered lunar orbit 11-2-74. Landed 11-6-74 at 13.5°N, 56.5°E in Mare Crisium (Sea of Crises). Problems with drill prevented sample of lunar soil from being inserted into return capsule. Return attempt canceled.
Helios 1	12-10-74 Cape Canaveral (USA)	Titan III E Centaur (TC-2)	Joint U.S.–West German probe to study solar phenomena. Entered solar orbit with perihelion of 0.31 AU, closest approach yet to Sun. Measurements of magnetic fields, solar wind, radio waves, plasmas, cosmic rays, dust, X rays.

NAME	LAUNCH DATE AND LOCATION	BOOSTER	REMARKS
Venera 9	6-8-75 Tyuratam (USSR)	Proton/ SL-12/ D-1-E	Combined Venus orbiter and landing mission. Lander separated from orbiter 10-20-75 during approach to Venus. Successful landing 10-22-75 near Beta Regio (32°N, 291°E), on day side of planet. Measurements of surface composition, wind speed. Returned a single photograph, the first ever taken on the surface of another planet. Operated on the surface for 53 min before failing. Main craft entered Venus orbit 10-22-75, becoming the first artificial satellite of that planet. Measurements of cloud structure and atmospheric composition.
Venera 10	6-14-75 Tyuratam (USSR)	Proton/ SL-12/ D-1-E	Combined Venus orbiter and landing mission. Lander touched down 10-25-75 near Beta Regio (16°N, 291°E) on day side of planet. Operated on surface for 65 min before failing. TV picture of surface returned. Main craft entered Venus orbit 10-25-75.
Viking 1	8-20-75 Cape Canaveral (USA)	Titan III E Centaur (TC-4)	Combined Mars orbiter and landing mission. Entered Mars orbit 6-19-76. Many photos of Martian surface and two moons returned. Remote-sensing studies of Martian surface and atmosphere. Lander touched down 7-20-76 in Chryse region of northern hemisphere of Mars (22.5°N, 48.0°W). First successful landing of spacecraft on Mars. TV pictures of surface. Measurements of soil composition, air pressure, temperature, wind velocity. Biochemical tests for life. Orbiter turned off 8-8-80 due to exhaustion of reaction gases. Lander computer failed 11-5-82.
Viking 2	9-9-75 Cape Canaveral (USA)	Titan III E Centaur (TC-3)	Combined Mars orbiter and lander. Entered Mars orbit 8-7-76. Photos of surface and two moons returned. Remote-sensing studies of Martian surface and atmosphere. Lander touched down 9-3-76 in Utopia region of Martian northern hemisphere (47.89°N, 225.86°W) 9-3-76. TV pictures of surface. Measurements of soil composition, air pressure, temperature, wind velocity. Biochemical tests for life. Orbiter shut off 7-25-78 due to exhaustion of reaction gases. Lander computer failed 4-12-80.
Helios 2	6-15-76 Cape Canaveral (USA)	Titan III E Centaur	Joint U.S.–West German probe to study solar phenomena. Entered solar orbit with perihelion of 0.29 AU, closest approach yet to Sun. Measurements of magnetic fields, solar wind, radio waves, plasmas, cosmic rays, dust, X rays.

NAME	LAUNCH DATE AND LOCATION	BOOSTER	REMARKS
Luna 24	8-9-76 Tyuratam (USSR)	Proton/ SL-12/ D-1-E	Lunar soft landing and soil sample return mission. Entered lunar orbit 8-14-76. Landed 8-18-76 at 12.75°N, 62.2°E in Mare Crisium (Sea of Crises). Surface drill collected 170-g soil sample from depth of 2 m and inserted it into return capsule. Capsule returned to Earth. Landed in USSR 8-22-76 and recovered.
Voyager 2	8-20-77 Cape Canaveral (USA)	Titan III E Centaur (TC-7)	Probe to outer planets of Solar System. Flew past Jupiter 7-9-79, coming within 650,180 km of cloud tops. Extensive studies of Jovian magnetosphere. Photographs of Jupiter and Galilean moons returned. Discovered a 14th Jovian moon. Flew past Saturn 8-26-81, coming within 100,800 km of cloud tops. Extensive studies of Saturnian magnetosphere. Photographs of Saturn, ring system, and moons. Flew past Uranus 1-24-86, passing within 107,080 km of planet. Study of Uranian magnetosphere. Photographs of Uranus, ring system, and moons returned. Passed within 29,000 km of Miranda. Discovered additional ring components and 10 new moons. Due at Neptune 8-24-89. Will pass within 7400 km of south pole of Neptune and within 10,000 km of Triton. Following Neptune encounter, will escape from Solar System into interstellar space. Carries recorded message for extraterrestrial intelligences.
Voyager 1	9-5-77 Cape Canaveral (USA)	Titan III E Centaur (TC-6)	Probe to outer planets of Solar System. Flew past Jupiter 3-5-79, coming within 277,840 km of cloud tops. 18,000 photos of Jupiter and Galilean moons returned. Discovered volcanoes of Io, two new Jovian moons, and a previously unknown ring system. Extensive survey of Jovian magnetosphere. Flew past Saturn 11-12-80, coming within 124,200 km of cloud tops. Returned 18,000 photos of Saturn, ring system, and moons. Discovered three previously unknown moons, extensive structure of rings. Passed within 4000 km of surface of Titan 11-11-80. After Saturn, entered trajectory that will expel it from Solar System into interstellar space. Carries recorded message for extraterrestrial intelligences.
Pioneer-Venus 1	5-20-78 Cape Canaveral (USA)	Atlas-Centaur	Venus orbiter mission. Entered near-polar Venus orbit 12-4-78. Measurements of Venusian ionosphere, ultraviolet study of clouds, plasma measurements, radiation studies. Equipped with

NAME	LAUNCH DATE AND LOCATION	BOOSTER	REMARKS
			radar altimeter that made map of virtually the entire Venusian surface with a resolution of about 75 km. Photographs of comet Halley returned 2-86. Craft still fully functional.
Pioneer-Venus 2	8-8-78 Cape Canaveral (USA)	Atlas-Centaur (AC-51)	Venus atmosphere "multiprobe." Bus spacecraft carried four atmospheric entry probes (one large *Sounder* probe and three smaller *North, Day,* and *Night* probes). All four probes entered atmosphere 12-9-78, and all operated from entry until surface impact. Impact points were: *Sounder*— 4.4°N, 304.0°E; *North*— 59.3°N, 4.8°E; *Day*— 31.3°S, 317°E; *Night*— 28.7°S, 56.7°E. *Day* probe survived on surface for 67 min before ceasing transmission. Bus spacecraft entered atmosphere 12 hours later (at 37.9°S, 290.9°E) and was destroyed. Studies of temperature, pressure, and composition of atmosphere and clouds.
ISEE-3 (ICE)	8-12-78 Cape Canaveral (USA)	Delta 2914	International Sun–Earth Explorer. Placed in "halo" orbit, located at stable point between Earth and Sun, where gravitational forces of the two bodies exactly balance. Arrived on station 11-28-78, 1.5×10^6 km from Earth. Performed long-term study of solar phenomena. Boosted out of halo orbit 6-82 and placed on trajectory to encounter Comet Giacobini–Zinner. Craft renamed International Comet Explorer. Passed through tail of Giacobini–Zinner 9-11-85, missing nucleus by 7870 km. First spacecraft encounter with a comet. Study of magentic field and plasma within coma. No TV cameras carried. Passed 3.0×10^7 km downstream of comet Halley 3-86. Will return to Earth in 2005 and may be recovered.
Venera 11	9-9-78 Tyuratam (USSR)	Proton/ SL-12/ D-1-E	Combined Venus flyby probe and lander. Lander touched down in Beta-Phoebe region of Venus 12-21-78 and operated 95 min on surface before failing. Performed chemical analyses of gases in atmosphere. May have failed in attempt to return surface photographs. Flyby bus passed within 35,000 km of Venus and continued on into solar orbit. Studies of gamma-ray bursts emitted by cosmic sources.
Venera 12	9-14-78 Tyuratam (USSR)	Proton/ SL-12/ D-1-E	Combined Venus flyby probe and lander. Landed 12-25-78 in Beta-Phoebe region of Venus and operated 110 min on surface before failing.

NAME	LAUNCH DATE AND LOCATION	BOOSTER	REMARKS
			Lander performed chemical analyses of gases in atmosphere. May have failed in attempt to return surface photographs. Flyby bus passed within 35,000 km of Venus and continued on into solar orbit. Studies of gamma-ray bursts emitted by cosmic sources.
Venera 13	10-30-81 Tyuratam (USSR)	Proton/ SL-12/ D-1-E	Combined Venus flyby probe and lander. Landed 3-2-82 at 7.5° S, 303.5° E, east of Beta-Phoebe region. Operated for 127 min on surface before failing. Color photos of surface, chemical analysis of soil sample. Flyby bus continued on into solar orbit.
Venera 14	11-4-81 Tyuratam (USSR)	Proton/ SL-12/ D-1-E	Combined Venus flyby probe and lander. Landed 3-5-82 at 13.25° S, 310° E, east of Beta-Phoebe region. Operated for 53 min on surface before failing. Color photos of surface, chemical analysis of soil sample. Flyby bus continued on into solar orbit.
Venera 15	6-2-83 Tyuratam (USSR)	Proton/ SL-12/ D-1-E	Venus orbiter and radar surface mapper. Entered Venus orbit 10-10-83. Equipped with side-looking synthetic-aperture radar to map the hidden surface to a resolution of 1 to 2 km. Region of northern hemisphere between 24°N and 33°N covered.
Venera 16	6-7-83 Tyuratam (USSR)	Proton/ SL-12/ D-1-E	Venus orbiter and radar surface mapper. Entered Venus orbit 10-14-83. Equipped with side-looking synthetic-aperture radar to map the hidden surface to a resolution of 1 to 2 km. Radar map covered the northern one-third of Venus.
Vega 1	12-15-84 Tyuratam (USSR)	Proton/ SL-12/ D-1-E	Combined Venus and Comet Halley probe. Initially dispatched toward Venus. Redirected to Halley after Venus flyby. Main body of craft carried television cameras as well as set of instruments for study of dust, gas, plasma, and magnetic fields within coma of Halley. Carried Venus surface and atmospheric probe for drop-off during Venus flyby phase of mission. Passed within 30,000 km of Venus 6-10-85. Venus probe detached from main body of probe and entered Venus atmosphere on night side of planet. On way down to surface, lander released instrumented balloon that drifted with the wind at altitude of 55 km for 2 days before batteries failed. Measurements of wind speeds, pressures, and temperatures. Venus lander set down within a

NAME	LAUNCH DATE AND LOCATION	BOOSTER	REMARKS
			lowland tract of Aphrodite Terra (7.3°N, 177.7°W). Operated on surface 56 min before failing. Measurements of atmospheric composition. Surface drilling experiment failed. No provision for surface photos. Following Venus encounter, deflected onto course intercepting orbit of Comet Halley. Encountered Halley 3-6-86, passing within 8890 km of nucleus. Study of composition of dust and gas in coma, TV image of nucleus.
Vega 2	12-21-84 Tyuratam (USSR)	Proton/ SL-12/ D-1-E	Combined Venus and Comet Halley probe. Identical in configuration and mission goals to *Vega 1*. Flyby of Venus 6-15-18, passing within 30,000 km of planet. Landing probe detached and entered Venus atmosphere on night side of planet. On way down to the surface, lander released instrumented balloon which drifted with the wind at altitude of 55 km for 2 days before batteries failed. Measurements of wind speeds, pressures, and temperatures. Venus lander set down within Atla Regio at eastern end of Aphrodite (6.6°S, 180.7°E). Operated 57 min on surface before failing. Measurements of composition of atmosphere and surface soil. No provision for surface photos. Following Venus encounter, deflected onto course intercepting orbit of Comet Halley. Encountered Halley 3-9-86, passing within 8030 km of nucleus. Study of composition of dust and gas in coma, TV image of nucleus.
MS-T5 Sakigake (Pioneer)	1-8-85 Uchinoura (Japan)	M-3S-2	Engineering test spacecraft for *Planet-A* mission to Comet Halley. Measurements of magnetic fields; studies of velocity, density, and temperature of solar wind. Approached within 7×10^6 km of Halley 3-11-86. Performed studies of interaction between comet and the solar wind.
Giotto	7-2-85 Kourou (French Guiana)	Ariane 1 (V14)	Probe to vicinity of Comet Halley. Equipped with television camera, neutral and ion mass spectrometers, dust-impact detector, dust mass spectrometer, plasma-analysis equipment, energetic-particle analyzer, magnetometer, optical probe. Passed through coma of Comet Halley (approaching to within 610 km of nucleus) 3-13-86. Studies of dust, gas, magnetic fields, electric fields, radiation, and plasmas within cometary coma. TV images of nucleus. Several instruments

NAME	LAUNCH DATE AND LOCATION	BOOSTER	REMARKS
			heavily damaged by dust impacts during passage through coma, but spacecraft survived. Retargeted for return to vicinity of Earth 7-2-90.
Suisei (Comet) Planet-A	8-19-85 Uchinoura (Japan)	M-3S-2	Probe to vicinity of Comet Halley. Equipped with ultraviolet imager and charged-particle energy analyzer. Passed within 151,000 km of nucleus of Halley 3-8-86. Performed ultraviolet studies of cometary coma; measurements of solar wind.
Mars Orbit/ Phobos Lander	1988 Tyuratam (USSR)	Proton/ SL-12/ D-1-E	Proposed Mars-orbit/Phobos-rendezvous mission. Two identical craft will be built. On way to Mars, craft will make solar observations. Will enter Mars orbit to make remote-sensing measurements of Martian surface and atmosphere. Craft will approach within 100 m of moon Phobos, performing laser and ion beam remote analysis of surface material. Landing craft will be released that will drop onto Phobos to provide direct observation of surface. Both fixed-site and "hopper" landers being considered. If first craft is successful, second craft may be diverted to Deimos.
Ulysses	September 1989 Cape Canaveral (USA)	Space Shuttle or Titan 34D7?	European-built solar polar mission. Initially aimed toward Jupiter. Upon encounter with Jupiter, craft will be thrown south of ecliptic plane. Perihelion will be 1.25 AU. Will make measurements of solar wind, solar flares, magnetic fields, cosmic rays, cosmic dust. First view of poles of Sun.
Galileo	December 1989 Cape Canaveral (USA)	Space Shuttle with IUS	Combined Jupiter orbiter/atmospheric entry probe. Will use Venus and Earth flybys to provide boost for flight to Jupiter. Will arrive in Jovian System 11-95. Entry probe will separate from main craft and enter Jovian atmosphere near equator on day side of planet. Physical and chemical measurements of upper atmosphere. Expected to reach depth where pressure is 10 to 20 atm before failing. Orbiter will then enter elliptical orbit around Jupiter to carry out extensive survey of magnetosphere and moons. May fly past asteroid on way to Jupiter.
Magellan	October 1989 Cape Canaveral (USA)	Space Shuttle or Titan 34D7?	Venus radar mapper. Will be placed in elliptical near-polar Venus orbit. Equipped with side-looking, synthetic-aperture radar capable of mapping most of surface at 0.2 to 0.7-km resolution.

NAME	LAUNCH DATE AND LOCATION	BOOSTER	REMARKS
Lunar Polar Orbiter	1989-90 Tyuratam (USSR)	?	Craft to be placed in polar lunar orbit to make geochemical map of entire surface.
Caesar	1989-94 Kourou (French Guiana)	Ariane	Comet Atmosphere and Earth Sample Return mission proposal by ESA. Will fly through coma of a comet and return sample of gas and dust to Earth orbit for recovery by space shuttle. Awaiting funding approval.
Mars Observer	1990 Cape Canaveral (USA)	Space Shuttle/ Transfer Orbit Stage	Craft to orbit Mars in year-long survey, gathering information on surface composition, magnetic field (if any), and seasonal cycles involving CO_2 and windblown dust.
Comet Rendezvous/ Asteroid Flyby	1991-92 Cape Canaveral (USA)	Space Shuttle	Proposed cometary-rendezvous/asteroid-flyby mission. Will rendezvous with comet and remain with it for most of active phase. Measurements of magnetic fields, plasmas, composition of gas and dust, photos of nucleus. During rendezvous, will drop penetrator into nucleus to study chemical composition and temperature. May be directed to fly past asteroid while en route to comet. Initial choice of target comet is Wild-2. Mission not yet funded.
Vesta	1992 Tyuratam (USSR)	Proton/ SL-12/ D-1-E	Joint Soviet – French Mars/asteroid mission. Two craft are to be built. They will be initially launched toward Mars, and will use a gravity-assist from that planet to throw them outward to the Asteroid Belt. Several possible encounters with asteroids or comets. Vesta encounter an important priority. During flyby, will drop lander onto the surface of Vesta to take TV pictures and make physical measurements.
Multicomet Sample Return	November 1992 Cape Canaveral (USA)	Space Shuttle	NASA proposal for return of samples of comet dust to Earth. Three craft are to be launched from Earth orbit. One is dedicated solar probe equipped with chronograph, X-ray telescope, and radio science equipment. Other two will be comet probes equipped with dust collectors. Each comet probe will pass through coma of a different comet, collecting dust and gas. Both comet probes will return to Earth orbit and will be retrieved by maneuvering vehicle dispatched from space station. Probes will be refurbished

NAME	LAUNCH DATE AND LOCATION	BOOSTER	REMARKS
			and then dispatched toward different comets. Samples from as many as four different comets may be collected during the years 1995–2005. Project not yet funded.
Cassini	1993-94 Cape Canaveral (USA)	Space Shuttle	Proposal for joint NASA–ESA Saturn orbiter/ Titan probe. Will arrive in Saturn system in year 2000. Will make orbital tour of Saturn and its moons, using Titan for gravity-assist. Measurement of upper Saturnian atmosphere and magnetosphere. Craft to be equipped with imaging radar for study of Titan surface. Will also carry lander that will descend into Titan atmosphere. Not yet funded.
Agora	Feb./March 1993 Kourou (French Guiana)	Ariane 44L	Asteroid, Gravity, Optical and Radar Analysis probe proposed by ESA. To be equipped with ion-drive propulsion system that will make it possible to rendezvous with several different asteroids. Measurements of magnetic fields, gamma-ray spectroscopy, microwave radiometry, studies of dust, TV photographs. First target is to be Vesta, with other asteroids possibly visited later. Not yet funded.
Lunar Polar Orbiter	1994 Uchinoura (Japan)	H-2	Japanese proposal for lunar polar orbiter. Will fire rocket-shaped penetrators into lunar far side to implant seimographs. Currently under study.
Kepler	Mid-1990s Kourou (French Guiana)	Ariane 3	ESA proposal for orbiter to study structure and composition of Martian atmosphere and magnetosphere. Not yet funded.
Io Probe	Mid-1990s Kourou (French Guiana)	Ariane 4	ESA proposal for probe to Jupiter's moon Io. May be impacted on the surface. Not yet approved.
Mercury Polar Orbiter	Mid-1990s Kourou (French Guiana)	Ariane 4	ESA proposal for probe to be placed in polar orbit around Mercury. May carry surface penetrators to provide surface data. Not yet approved.
Venus Orbiter	Mid-1990s Uchinoura (Japan)	H-2	Japanese proposal for Venus orbiter to gather magnetospheric data. Currently under study.
Mars Lander/ Rover	Mid-1990s Tyuratam (USSR)	?	Mission to place automated rover vehicle on Martian surface. In early definition phase.
Mars Sample Return	Late-1990s Tyuratam (USSR)	?	Mission to return sample of Martian surface to Earth. In early definition phase.

NAME	LAUNCH DATE AND LOCATION	BOOSTER	REMARKS
Mars Aeronomy Orbiter	Late 1990s Cape Canaveral (USA)	Space Shuttle	NASA proposal for orbiter to study region of space near Mars and to sample upper atmosphere. Not yet funded.

SOURCE: Table derived from information published in *Aviation Week and Space Technology, Spaceflight, Journal of the British Interplanetary Society, Missiles and Rockets.*

Appendix B

ELEMENTAL COMPOSITION OF SOLAR ATMOSPHERE

ATOMIC NUMBER	ELEMENT	ATOMIC PERCENT	WEIGHT PERCENT	ATOMIC NUMBER	ELEMENT	ATOMIC PERCENT	WEIGHT PERCENT
1	Hydrogen	93.96	77.82	47	Silver	6.7×10^{-10}	6.0×10^{-8}
2	Helium	5.92	19.61	48	Cadmium	6.7×10^{-9}	6.2×10^{-7}
3	Lithium	9.39×10^{-8}	5.4×10^{-7}	49	Indium	4.2×10^{-9}	4.0×10^{-7}
4	Beryllium	1.32×10^{-9}	9.8×10^{-9}	50	Tin	9.4×10^{-9}	9.2×10^{-7}
5	Boron	1.2×10^{-8}	1.1×10^{-7}	51	Antimony	9.4×10^{-10}	9.5×10^{-8}
6	Carbon	0.039	0.39	52	Tellurium	—	—
7	Nitrogen	0.0082	0.095	53	Iodine	—	—
8	Oxygen	0.064	0.859	54	Xenon	—	—
9	Fluorine	3.4×10^{-6}	5.3×10^{-5}	55	Cesium	$<7.4 \times 10^{-9}$	$<8.2 \times 10^{-7}$
10	Neon	0.0035	0.058	56	Barium	1.1×10^{-8}	1.3×10^{-8}
11	Sodium	1.79×10^{-5}	0.0034	57	Lanthanum	1.2×10^{-9}	1.4×10^{-7}
12	Magnesium	0.0038	0.076	58	Cerium	3.3×10^{-9}	3.8×10^{-7}
13	Aluminum	3.1×10^{-4}	0.0069	59	Praseodymium	4.3×10^{-10}	5.0×10^{-8}
14	Silicon	0.0042	0.098	60	Neodymium	1.6×10^{-9}	1.9×10^{-7}
15	Phosphorus	3.0×10^{-6}	7.7×10^{-4}	61	Promethium	—	—
16	Sulfur	0.0015	0.040	62	Samarium	4.9×10^{-10}	6.1×10^{-8}
17	Chlorine	3.0×10^{-5}	8.8×10^{-4}	63	Europium	4.7×10^{-10}	5.9×10^{-8}
18	Argon	9.4×10^{-5}	0.0031	64	Gadolinium	1.2×10^{-9}	1.6×10^{-7}
19	Potassium	1.31×10^{-5}	4.3×10^{-4}	65	Terbium	—	—
20	Calcium	2.1×10^{-4}	0.0069	66	Dysprosium	1.0×10^{-9}	1.4×10^{-7}
21	Scandium	1.03×10^{-7}	3.9×10^{-6}	67	Holmium	—	—
22	Titanium	1.03×10^{-5}	4.1×10^{-4}	68	Erbium	5.5×10^{-10}	7.6×10^{-8}
23	Vanadium	9.4×10^{-7}	4.0×10^{-5}	69	Thulium	1.6×10^{-10}	2.4×10^{-8}

Number	Element		
24	Chromium	4.8×10^{-5}	0.0021
25	Manganese	2.4×10^{-5}	0.0011
26	Iron	0.0030	0.139
27	Cobalt	7.4×10^{-6}	3.6×10^{-4}
28	Nickel	1.7×10^{-4}	0.0089
29	Copper	1.03×10^{-6}	5.4×10^{-5}
30	Zinc	2.6×10^{-6}	1.4×10^{-4}
31	Gallium	5.9×10^{-8}	3.4×10^{-6}
32	Germanium	3.0×10^{-7}	1.8×10^{-5}
33	Arsenic	—	—
34	Selenium	—	—
35	Bromine	—	—
36	Krypton	—	—
37	Rubidium	3.8×10^{-8}	2.7×10^{-6}
38	Strontium	7.4×10^{-8}	5.4×10^{-6}
39	Yttrium	1.2×10^{-8}	9.0×10^{-7}
40	Zirconium	5.3×10^{-8}	4.0×10^{-6}
41	Niobium	7.4×10^{-9}	5.7×10^{-7}
42	Molybdenum	1.3×10^{-8}	1.0×10^{-6}
43	Technetium	—	—
44	Rubidium	6.4×10^{-9}	5.3×10^{-7}
45	Rhodium	2.3×10^{-9}	2.0×10^{-7}
46	Palladium	3.0×10^{-9}	2.7×10^{-7}
70	Ytterbium	7.4×10^{-10}	1.1×10^{-7}
71	Lutetium	5.5×10^{-10}	7.9×10^{-8}
72	Hafnium	5.9×10^{-10}	8.8×10^{-8}
73	Tantalum	—	—
74	Tungsten	4.7×10^{-9}	7.2×10^{-7}
75	Rhenium	4.7×10^{-11}	$<7.2 \times 10^{-9}$
76	Osmium	4.7×10^{-10}	9.0×10^{-8}
77	Iridium	6.7×10^{-10}	1.1×10^{-7}
78	Platinum	5.3×10^{-9}	8.5×10^{-7}
79	Gold	5.3×10^{-10}	8.6×10^{-8}
80	Mercury	$<1.2 \times 10^{-8}$	$<2.0 \times 10^{-6}$
81	Thallium	7.4×10^{-10}	1.3×10^{-7}
82	Lead	8.0×10^{-9}	1.4×10^{-6}
83	Bismuth	$<7.4 \times 10^{-9}$	$<1.3 \times 10^{-6}$
84	Polonium	—	—
85	Astatine	—	—
86	Radon	—	—
87	Francium	—	—
88	Radium	—	—
89	Actinium	—	—
90	Thorium	1.5×10^{-10}	2.9×10^{-8}
91	Protactinium	—	—
92	Uranium	$<3.8 \times 10^{-10}$	$<7.4 \times 10^{-8}$

SOURCE: Adapted from Pasachoff and Kutner (1978).
A dash indicates that the element has not yet been detected in the solar atmosphere.

Appendix C

PROPERTIES OF THE SUN

Mass	1.991×10^{30} kg
Mean Earth–Sun distance	1.49×10^{11} = 1 astronomical unit (AU)
Radius	6.960×10^8 m
Density	1.41 g/cm³ (average)
	150 g/cm³ (core)
Rotational period	24 days, 16 hours (at equator)
Rotational velocity	2.06 km/sec (at equator)
Escape velocity from surface	617 km/sec
Gravitational acceleration at surface	273.7 m/sec² (27.9 g)
Angle of inclination between equator and ecliptic plane	7°
Oblateness	None observable
Solar luminosity ($L_\odot$)	3.9×10^{26} W
Average power emitted per unit surface area	6.44×10^7 W/m²
Solar constant (sunlight intensity at Earth's orbit)	1353 W/m² ($\pm$ 2 percent)
Effective surface temperature	5500°C
Composition (atomic percent)	93.9 percent H; 5.9 percent He; rest heavier elements

Appendix D

TABLE OF PLANETARY PROPERTIES

	MERCURY	VENUS	EARTH
Semimajor axis (AU)[a]	0.39	0.72	1.00
Eccentricity	0.206	0.007	0.017
Inclination w.r.t. ecliptic (°)	7.0	3.39	0.00
Sidereal orbital period (Earth years)	0.241	0.615	1.00
Longitude of perihelion (°)	75.90	130.15	101.22
Longitude of ascending node (°)	47.15	75.78	—
Sidereal rotation period (Earth days)	58.65	243.01	0.997
Inclination of equator to orbital plane (°)	7.0	177.4	23.5
Equatorial radius (km)	2440	6070	6378
Oblateness	0	0	0.003
Mass (Earth masses)[d]	0.056	0.82	1.00
Average density (g/cm^3)	5.43	5.25	5.50
Effective temperature (°C)	430	−29	−20
Average surface temperature (°C)	430(d)/−170(n)	460	15–20
Surface pressure (atm)	—	90	1.0
Magnetic moment (gauss cm^3)	2.4×10^{22} – 6.0×10^{22}	$<10^{22}$	7.91×10^{25}
Atmospheric constituents (by weight percent)	none	97% CO_2; 3% N_2; H_2O; Ar, CO	78% N_2; 21% O_2; 1% Ar; CO_2; H_2O

[a] 1 AU $= 1.497 \times 10^m$.
[b] Radio period.
[c] Retrograde orbit.
[d] Mass of Earth $= 5.98 \times 10^{24}$ kg.

MARS	JUPITER	SATURN	URANUS	NEPTUNE	PLUTO
1.52	5.20	9.5	19.2	30.1	39.72
0.093	0.049	0.054	0.047	0.009	0.25
1.85	1.30	2.49	0.77	1.77	17.2
1.88	11.86	29.46	84.01	164.79	250.3
334.22	12.72	91.10	171.53	46.33	223
48.79	99.44	112.79	73.48	130.68	109.7
1.026	0.415^b	0.445^b	$0.718^{b,c}$	0.75 – 1.0	6.4
23.98	3.08	26.73	97.92	28.8	50 (?)
3389	71,540	60,330	26,145	25,000 (?)	1500 – 1800
0.009	0.064	0.10	0.024	0.02 (?)	?
0.108	318	95.1	14.5	17.2	0.002
3.93	1.33	0.71	1.24	1.67	0.4 – 1.0
−33	−149	−179	−213	−218	−220(d)
−33(d)/ −85(n)	no surface	no surface	no surface	no surface	−229 (d)/ −270 (n)
0.008	no surface	no surface	no surface	no surface	0.0001 – 0.05
$<2.5 \times 10^{21}$	1.55×10^{30}	4.7×10^{28}	4.0×10^{27}	?	?
95% CO_2; 2.5% N_2; 1.5% Ar	81% H_2; 19% He	89% H_2; 11% He	74% H_2; 26% He; CH_4	H_2; He; CH_4	CH_4

Appendix E

THE MOONS OF THE SOLAR SYSTEM

NAME	SEMIMAJOR AXIS (km)	ECCENTRICITY	ORBITAL INCLINATION W.R.T. EQUATOR OF PRIMARY (°)
Earth			
Moon	384,400	0.055	varies
Mars			
Phobos	9378	0.015	1.02
Deimos	23,459	0.00005	1.82
Jupiter			
Metis (1979J3/J16)	127,960	0.004(?)	0.000
Adrastea (1979J1/J14)	128,980	0.000	0.000
Amalthea (J5)	161,140	0.003	0.46
Thebe (1979J2/J15)	222,330	0.015	0.8
Io (J1)	421,600	0.000	0.03
Europa (J2)	670,900	0.000	0.47
Ganymede (J3)	1,070,000	0.001	0.18
Callisto (J4)	1,883,000	0.007	0.25
Leda (J13)	11,110,000	0.147	27
Himalia (J6)	11,476,000	0.158	28
Lysithea (J10)	11,700,000	0.12	29
Elara (J7)	11,737,000	0.207	26
Anake (J12)	21,200,000	0.169	147
Carme (J11)	22,600,000	0.207	163
Pasiphae (J8)	23,500,000	0.40	147
Sinope (J9)	23,600,000	0.275	156
Saturn			
Atlas (1980S28/S17)	137,670	0.002	0.3
Prometheus (1980S27/S16)	139,400	0.003	0.0
Pandora (1980S26/S15)	141,700	0.004	0.05
Epimetheus (1980S3/S11)	151,422	0.009	0.34
Janus (1980S1/S10)	151,472	0.007	0.14

THE MOONS OF THE SOLAR SYSTEM *(continued)*

SIDEREAL ROTATION PERIOD (EARTH DAYS)	SIDEREAL ORBITAL PERIOD (EARTH DAYS)[a]	RADIUS (km)	MASS (EARTH'S MOON = 1.00)[b]	DENSITY (g/cm²)
27.32	27.32	1738	1.00	3.34
0.319	0.319	$9.0 \times 10.5 \times 13.5$	1.0×10^{-6}	~2.0
1.262	1.262	$7.5 \times 6.0 \times 5.0$	1.3×10^{-7}	~1.9
0.295	0.295	20	?	?
0.297	0.297	12	?	?
0.418	0.418	$75 \times 85 \times 135$	?	?
0.675	0.675	35 – 40	?	?
1.769	1.769	1815	1.21	3.53
3.551	3.551	1570	0.66	3.03
7.155	7.155	2630	2.03	1.93
16.689	16.689	2400	1.45	1.79
?	240	3 – 8	?	?
0.5(?)	250.6	50 – 60	?	?
?	259.2	7 – 8	?	?
?	259.7	12 – 20	?	?
?	630(R)	6 – 8	?	?
?	692(R)	7 – 10	?	?
?	739(R)	6 – 10	?	?
?	758(R)	7 – 10	?	?
0.602	0.602	$20 \times 15 \times ?$	?	?
0.613	0.613	$70 \times 50 \times 40$	?	?
0.629	0.629	$55 \times 45 \times 35$	?	?
0.694	0.694	$70 \times 60 \times 50$	?	?
0.695	0.695	$110 \times 90 \times 80$	?	?

NAME	SEMIMAJOR AXIS (km)	ECCENTRICITY	ORBITAL INCLINATION W.R.T. EQUATOR OF PRIMARY (°)
Mimas (S1)	188,224	0.0201	1.52
1980S12(?)	188,200	?	?
Enceladus (S2)	240,192	0.0044	0.023
Tethys (S3)	296,563	0.000	1.09
1980S34(?)	?	?	?
Telesto (1980S13/S13)	296,560	?	?
Calypso (1980S25/S14)	296,560	?	?
1981S6(?)	296,560	?	?
1981S10(?)	350,000	?	?
1980S6 (S12)	378,600	0.005	0.15
1981S7(?)	378,600	?	?
Dione (S4)	379,074	0.0022	0.0023
1981S9(?)	470,000	?	?
Rhea (S5)	527,828	0.0010	0.35
Titan (S6)	1,221,432	0.0290	0.33
Hyperion (S7)	1,502,275	0.1042	0.28–0.93
Iapetus (S8)	3,559,400	0.0283	115[c]
Phoebe (S9)	12,900,000	0.1633	150
Uranus			
1986U7	49,700	0	0
1986U8	53,800	0	0
1986U9	59,200	0	0
1986U3	61,750	0	0
1986U6	62,700	0	0
1986U2	64,600	0	0
1986U1	66,090	0	0
1986U4	69,920	0	0
1986U5	75,300	0	0
1985U1	85,890	0	0
Miranda	129,780	0.000	4
Ariel	191,240	0.003	0
Umbriel	265,970	0.004	0
Titania	435,840	0.002	0
Oberon	582,600	0.001	0

SIDEREAL ROTATION PERIOD (EARTH DAYS)	SIDEREAL ORBITAL PERIOD (EARTH DAYS)[a]	RADIUS (km)	MASS (EARTH'S MOON = 1.00)[b]	DENSITY (g/cm^2)
0.964	0.964	195	0.00051	~1.2
?	0.964	5	?	?
1.389	1.389	250	0.0010	~1.1
1.906	1.906	525	0.0083	~1.0
?	?	5	?	?
?	1.906	? $\times$ 12 $\times$ 11	?	?
?	1.906	17 $\times$ 11 $\times$ 11	?	?
?	1.906	?	?	?
?	2.44	5	?	?
?	2.739	18 $\times$ 16 $\times$ 15	?	?
?	?	5	?	?
2.756	2.756	560	0.0140	~1.4
?	3.8	5	?	?
4.528	4.528	765	0.033	~1.3
?	15.938	2570	1.85	1.92
13(?)[c]	21.739	115 $\times$ 145 $\times$ 190	?	?
79.243	79.243	720	0.026	~1.2
0.40	550.4(R)	110	?	?
0.330(R)	0.330(R)	20	?	?
0.372(R)	0.372(R)	25	?	?
0.433(R)	0.433(R)	25	?	?
0.463(R)	0.463(R)	30	?	?
0.475(R)	0.475(R)	30	?	?
0.493(R)	0.493(R)	40	?	?
0.513(R)	0.513(R)	40	?	?
0.558(R)	0.558(R)	30	?	?
0.622(R)	0.622(R)	30	?	?
0.761(R)	0.761(R)	85	?	?
1.414(R)	1.414(R)	240	~0.001	~1.3
2.520(R)	2.520(R)	580	~0.020	~1.7
4.144(R)	4.144(R)	595	~0.020	~1.4
8.706(R)	8.706(R)	805	0.048	~1.6
13.463(R)	13.463(R)	775	0.040	~1.5

THE MOONS OF THE SOLAR SYSTEM *(continued)*

NAME	SEMIMAJOR AXIS (km)	ECCENTRICITY	ORBITAL INCLINATION W.R.T. EQUATOR OF PRIMARY (°)
Neptune			
Triton	355,000	0.0(?)	160
Nereid	5,562,000	0.75	28
Pluto			
Charon	17,000	?	?

[a](R) = Retrograde.
[b]Mass of Earth's Moon = 7.354×10^{22} kg.
[c]Variable.
SOURCE: Adapted from Mitton (1977).

SIDEREAL ROTATION PERIOD (EARTH DAYS)	SIDEREAL ORBITAL PERIOD (EARTH DAYS)[a]	RADIUS (km)	MASS (EARTH'S MOON = 1.00)[b]	DENSITY (g/cm^2)
5.876(R)	5.876(R)	1600 – 2640	1.9(?)	?
?	359.88	140 – 300	0.0007(?)	?
6.4(?)	6.4	675 – 900	0.02(?)	?

Appendix F

ISOTOPIC ABUNDANCE RATIOS IN THE ATMOSPHERES OF THE TERRESTRIAL PLANETS[a,b]

	SUN	VENUS	EARTH	MARS
$^{12}C/^{13}C$	0.011	0.012	0.0112	0.0118
$^{14}N/^{15}N$	260	160	280	160
$^{17}O/^{16}O$	?	?	3.71×10^{-4}	3.91×10^{-4}
$^{18}O/^{16}O$	?	2.0×10^{-3}	2.05×10^{-3}	2.06×10^{-3}
$^{22}Ne/^{20}Ne$	?	?	0.097	0.10
$^{20}Ne/^{36}Ar$	31	0.5	0.6	0.5
$^{36}Ar/^{40}Ar$	?	0.80	3.4×10^{-3}	3.0×10^{-4}
$^{38}Ar/^{40}Ar$	?	0.14	6.4×10^{-4}	6.0×10^{-5}
$^{38}Ar/^{36}Ar$	?	0.18	0.19	0.20
$^{36}Ar/^{12}C$	8.4×10^{-3}	1×10^{-5} to 6×10^{-5}	3×10^{-7}	6×10^{-6}
$^{84}Kr/^{36}Ar$	3.0×10^{-4}	0.02	0.04	0.03
$^{129}Xe/^{132}Xe$		—	0.983	2.56
$^{36}Ar/^{132}Xe$			$>3 \times 10^3$	1.3×10^3

SOURCE: Adapted from Rasool, Hunten, and Kaula (1977); Cameron (1983); Kerridge (1980); Arvidson, Goettel, Hohenberg (1980); Gautier and Owen (1983); Pollack and Black (1979).
[a]Ratios of numbers of atoms.
[b]The isotopes ^{20}Ne, ^{36}Ar, ^{38}Ar, ^{84}Kr, ^{132}Xe are primordial; ^{40}Ar, ^{129}Xe are radiogenic.

Appendix G

ABUNDANCE RATIOS IN THE ATMOSPHERES OF THE SUN AND THE JOVIAN PLANETS[a]

	SUN	JUPITER	SATURN	URANUS	NEPTUNE
He/H	0.06	0.05	0.03	0.06	?
C/H	4.7×10^{-4}	0.0011	9.9×10^{-4}	0.0094	0.012
N/H	9.8×10^{-5}	9.8×10^{-5}	2.4×10^{-4}	$<9.8 \times 10^{-5}$	$<9.8 \times 10^{-5}$
O/H	6.8×10^{-4}	2.3×10^{-5}	?	?	?
P/H	2.4×10^{-7}	2.4×10^{-7}	7.9×10^{-7}	?	?
D/H	2.5×10^{-5}	3.2×10^{-5}	2.4×10^{-5}	4.8×10^{-5}	?
$^{13}C/^{12}C$	0.011	0.0062	0.011	?	?
$^{14}N/^{15}N$	0.0026	0.0016	?	?	?

SOURCE: Adapted from Rasool, Hunten, and Kaula (1977); Cameron (1983); Kerridge (1980); Arvidson, Goettel, and Hohenberg (1980); Gautier and Owen (1983); Pollack and Black (1979).
[a]Ratios of numbers of atoms.

Appendix H

INVENTORY OF VOLATILES ON THE TERRESTRIAL PLANETS[a,b]

	VENUS	EARTH	MARS
H_2O	1×10^{-7} (atmosphere)	2.6×10^{-4} (oceans)	1×10^{-7} (north polar cap) 2.1×10^{-12} (atmosphere) 2×10^{-6} to 5×10^{-6} (surface) 2×10^{-6}–5×10^{-6} (total)
N	2×10^{-6} (atmosphere)	6.6×10^{-7} (atmosphere) 2.04×10^{-10} (fixed) 2.34×10^{-7} (sediments) 8.90×10^{-7} (total)	6.2×10^{-10} (atmosphere) 2.4×10^{-8} (surface rocks) 2.4×10^{-8} (total) (1.3×10^{-7} initial outgassing)
CO_2	9.5×10^{-5} (atmosphere)	4.1×10^{-10} (atmosphere) 1.47×10^{-9} (life) 4.4×10^{-9} (fossil fuel) 2.4×10^{-8} (oceans) 9.2×10^{-6} (sediments) 3.68×10^{-5} (rocks) 4.60×10^{-5} (total)	5.55×10^{-8} (atmosphere) 1×10^{-6} (absorbed on surface) 6×10^{-6}–2×10^{-5} (in soil) 7×10^{-6}–2.1×10^{-5} (total)
^{20}Ne	1×10^{-9}	9.6×10^{-12}	6.0×10^{-14}
^{36}Ar	2.5×10^{-9}	3.50×10^{-11}	2.1×10^{-13}
^{38}Ar	4.5×10^{-10}	6.7×10^{-12}	4.2×10^{-14}
^{40}Ar	2.7×10^{-8}	1.06×10^{-7}	6.6×10^{-9}
^{84}Kr	4.0×10^{-10}	3.0×10^{-12}	1.9×10^{-14}
^{129}Xe	—	9.0×10^{-14}	1.1×10^{-14}
^{132}Xe	—	3.7×10^{-12}	1.0×10^{-14}

SOURCE: Adapted from Cameron (1983); Phillips et al. (1981); Arvidson, Goettel, and Hohenberg (1980); Fanale and Cannon (1979); Pollack and Black (1979).
[a]The planet Mercury is not believed to possess any volatiles.
[b]Mass of volatile/mass of planet.

426

Bibliography

ABELL, G. O. (1975). *Exploration of the Universe.* 3d ed. Holt, Rinehart & Winston, New York.

ACUÑA, M. H., AND N. F. NESS (1980). "The Magnetic Field of Saturn: Pioneer 11 Observations," *Science,* **207,** 444–446.

AKSNES, K., AND F. A. FRANKLIN (1978). "The Evidence for Faint Satellites of Saturn Reexamined," *Icarus,* **36,** 107–118.

ALEXANDROV, YU. N., ET AL. (1986). "Venus: Detailed Mapping of Maxwell Montes Region," *Science,* **231,** 1271–1273.

ALLENBY, R. J. (1970). "Lunar Orbital Science," *Space Science Reviews,* **11,** 5–53.

ALVAREZ, L. W., ET AL. (1980). "Extraterrestrial Cause for the Cretaceous–Tertiary Extinction," *Science,* **208,** 1095–1108.

ARVIDSON, R. E., A. B. BINDER, AND K. L. JONES (1978). "The Surface of Mars," *Scientific American,* **238**(3), 76–89.

ARVIDSON, R. E., K. A. GOETTEL, AND C. M. HOHENBERG (1980). "A Post-Viking View of Martian Geologic Evolution," *Reviews of Geophysics and Space Physics,* **18,** 565–603.

ATREYA, S. K., AND T. M. DONAHUE (1982). "The Atmosphere and Ionosphere of Jupiter," *Vistas in Astronomy,* **25,** 315–335.

ATREYA, S. K., T. M. DONAHUE, AND M. C. FESTOU (1981). "Jupiter: Structure and Composition of the Upper Atmosphere," *Astrophysical Journal,* **247,** L43–L47.

BAKER, D. (1974–75). "Report from Jupiter, Parts 1 and 2," *Spaceflight,* **16,** 140–144; **17,** 102–107.

BAKER, D. (1975). "Mariner-Venus–Mercury 1973 Project History," *Spaceflight,* **17,** 131–133, 191–194, 298–301.

BALSIGER, H., ET AL. (1986). "Ion Composition and Dynamics at Comet Halley," *Nature,* **321,** 330–334.

BAMBACH, R. K., C. R. SCOTESE, AND A. M. ZIEGLER (1980). "Before Pangaea: The Geographies of the Paleozoic World," *American Scientist,* **68,** 26–38.

BARATH, F. T., ET AL. (1963). "Mariner II: Preliminary Reports on Measurements of Venus, Microwave Radiometer," *Science,* **139,** 908–909.

BAUGHER, J. F. (1985). *On Civilized Stars: The Search for Intelligent Life in Outer Space.* Prentice–Hall, Englewood Cliffs, N.J.

BEATTY, J. K. (1985). "A Radar Tour of Venus," *Sky and Telescope,* **69,** 507–510.

BEATTY, J. K. (1986). "A Place Called Uranus," *Sky and Telescope,* **71,** 333–337.

BEATTY, J. K., B. O'LEARY, AND A. CHAIKIN, EDS. (1981). *The New Solar System.* Sky Publishing, Cambridge, Mass.

BEEBE, R. (1983). "Planetary Atmospheres," *Reviews of Geophysics and Space Physics,* **21,** 143–151.

BERRY, R. (1986). "Voyager: Discovery at Uranus," *Astronomy,* **14**(5), 6–22.

BERRY, R., AND R. TALCOTT (1986). "What Have We Learned from Comet Halley?" *Astronomy,* **14**(9), 6–22.

BIEMANN, K., ET AL. (1976). "The Atmosphere of Mars Near the Surface: Isotope Ratios and Upper Limits on Noble Gases," *Science,* **194,** 70–72.

BINDER, A. B. (1982). "The Moon: Its Figure and Orbital Evolution," *Geophysical Research Letters,* **9,** 33–36.

BINDER, A. B., ET AL. (1977). "The Geology of the Viking Lander 1 Site," *Journal of Geophysical Research,* **82,** 4439–4451.

BINZEL, R. P. (1984). "The Rotation of Small Asteroids," *Icarus,* **57,** 294–306.

BINZEL, R. P., ET AL. (1985). "The Detection of Eclipses in the Pluto–Charon System," *Science,* **228,** 1193–1195.

BIRMINGHAM, T. J. (1983). "The Jovian Magnetosphere," *Reviews of Geophysics and Space Physics,* **21,** 375–389.

BLAMONT, J. E., ET AL. (1986). "Implications of the VEGA Balloon Results for Venus Atmospheric Dynamics," *Science,* **231,** 1422–1425.

BLASIUS, K. R., ET AL. (1977). "Geology of the Valles Marineris: First Analysis of Imaging from the Viking I Orbiter Primary Mission," *Journal of Geophysical Research,* **82,** 4067–4091.

BOSS, A. P. (1986). "The Origin of the Moon," *Science,* **231,** 341–345.

BRATT, S. R., AND S. C. SOLOMON (1985). "The Evolution of Impact Basins: Cooling, Subsidence, and Thermal Stress," *Journal of Geophysical Research,* **90**(B14), 12,415–12,433.

BRIDGE, H. S., ET AL. (1979). "Plasma Observations Near Jupiter: Inital Results from Voyager 1," *Science,* 204, 987–991.

BRIDGE, H. S., ET AL. (1986). "Plasma Observations Near Uranus: Initial Results from Voyager 2," *Science,* **233,** 89–93.

BROADFOOT, A. L., ET AL. (1979). "Extreme Ultraviolet Observations from Voyager 1 Encounter with Jupiter," *Science,* **204,** 979–982.

BROADFOOT, A. L., ET AL. (1981). "Extreme Ultraviolet Observations from Voyager 1 Encounter with Saturn," *Science,* **212,** 206–211.

BROADFOOT, A. L., ET AL. (1986). "Ultraviolet Spectrometer Observations of Uranus," *Science,* **233,** 74–79.

BROWN, J. C., AND D. W. HUGHES (1977). "Tunguska's Comet and Non-Thermal ^{14}C Production in the Atmosphere," *Nature,* **268,** 512–514.

BROWN, R. H. (1983). "The Uranian Satellites and Hyperion: New Spectrophotometry and Compositional Implications," *Icarus,* **56,** 414–425.

BROWN, R. H., AND R. N. CLARK (1984). "Surface of Miranda: Identification of Water Ice," *Icarus,* **58,** 288–292.

BROWN, R. H., AND D. P. CRUIKSHANK (1985). "The Moons of

Uranus, Neptune, and Pluto," *Scientific American*, **253**(1), 38–47.

BROWN, R. H., D. P. CRUIKSHANK, AND D. MORRISON (1982). "Diameters and Albedos of Satellites of Uranus," *Nature*, **300**, 423–425.

BROWN, W. L. (1976). "Possible Radio Emission from Uranus at 0.5 MHz," *Astrophysical Journal*, **207**, L209–L212.

BURY, J. S. (1979). "The Planet Venus," *Journal of the British Interplanetary Society*, **32**, 122–155.

BUTTERWORTH, P. S. (1984). "Physical Properties of Comets," *Vistas in Astronomy*, **27**, 361–419.

CAMERON, A. G. W. (1962). "The Formation of the Sun and Planets," *Icarus*, **1**, 13–69.

CAMERON, A. G. W. (1975). "The Origin and Evolution of the Solar System," *Scientific American*, **233**(3), 33–41.

CAMERON, A. G. W. (1983). "Origin of the Atmospheres of the Terrestrial Planets," *Icarus*, **56**, 195–201.

CAMERON, A. G. W., AND J. W. TRURAN (1977). "The Supernova Trigger for Formation of the Solar System," *Icarus*, **30**, 447–461.

CAMPBELL, D. B., ET AL. (1984). "Venus: Volcanism and Rift Formation in Beta Regio," *Science*, **226**, 167–170.

CAMPINS, H., ET AL. (1982). "A Search for Frosts in Comet Bowell (1980b)," *Astronomical Journal*, **87**, 1867–1873.

CAMPINS, H., G. H. RIEKE, AND M. J. LEBOFSKY (1983). "Ice in Comet Bowell," *Nature*, **301**, 405–406.

CANUTO, V. M., ET AL. (1982). "UV Radiation from the Young Sun and Oxygen and Ozone Levels in the Prebiological Palaeoatmosphere," *Nature*, **296**, 816–820.

CANUTO, V. M., ET AL. (1983). "The Young Sun and the Atmosphere and Photochemistry of the Early Earth," *Nature*, **305**, 281–286.

CARR, M. H. (1976). "The Volcanoes of Mars," *Scientific American*, **234**(1), 33–43.

CARR, M. H. (1983). "The Geology of the Terrestrial Planets," *Reviews of Geophysics and Space Physics*, **21**, 160–172.

CARR, M. H., ET AL. (1976). "Preliminary Results from the Viking Orbiter Imaging Experiment," *Science*, **193**, 766–776.

CARR, M. H., ET AL. (1977). "Martian Impact Craters and Emplacement of Ejecta by Surface Flows," *Journal of Geophysical Research*, **82**, 4055–4065.

CARR, M. H., ET AL. (1979). "Volcanic Features of Io," *Nature*, **280**, 729–733.

CARRIGAN, C. R., AND D. GUBBINS (1979). "The Source of the Earth's Magnetic Field," *Scientific American*, **240**(2), 118–130.

CARVER, J. H. (1981). "Prebiotic Atmosphere Oxygen Levels," *Nature*, **292**, 136–138.

CASSEN, P., R. T. REYNOLDS, AND S. J. PEALE (1979). "Is There Liquid Water on Europa?" *Geophysical Research Letters*, **6**, 731–734.

CHAIKIN, A. (1986). "Voyager Among the Ice Worlds," *Sky and Telescope*, **71**, 338–343.

CHAPMAN, C. R. (1982). *Planets of Rock and Ice: From Mercury to the Moons of Saturn*. Scribner's, New York.

CHAPMAN, C. R. (1983). "Asteroids and Comets," *Reviews of Geophysics and Space Physics*, **21**, 196–206.

CHAPMAN, R. D. (1978). *Discovering Astronomy*. Freeman, San Francisco.

CHASE, S. C., L. D. KAPLAN, AND G. NEUGEBAUER (1963).

"Mariner II: Preliminary Reports on Measurements of Venus: Infrared Radiometer," *Science*, **139**, 907–908.

CHRISTY, J. W., AND R. S. HARRINGTON (1978). "The Satellite of Pluto," *Astronomical Journal*, **83**, 1005–1008.

CHRISTY, J. W., AND R. S. HARRINGTON (1980). "The Discovery and Orbit of Charon," *Icarus*, **44**, 38–40.

CHURMS, J., J. L. ELLIOT, AND E. DUNHAM (1979). "Structure of the Uranian Upper Atmosphere," *Nature*, **282**, 195–196.

CLARK, B. C., ET AL. (1976). "Argon Content of the Martian Atmosphere at the Viking 1 Landing Site: Analysis by X-Ray Fluorescence Spectroscopy," *Science*, **193**, 804–805.

CLARK, B. S., ET AL. (1976). "Inorganic Analyses of Martian Surface Samples at the Viking Landing Sites," *Science*, **194**, 1329–1337.

CLARK, R. N., AND T. B. MCCORD (1982). "Mars Residual North Polar Cap: Earth-Based Spectroscopic Confirmation of Water Ice as a Major Constituent and Evidence for Hydrated Minerals," *Journal of Geophysical Research*, **87**, 367–370.

CLOUD, P. (1983). "The Biosphere," *Scientific American*, **249**(3), 176–189.

CLUBE, S. V. M, AND W. M. NAPIER (1984). "Terrestrial Catastrophism—Nemesis or Galaxy?" *Nature*, **311**, 635–636.

COLE, G. H. A. (1980). "The Internal Structure and Early History of the Moon," *Journal of the British Astronomical Association*, **90**, 539–559.

COLE, G. H. A. (1981). "Aspects of the Physics of Planetary Interiors," *Contemporary Physics*, **22**, 397–424.

COLLINS, S. A., ET AL. (1980). "First Voyager View of the Rings of Saturn," *Nature*, **288**, 439–442.

COMBES, M., ET AL. (1986). "Infrared Sounding of Comet Halley from Vega 1," *Nature*, **321**, 266–268.

CONSOLMAGNO, G. G. (1979). "Sulfur Volcanoes on Io," *Science*, **205**, 397–398.

COOK, A. F., T. C. DUXBURY, AND G. E. HUNT (1979). "First Results on Jovian Lightning," *Nature*, **280**, 794.

COVAULT, C. (1985). "Soviets in Houston Reveal New Lunar, Mars, Asteroid Flights," *Aviation Week and Space Technology*, **122**(13), 18–20.

COVAULT, C. (1985). "NASA Defining Mission to Return Cometary Matter to Earth in 1990s." *Aviation Week and Space Technology*, **123**(23), 115–117.

COWAN, C., C. R. ATLURI, AND W. F. LIBBY (1965). "Possible Anti-Matter Content of the Tunguska Meteor of 1908," *Nature*, **206**, 861–865.

CRAIG, R. A., L. COLIN, AND R. O. FIMMEL (1984). "Pioneer to Venus: The Multiprobe and Orbiter Missions," *Journal of the British Interplanetary Society*, **37**, 453–466.

CROWLEY, T. J. (1983). "The Geologic Record of Climatic Change," *Reviews of Geophysics and Space Physics*, **21**, 828–877.

CRUIKSHANK, D. P. (1981). "Near-Infrared Studies of the Satellites of Saturn and Uranus," *Icarus*, **41**, 246–258.

CRUIKSHANK, D. P., AND J. APT (1984). "Methane on Triton: Physical State and Distribution," *Icarus*, **58**, 306–311.

CRUIKSHANK, D. P., AND R. H. BROWN (1981). "The Uranian Satellites: Water Ice on Ariel and Umbriel," *Icarus*, **45**, 605–611.

CRUIKSHANK, D. P., R. H. BROWN, AND R. N. CLARK (1984). "Nitrogen on Triton," *Icarus,* **58,** 293–305.

CRUIKSHANK, D. P., AND P. M. SILVAGGIO (1979). "Triton: A Satellite with an Atmosphere," *Astrophysical Journal,* **233,** 1016–1020.

CRUIKSHANK, D. P., ET AL. (1979). "The Diameter and Reflectance of Triton," *Icarus,* **40,** 104–114.

CUTTING, E., J. H. KWOK, AND S. N. MOHAN (1984). "The Venus Radar Mapper Mission," *Journal of the British Interplanetary Society,* **37,** 443–452.

CUTTS, J. A., ET AL. (1976). "North Polar Regions of Mars: Imaging Results from Viking 2," *Science,* **194,** 1329–1337.

CUZZI, J., (1983). "Planetary Ring Systems," *Reviews of Geophysics and Space Physics,* **21,** 173–186.

DAVIES, J. K., ET AL. (1982). "The Classification of Asteroids," *Vistas in Astronomy,* **26,** 243–251.

DAVIS, M., P. HUT, AND R. A. MULLER (1984). "Extinction of Species by Periodic Comet Showers," *Nature,* **308,** 715–717.

DERMOTT, S. F., ED. (1978). *The Origin of the Solar System.* Wiley, New York.

DERMOTT, S. F. (1979). "Shapes and Gravitational Moments of Satellites and Asteroids," *Icarus,* **37,** 575–586.

DERMOTT, S. F., AND T. GOLD (1977). "The Rings of Uranus: Theory," *Nature,* **267,** 590–593.

DERMOTT, S. F., T. GOLD, AND A. T. SINCLAIR (1979). "The Rings of Uranus: Nature and Origin," *Astronomical Journal,* **84,** 1225–1234.

DERMOTT, S. F., A. W. HARRIS, AND C. D. MURRAY (1984). "Asteroid Rotation Rates," *Icarus,* **57,** 14–34.

DERMOTT, S. F., AND C. D. MURRAY (1982). "Asteroid Rotation Rates Depend on Diameter and Type," *Nature,* **296,** 418–421.

DERMOTT, S. F., AND C. D. MURRAY (1983). "Nature of the Kirkwood Gaps in the Asteroid Belt," *Nature,* **301,** 201–205.

DICKERSON, R. E. (1978). "Chemical Evolution and the Origin of Life," *Scientific American,* **239**(3), 70–86.

DOLLFUS, A. (1968). "The Discovery of the Tenth Satellite of Saturn," *L'Astronomie,* **82,** 253–262.

DONAHUE, T. M. (1979). "Pioneer Venus Results: An Overview," *Science,* **208,** 41–44.

DONAHUE, T. M., ET AL. (1982). "Venus Was Wet: A Measurement of the Ratio of Deuterium to Hydrogen," *Science,* **216,** 630–633.

DORMAND, J. R., AND M. M. WOOLFSON (1980). "The Origin of Pluto," *Monthly Notices of the Royal Astronomical Society,* **193,** 171–174.

DOTT, R. H., AND R. L. BATTEN (1971). *Evolution of the Earth.* McGraw–Hill, New York.

DREIBUS, G., AND H. WANKE (1985). "Mars, a Volatile-Rich Planet," *Meteoritics,* **44,** 367–381.

DUNCOMBE, R. L., AND P. K. SEIDELMANN (1980). "A History of the Determination of Pluto's Mass," *Icarus,* **44,** 12–18.

DUNHAM, E., J. L. ELLIOT, AND P. J. GIERASCH (1980). "The Upper Atmosphere of Uranus: Mean Temperature and Temperature Variations," *Astrophysical Journal,* **235,** 274–284.

EBERHART, J. (1986). "Voyager 2's Uranus: 'Totally Different'," *Science News,* **129,** 72–73.

EDDY, J. A. (1977). "The Case of the Missing Sunspots," *Scientific American,* **236**(5), 80–92.

EDENHOFER, P., ET AL. (1986). "First Results from the Giotto Radio Science Experiment," *Nature,* **321,** 355–357.

ELLIOT, J. L., AND E. DUNHAM (1979). "Temperature Structure of the Uranian Upper Atmosphere," *Nature,* **279,** 307–308.

ELLIOT, J. L., E. DUNHAM, AND D. J. MINK (1980). "The Radius and Ellipticity of Uranus from its Occultation of SAO 158687," *Astrophysical Journal,* **236,** 1026–1030.

ELLIOT, J. L., ET AL. (1981). "No Evidence of Rings Around Neptune," *Nature,* **294,** 526–529.

ELSON, B. M. (1973). "Mission Dictates New Mariner Features," *Aviation Week and Space Technology,* **99**(15), 50–56.

ESHLEMAN, V. R., G. F. LINDAL, AND G. L. TYLER (1983). "Is Titan Wet or Dry?" *Science,* **221,** 53–55.

ESHLEMAN, V. R., ET AL. (1979). "Radio Science with Voyager 1 at Jupiter: Preliminary Profiles of the Atmosphere and Ionosphere," *Science,* **204,** 976–978.

ESHLEMAN, V. R., ET AL. (1979). "Radio Science with Voyager at Jupiter: Initial Voyager 2 Results and a Voyager 1 Measure of the Io Torus," *Science,* **206,** 959–962.

ESPOSITO, L. W. (1984). "Sulfur Dioxide: Episodic Injection Shows Evidence for Active Venus Volcanism," *Science,* **223,** 1072–1074.

FANALE, F. P., AND W. A. CANNON (1979). "Mars: CO_2 Adsorption and Capillary Condensation on Clays—Significance for Volatile Storage and Atmospheric History," *Journal of Geophysical Research,* **84**(B14), 8404–8414.

FANALE, F. P., ET AL. (1979). "Significance of Absorption Features in Io's IR Reflectance Spectrum," *Nature,* **280,** 761–763.

FARMER, C. B., D. W. DAVIES, AND D. D. LaPORTE (1976). "Viking: Mars Atmospheric Water Vapor Mapping Experiment—Preliminary Report of Results," *Science,* **193,** 776–780.

FARMER, C. B., D. W. DAVIES, AND D. D. LaPORTE (1976). "Mars: Northern Summer Ice Caps—Water Vapor Observations from Viking 2," *Science,* **194,** 1339–1341.

FARMER, C. B., ET AL. (1977). "Mars: Water Vapor Observations from the Viking Orbiters," *Journal of Geophysical Research,* **82,** 4225–4248.

FECHTIG, H., AND J. RAHE (1984). "Comets and ESA's Space Mission GIOTTO to Halley's Comet," *Naturwissenschaften,* **71,** 275–293.

FILLIUS, W., W. H. IP, AND C. E. McILWAIN (1980). "Trapped Radiation Belts of Saturn: First Look," *Science,* **207,** 425–431.

FINK, U., ET AL. (1980). "Detection of a CH_4 Atmosphere on Pluto," *Icarus,* **44,** 62–71.

FLASAR, F. M. (1983). "Oceans on Titan?" *Science,* **221,** 55–57.

FOUNTAIN, J. W., AND S. M. LARSON (1978). "Saturn's Ring and Nearby Faint Satellites," *Icarus,* **36,** 92–106.

FOX, J. L., AND A. DALGARNO (1983). "Nitrogen Escape from Mars," *Journal of Geophysical Research,* **88**(A11), 9027–9032.

FRENCH, R. G., J. L. ELLIOT, AND D. A. ALLEN (1982). "Inclination of the Uranian Rings," *Nature,* **298,** 827–829.

FROUDE, D. O., ET AL. (1983). "Ion Microprobe Identification of 4100–4200Myr-old Terrestrial Zircons," *Nature,* **304,** 616–618.

GARWIN, J. B., ET AL. (1984). "Venus: The Nature of the Surface from Venera Panoramas," *Journal of Geophysical Research,* **89**(B5), 3381–3399.

GAUTIER, D., AND T. OWEN (1983). "Cosmogonical Implications of Elemental and Isotopic Abundances in Atmospheres of the Giant Planets," *Nature,* **304,** 691–694.

GENTRY, R. V. (1966). "Anti-Matter Content of the Tunguska Meteor," *Nature,* **211,** 1071–1072.

GINGERICH, O. (1978). "The Discovery of the Satellites of Mars," *Vistas in Astronomy,* **22,** 127–132.

GOLDSTEIN, R. M., AND R. L. CARPENTER (1963). "Rotation of Venus: Period Estimated from Radar Measurements," *Science,* **139,** 910–911.

GRADIE, J., AND E. TEDESCO (1982). "Compositional Structure of the Asteroid Belt," *Science,* **216,** 1405–1407.

GRARD, R. (1986). "Observations of Waves and Plasma in the Environment of Comet Halley," *Nature,* **321,** 290–291.

GRASSHOFF, R. (1985). "Comets, 'Death Stars,' and Extinctions—The Rise and Fall of a Scientific Theory," *Astronomy,* **13**(3), 18–22.

GREELEY, R., AND P. D. SPUDIS (1981). "Volcanism on Mars," *Reviews of Geophysics and Space Physics,* **19,** 13–41.

GUEST, J., ET AL. (1979). *Planetary Geology.* Wiley, New York.

HALLAM, A. (1975). "Alfred Wegener and the Hypothesis of Continental Drift," *Scientific American,* **232**(2), 88–97.

HANEL, R. A., ET AL. (1979). "Infrared Observations of the Jovian System from Voyager 2," *Science,* **206,** 952–956.

HANEL, R. A., ET AL. (1981). "Albedo, Internal Heat, and Energy Balance of Jupiter: Preliminary Results of the Voyager Infrared Experiment," *Journal of Geophysical Research,* **86**(A10), 8705–8712.

HANEL, R. A., ET AL. (1981). "Infrared Observation of the Saturnian System From Voyager 1," *Science,* **212,** 192–200.

HANEL, R. A., ET AL. (1982). "Infrared Observation of the Saturn System From Voyager 2," *Science,* **215,** 544–548.

HANEL, R. A., ET AL (1986). "Infrared Observation of the Uranian System," *Science,* **233,** 70–74.

HARRINGTON, R. S., AND J. W. CHRISTY (1980). "The Satellite of Pluto. II," *Astronomical Journal,* **85,** 168–170.

HARRINGTON, R. S., AND J. W. CHRISTY (1981). "The Satellite of Pluto. III," *Astronomical Journal,* **86,** 442–443.

HARRINGTON, R. S., AND T. C. VAN FLANDERN (1979). "The Satellites of Neptune and the Origin of Pluto," *Icarus,* **39,** 131–136.

HART, M. H. (1978). "The Evolution of the Atmosphere of the Earth," *Icarus,* **33,** 23–39.

HARTMANN, W. K. (1975). "The Smaller Bodies of the Solar System," *Scientific American,* **233**(3), 143–159.

HARTMANN, W. K. (1977). "Cratering in the Solar System," *Scientific American,* **236**(1), 84–99.

HARTMANN, W. K. (1978). *Astronomy—The Cosmic Journey.* Wadsworth, Belmont, Calif.

HARTMANN, W. K. (1983). *Moons and Planets.* 2d ed. Wadsworth, Belmont, Calif.

HARTMANN, W. K., R. MILLER, AND P. LEE (1984). *Out of the Cradle—Exploring the Frontiers Beyond Earth.* Workman, New York.

HAYES, S. H., AND M. J. S. BELTON (1977). "The Rotational Periods of Uranus and Neptune," *Icarus,* **32,** 383–401.

HAYS, J. D., J. IMBRIE, AND N. J. SHACKLETON (1976). "Variations in the Earth's Orbit: Pacemaker of the Ice Ages," *Science,* **194,** 1121–1132.

HEAD, J. W. (1976). "Lunar Volcanism in Space and Time," *Reviews of Geophysics and Space Physics,* **14,** 265–300.

HEAD, J. W., C. A. WOOD, AND T. A. MUTCH (1977). "Geologic Evolution of the Terrestrial Planets," *American Scientist,* **65,** 21–29.

HEAD, J. W., S. E. YUTER, AND S. C. SOLOMON (1981). "Topography of Venus and Earth: A Test for the Presence of Plate Tectonics," *American Scientist,* **69,** 614–623.

HENDERSON-SELLERS, A., AND J. G. COGLEY (1982). "The Earth's Early Hydrosphere," *Nature,* **298,** 832–835.

HERTZ, H. G. (1968). "Mass of Vesta," *Science,* **160,** 299–300.

HESS, S. L., ET AL. (1976). "Mars Climatology from Viking 1 After 20 Sols," *Science,* **194,** 78–81.

HILLS, J. G. (1984). "Dynamical Constraints on the Mass and Perihelion Distance of Nemesis and the Stability of its Orbit," *Nature,* **311,** 636–638.

HIRAO, K., AND T. ITOH (1986). "The Planet-A Halley Encounters," *Nature,* **321,** 294–297.

HODGE, P. W. (1974). *Concepts of Contemporary Astronomy.* McGraw–Hill, New York.

HOOD, L. L., C. T. RUSSELL, AND P. J. COLEMAN (1981). "Contour Maps of Lunar Remanent Magnetic Fields," *Journal of Geophysical Research,* **86**(B2), 1055–1069.

HOROWITZ, N. H. (1977). "The Search for Life on Mars," *Scientific American,* **237**(5), 52–61.

HOWARD, R. (1975). "The Rotation of the Sun," *Scientific American,* **232**(4), 106–114.

HOWARD, R. (1978). "The Rotation of the Sun," *Reviews of Geophysics and Space Physics,* **16,** 721–732.

HUBBARD, W. B. (1980). "Intrinsic Luminosities of the Jovian Planets," *Reviews of Geophysics and Space Physics,* **18,** 1–9.

HUBBARD, W. B. (1981). "Constraints on the Origin and Interior Structure of the Major Planets," *Philosophical Transactions of the Royal Society of London,* **A303,** 315–326.

HUBBARD, W. B. (1981). "Interiors of the Giant Planets," *Science,* **214,** 145–149.

HUBBARD, W. B., ET AL. (1980). "Interior Structure of Saturn Inferred From Pioneer 11 Gravity Data," *Journal of Geophysical Research,* **85**(A11), 5909–5916.

HUGHES, D. W. (1982). "Comets," *Contemporary Physics,* **23,** 257–283.

HUNT, B. G. (1979). "The Effects of Past Variations of the Earth's Rotation Rate on Climate," *Nature,* **281,** 188–191.

HUNTEN, D. M. (1975). "The Outer Planets," *Scientific American,* **233**(3), 131–140.

HUT, P. (1984). "How Stable is an Astronomical Clock That Can Trigger Mass Extinctions on Earth?" *Nature,* **311,** 638–641.

HUTCHISON, R. (1983). *The Search for Our Beginning.* Oxford Univ. Press, London/New York.

INGERSOLL, A. P. (1976). "The Atmosphere of Jupiter," *Space Science Reviews,* **18,** 603–639.

INGERSOLL, A. P. (1976). "The Meteorology of Jupiter," *Scientific American,* **234**(3), 46–56.

INGERSOLL, A. P. (1981). "Jupiter and Saturn," *Scientific American*, **245**(6), 90–108.

INGERSOLL, A. P. (1983). "The Atmosphere," *Scientific American*, **249**(3), 162–174.

INGERSOLL, A. P., A. R. DOBROVOLSKIS, AND B. M. JAKOSKY (1979). "Planetary Atmospheres," *Reviews of Geophysics and Space Physics*, **17**, 1722–1735.

INGERSOLL, A. P., ET AL. (1980). "Pioneer Saturn Infrared Radiometer: Preliminary Results," *Science*, **207**, 439–443.

JACKSON, A. A., AND M. P. RYAN (1973). "Was the Tungus Event Due to a Black Hole?" *Nature*, **245**, 88–89.

JAKOSKY, B. M. (1985). "The Seasonal Cycle of Water on Mars," *Space Science Reviews*, **41**, 131–200.

JAKOSKY, B. M., AND C. B. FARMER (1982). "The Seasonal and Global Behavior of Water Vapor in the Mars Atmosphere: Complete Global Results of the Viking Atmospheric Water Detector Experiment," *Journal of Geophysical Research*, **87**(B4), 2999–3019.

JASTROW, R. (1979). *Until the Sun Dies*. Norton, New York.

JASTROW, R., AND M. H. THOMPSON (1977). *Astronomy: Fundamentals and Frontiers*. Wiley, New York.

JEWITT, D. C., G. E. DANIELSON, AND S. J. SYNNOTT (1979). "Discovery of a New Jupiter Satellite," *Science*, **206**, 951.

JOHNSON, F. S. (1965). "Atmosphere of Mars," *Science*, **150**, 1445–1448.

JOHNSON, T. V., AND L. A. SODERBLOM (1983). "Io," *Scientific American*, **249**(6), 56–67.

JOHNSON, T. V., ET AL. (1979). "Volcanic Resurfacing Rates and Implications for Volatiles on Io," *Nature*, **280**, 746–750.

JOHNSTONE, A., ET AL. (1986). "Ion Flow at Comet Halley," *Nature*, **321**, 344–347.

JUDGE, D. L., AND F. M. WU (1980). "Ultraviolet Photometer Observations of the Saturnian System," *Science*, **207**, 431–434.

KAMOUN, P. G. (1982). "Comet Encke: Radar Detection of Nucleus," *Science*, **216**, 293–295.

KANEDA, E., ET AL. (1986). "Observations of Comet Halley by the Ultraviolet Imager of Suisei," *Nature*, **321**, 297–299.

KASTING, J. F., AND T. M. DONAHUE (1980). "The Evolution of Atmospheric Ozone," *Journal of Geophysical Research*, **85**(C6), 3255–3263.

KASTING, J. F., S. C. LIU, AND T. M. DONAHUE (1979). "Oxygen Levels in the Prebiological Atmosphere," *Journal of Geophysical Research*, **84**(C6), 3097–3107.

KASTING, J. F., J. B. POLLACK, AND T. P. ACKERMAN (1984). "Response of Earth's Atmosphere to Increases in Solar Flux and Implications for Loss of Water from Venus," *Icarus*, **57**, 335–355.

KASTING, J. F., AND J. C. G. WALKER (1981). "Limits on Oxygen Concentration in the Prebiological Atmosphere and the Rate of Abiotic Fixation of Nitrogen," *Journal of Geophysical Research*, **86**(C2), 1147–1158.

KAUFMAN, W. J. (1977). *Astronomy: The Structure of the Universe*. MacMillan, New York.

KAUFMANN, W. K. (1979). *Planets and Moons*. Freeman, San Francisco.

KELLER, H. U., ET AL. (1986). "First Halley Multicolour Camera Imaging Results from Giotto," *Nature*, **321**, 320–326.

KEPPER, E. (1986). "Neutral Gas Measurements of Comet Halley from Vega 1," *Nature*, **321**, 273–274.

KERR, R. A. (1986). "Voyager Finds Uranian Shepherds and a Well-Behaved Flock of Rings," *Science*, **231**, 793–796.

KERRIDGE, J. F. (1980). "Accretion of Nitrogen During the Growth of Planets," *Nature*, **283**, 183–184.

KIEFFER, H. H. (1976). "Soil and Surface Temperatures at the Viking Landing Sites," *Science*, **194**, 1344–1346.

KIEFFER, H. H., ET AL. (1976). "Infrared Thermal Mapping of the Martian Surface and Atmosphere: First Results," *Science*, **193**, 780–786.

KIEFFER, H. H., ET AL. (1976). "Martian North Pole Summer Temperatures: Dirty Water Ice," *Science*, **194**, 1341–1344.

KIEFFER, H. H., ET AL. (1976). "Temperature of the Martian Surface and Atmosphere: Viking Observation of Diurnal and Geometric Variations," *Science*, **194**, 1346–1351.

KISSEL, J., ET AL. (1986). "Composition of Comet Halley Dust Particles From Vega Observations," *Nature*, **321**, 280–282.

KISSEL, J., ET AL. (1986). "Composition of Comet Halley Dust Particles From Giotto Observations," *Nature*, **321** 336–337.

KLEIN, H. P. (1979). "The Viking Mission and the Search for Life on Mars," *Reviews of Geophysics and Space Physics*, **17**, 1655–1662.

KLIMOV, S., ET AL. (1986). "Extremely-Low-Frequency Plasma Waves in the Environment of Comet Halley," *Nature*, **321**, 292–293.

KLINGER, J. (1983). "Extraterrestrial Ice — A Review," *Journal of Physical Chemistry*, **87**, 4209-4214.

KLINGER, J. (1982). "A Possible Resurfacing Mechanism for Icy Satellites," *Nature*, **299**, 41.

KOLODNY, Y., J. F. KERRIDGE, AND I. R. KAPLAN (1980). "Deuterium in Carbonaceous Chondrites," *Earth and Planetary Science Letters*, **46**, 149–158.

KOPAL, Z. (1979). *The Realm of the Terrestrial Planets*. Wiley, New York.

KOPAL, Z. (1980). "Planetary Exploration by Spacecraft," *Contemporary Physics*, **21**, 359–380.

KORTH, A., ET AL. (1986). "Mass Spectra of Heavy Ions Near Comet Halley," *Nature*, **321**, 335–336.

KOTEL'NIKOV, V. A., ET AL. (1984). "The Maxwell Montes Region, Surveyed by the Venera 15, Venera 16 Orbiters," *Soviet Astronomy Letters*, **10**, 369–373. Translated from *Pis'ma Astronomicheskii Zhurnal*, **10**, 883–889.

KRANKOWSKY, D., ET AL. (1986). "In-Situ Gas and Ion Measurements at Comet Halley," *Nature*, **321**, 326–329.

KRASNOPOLSKY, V. A., ET AL. (1986). "Spectroscopic Study of Comet Halley by the Vega 2 Three-Channel Spectrometer," *Nature*, **321**, 269–271.

KRIMIGIS, S. M. (1981). "A Post-Voyager View of Jupiter's Magnetosphere," *Endeavour*, **5**, 50–60.

KRIMIGIS, S. M., ET AL. (1979). "Low-Energy Charged-Particle Environment at Jupiter: A First Look," *Science*, **204**, 998–1003.

KRIMIGIS, S. M., ET AL. (1979). "Hot Plasma Environment at Jupiter: Voyager 2 Results," *Science*, **206**, 977–984.

KRIMIGIS, S. M., ET AL. (1986). "The Magnetosphere of Uranus: Hot Plasma and Radiation Environment," *Science*, **233**, 97–102.

KUMAR, S. (1979). "The Stability of an SO_2 Atmosphere on Io," *Nature*, **280**, 758–760.

LANE, A. L., ET AL. (1982). "Photopolarimetry from Voyager 2: Preliminary Results on Saturn, Titan, and the Rings," *Science*, **215**, 537–543.

LANE, A. L, ET AL. (1986). "Photometry from Voyager 2: Initial Results from the Uranian Atmosphere, Satellites, and Rings," *Science*, **233**, 65–70.

LARSON, H. P., ET AL. (1979). "Remote Spectroscopic Identification of Carbonaceous Chondrite Mineralogies: Applications to Ceres and Pallas," *Icarus*, **39**, 257–271.

LAWTON, A. T. (1980). "Asteroid Chiron — The First of A Few?" *Spaceflight*, **20**, 312–313.

LECACHEUX, J., ET AL. (1980). "A New Satellite of Saturn: Dione B," *Icarus*, **43**, 111–115.

LEE, T. (1979). "New Isotopic Clues to Solar System Formation," *Reviews of Geophysics and Space Physics*, **17**, 1591–1611.

LEIGHTON, R. B., ET AL. (1965). "Mariner 4 Photography of Mars: Initial Results," *Science*, **149**, 627–630.

LEIGHTON, R. B., ET AL. (1969). "Mariner 6 and 7 Television Pictures: Preliminary Analysis," *Science*, **166**, 49–67.

LENOROVITZ, J. M. (1986). "Europe Proposes Comet Sample Return Mission for 1990s," *Aviation Week and Space Technology*, **124**(1), 57–59.

LENOROVITZ, J. M. (1986). "Soviets Urge International Effort Leading to Manned Mars Mission," *Aviation Week and Space Technology*, **124**(12), 76–77.

LEOVY, C. B. (1977). "The Atmosphere of Mars," *Scientific American*, **237**(1), 34–43.

LERMAN, J. C., W. G. MOOK, AND J. C. VOGEL (1967). "Effect of the Tunguska Meteor and Sunspots on Radiocarbon in Tree Rings," *Nature*, **216**, 990–991.

LEVASSEUR-REGOURD, A. C., ET AL. (1986). "Optical Probing of Comet Halley From the Giotto Spacecraft," *Nature*, **321**, 341–344.

LEVINE, J. S. (1982). "The Photochemistry of the Paleoatmosphere," *Journal of Molecular Evolution*, **18**, 161–172.

LEVINE, J. S., AND F. ALLARIO (1982). "The Global Troposphere: Biogeochemical Cycles, Chemistry, and Remote Sensing," *Environmental Monitoring and Assessment*, **1**, 263–306.

LINKIN, V. M., ET AL. (1986). "VEGA Balloon Dynamics and Vertical Winds in the Venus Middle Cloud Region," *Science*, **231**, 1417–1419.

LINKIN, V. M., ET AL. (1986). "Thermal Structure of the Venus Atmosphere in the Middle Cloud Layer," *Science*, **231**, 1420–1422.

MCCAULEY, J. F., B. A. SMITH, AND L. A. SODERBLOM (1979). "Erosional Scarps on Io," *Nature*, **280**, 736–738.

MCCORD, T. B., J. B. ADAMS, AND T. V. JOHNSON (1970). "Asteroid Vesta: Spectral Reflectivity and Compositional Implications," *Science*, **168**, 1445–1447.

MCDONNELL, J. A. M., ET AL. (1986). "Dust Density and Mass Distribution Near Comet Halley From Giotto Observations," *Nature*, **321**, 338–341.

MCELROY, M. B., T. Y. KONG, AND Y. L. YUNG (1977). "Photochemistry and Evolution of Mars's Atmosphere: A Viking Perspective," *Journal of Geophysical Research*, **82**, 4379–4388.

MCELROY, M. B., AND M. J. PRATHER (1981). "Noble Gases in the Terrestrial Planets," *Nature*, **293**, 535–539.

MCELROY, M. B., AND Y. L. YUNG (1976). "Isotopic Composition of the Martian Atmosphere," *Science*, **194**, 68–70.

MCELROY, M. B., Y. L. YUNG, AND A. O. NIER (1976). "Isotopic Composition of Nitrogen: Implications for the Past History of Mars's Atmosphere," *Science*, **194**, 70–72.

MCEWEN, A. S., AND L. A. SODERBLOM (1983). "Two Classes of Volcanic Plumes on Io," *Icarus*, **55**, 191–217.

MCKENNA-LAWLOR, S., ET AL. (1986). "Energetic Ions in the Environment of Comet Halley," *Nature*, **321**, 347–349.

MCKINNON, W. B. "Geology of Icy Satellites." In J. Klinger (ed.), *Ices in the Solar System*. D. Reidel: Dordrecht, 1985.

MCNUTT, R. L., ET AL. (1979). "Departure from Rigid Co-Rotation of Plasma in Jupiter's Dayside Magnetosphere," *Nature*, **280**, 803.

MCSWEEN, H. Y., AND E. M. STOLPER (1980). "Basaltic Meteorites," *Scientific American*, **242**(6), 54–63.

MARGULIS, L., AND J. E. LOVELOCK (1974). "Biological Modulation of the Earth's Atmosphere," *Icarus*, **21**, 471–489.

MARSDEN, B. G. (1980). "Saturn's Satellite Situation," *Journal of Geophysical Research*, **85**(A11), 5957–5958.

MASSON, P. (1984). "Comparative Geology of the Satellites of the Giant Planets," *Space Science Reviews*, **38**, 281–324.

MASURSKY, H. (1982). "The Moon After Apollo," *Endeavour*, **6**, 48–58.

MASURSKY, H., ET AL. (1977). "Classification and Time of Formation of Martian Channels Based on Viking Data," *Journal of Geophysical Research*, **82**, 4016–4038.

MASURSKY, H., ET AL. (1979). "Preliminary Geological Mapping of Io," *Nature*, **280**, 725–729.

MAYR, E. (1978). "Evolution," *Scientific American*, **239**(3), 47–55.

MAZETS, E. P., ET AL. (1986). "Comet Halley Dust Environment from SP-2 Detector Measurements," *Nature*, **321**, 276–278.

MAZUR, P., ET AL. (1978). "Biological Implications of the Viking Mission to Mars," *Space Science Reviews*, **22**, 3–34.

MENZEL, D. H., F. L. WHIPPLE, AND G. DE VAUCOULEURS (1970). *Survey of the Universe*. Prentice–Hall, Englewood Cliffs, N.J.

MILLER, R., AND W. K. HARTMANN (1981). *The Grand Tour: A Traveler's Guide to the Solar System*. Workman, New York.

MITTON, S. (1977). *The Cambridge Encyclopaedia of Astronomy*. Crown, New York.

MOORE, J. M. (1984). "The Tectonic and Volcanic History of Dione," *Icarus*, **59**, 205–220.

MOORE, J. M, V. M. HORNER, AND R. GREELEY (1985). "The Geomorphology of Rhea: Implications for Geologic History and Surface Processes," *Journal of Geophysical Research*, **90**, C785–C795.

MOORE, P. (1976). *New Guide to the Moon*. Norton, New York.

MOORE, P., AND G. HUNT (1983). *Atlas of the Solar System*. Rand McNally, Chicago.

MORABITO, L. A., ET AL. (1979). "Discovery of Currently-Active Extraterrestrial Volcanism," *Science*, **204**, 972.

MOREELS, G., ET AL. (1986). "Near-Ultraviolet and Visible Spectrophotometry of Comet Halley from Vega 2," *Nature*, **321**, 271–273.

MORRISON, D. (1977). "Asteroid Sizes and Albedos," *Icarus*, **21**, 185–220.

MUHLEMAN, D. O., G. S. ORTON, AND G. L. BERGE (1979). "A Model of the Venus Atmosphere From Radio, Radar, and

Occultation Observations," *Astrophysical Journal,* **234,** 733–745.

MUKAI, T., ET AL. (1986). "Plasma Observations by Suisei of Solar-Wind Interaction with Comet Halley," *Nature,* **321,** 299–303.

MURRAY, B. C. (1975). "Mercury," *Scientific American,* **233**(3), 59–68.

MUTCH, T. A., AND J. W. HEAD (1975). "The Geology of Mars: A Brief Review of Some Recent Results," *Reviews of Geophysics and Space Physics,* **13,** 411–416.

MUTCH, T. A., AND R. S. SAUNDERS (1976). "The Geologic Development of Mars: A Review," *Space Science Reviews,* **19,** 3–57.

MUTCH, T. A., ET AL. (1976). "The Surface of Mars: The View from the Viking 1 Lander," *Science,* **193,** 791–801.

MUTCH, T. A., ET AL. (1976). "The Surface of Mars: The View from the Viking 2 Lander," *Science,* **194,** 1277–1283.

NESS, N. F. (1978). "Mercury: Magnetic Field and Interior," *Space Science Reviews,* **21,** 527–553.

NESS, N. F., ET AL. (1979). "Magnetic Field Studies at Jupiter by Voyager 1: Preliminary Results," *Science,* **204,** 982–987.

NESS, N. F., ET AL. (1979). "Magnetic Field Studies at Jupiter by Voyager 2: Preliminary Results," *Science,* **206,** 966–972.

NESS, N. F., ET AL. (1982). "Magnetic Field Studies by Voyager 2: Preliminary Results at Saturn," *Science,* **215,** 558–563.

NESS, N. F., ET AL. (1986). "Magnetic Fields at Uranus," *Science,* **233,** 85–89.

NEUBAUER, F. M., ET AL. (1986). "First Results from the Giotto Magnetometer Experiment at Comet Halley," *Nature,* **321,** 352–355.

NEUKUM, G. (1985). "Cratering Records of the Satellites of Jupiter and Saturn," *Advances in Space Research,* **5**(8), 107–116.

NEUKUM, G., AND K. HILLER (1981). "Martian Ages," *Journal of Geophysical Research,* **86**(B4), 3097–3121.

NEUKUM, G., AND D. U. WISE (1976). "Mars: A Standard Crater Curve and Possible New Time Scale," *Science,* **194,** 1381–1387.

NEWMAN, M. J., AND R. T. ROOD (1977). "Implications of Solar Evolution for the Earth's Early Atmosphere," *Science,* **198,** 1035–1037.

NIER, A. O., ET AL. (1976). "Composition and Structure of the Martian Atmosphere: Preliminary Results from Viking 1," *Science,* **193,** 786–788.

OBERG, J. E. (1975). "Russia Meant to Win the 'Moon Race'," *Spaceflight,* **17,** 163–171.

OPP, A. G. (1980). "Scientific Results from the Pioneer Saturn Encounter: Summary," *Science,* **207,** 401–403.

OWEN, T. (1982). "Titan," *Scientific American,* **246**(2), 98–109.

OWEN, T., AND K. BIEMANN (1976). "Composition of the Atmosphere at the Surface of Mars: Detection of Argon 36 and Preliminary Analysis," *Science,* **193,** 801–803.

OWEN, T., AND R. J. TERRILE (1981). "Colors on Jupiter," *Journal of Geophysical Research,* **86**(A10), 8797–8814.

OWEN, T., ET AL. (1976). "The Atmosphere of Mars: Detection of Krypton and Xenon," *Science,* **194,** 1293–1295.

OWEN, T., ET AL. (1979). "Jupiter's Rings," *Nature,* **281,** 442–446.

OYA, H., ET AL. (1986). "Discovery of Cometary Kilometric Radiations and Plasma Waves at Comet Halley," *Nature,* **321,** 307–310.

OYAMA, K., ET AL. (1986). "Was the Solar Wind Decelerated by Comet Halley?" *Nature,* **321,** 310–313.

PANG, K. D., ET AL. (1980). "Spectral Evidence for a Carbonaceous Chondrite Surface Composition on Deimos," *Nature,* **283,** 277–278.

PARKER, E. N. (1975). "The Sun," *Scientific American,* **233**(3), 43–50.

PARMENTIER, E. M., ET AL. (1982). "The Tectonics of Ganymede," *Nature,* **295,** 290–293.

PASACHOFF, J. M. (1973). *Astronomy Now.* Saunders, Philadelphia.

PASACHOFF, J. M., AND M. L. KUTNER (1978). *University Astronomy.* Saunders, Philadelphia.

PEALE, S. J., P. CASSEN, AND R. T. REYNOLDS (1979). "Melting of Io by Tidal Dissipation," *Science,* **203,** 892–894.

PEARL, J., ET AL. (1979). "Identification of Gaseous SO_2 and New Upper Limits for Other Gases on Io," *Nature,* **280,** 755–758.

PETTENGILL, G. H., D. B. CAMPBELL, AND H. MASURSKY (1980). "The Surface of Venus," *Scientific American,* **243**(2), 54–65.

PHILLIPS, R. J., AND M. C. MALIN (1980). "Ganymede: A Relationship Between Thermal History and Crater Statistics," *Science,* **210,** 185–188.

PHILLIPS, R. J., ET AL. (1981). "Tectonics and Evolution of Venus," *Science,* **212,** 879–887.

PIETERS, C. M., ET AL. (1985). "The Nature of Crater Rays: The Copernicus Example," *Journal of Geophysical Research,* **90**(B14), 12,393–12,413.

PLESCIA, J. B., AND J. M. BOYCE (1982). "Crater Densities and Geological Histories of Rhea, Dione, Mimas, and Tethys," *Nature,* **295,** 285–290.

PLESCIA, J. B., AND J. M. BOYCE (1983). "Crater Numbers and Geological Histories of Iapetus, Enceladus, Tethys, and Hyperion," *Nature,* **301,** 666–670.

POIRIER, J. P. (1982). "Rheology of Ices: A Key to the Tectonics of the Ice Moons of Jupiter and Saturn," *Nature,* **299,** 638–640.

POLLACK, J. B. (1975). "Mars," *Scientific American,* **233**(3), 107–117.

POLLACK, J. B. (1978). "The Rings of Saturn," *American Scientist,* **66,** 30–37.

POLLACK, J. B., AND D. C. BLACK (1979). "Implications of the Gas Compositional Measurements of Pioneer Venus for the Origin of Planetary Atmospheres," *Science,* **205,** 56–59.

POLLACK, J. B., J. A. BURNS, AND M. E. TAUBER (1979). "Gas Drag in Primordial Circumplanetary Envelopes: A Mechanism for Satellite Capture," *Icarus,* **37,** 587–611.

POLLACK, J. B., AND J. N. CUZZI (1981). "Rings in the Solar System," *Scientific American,* **245**(5), 104–129.

PRESTON, R. A., ET AL. (1986). "Determination of Venus Winds by Ground-Based Radio Tracking of the VEGA Balloons," *Science,* **231,** 1414–1416.

PROTHEROE, W. M., E. R. CAPRIOTTI, AND G. H. NEWSON (1979). *Exploring the Universe.* Merrill, Columbus, Ohio.

RAMPINO, M. R., AND R. B. STOTHERS (1984). "Terrestrial Mass Extinctions, Cometary Impacts, and the Sun's Motion Perpendicular to the Galactic Plane," *Nature,* **308,** 709–712.

RANDOLPH, A. (1984). "JPL Set to Integrate Galileo Jupiter Probe with Orbiter," *Aviation Week and Space Technology,* **120**(9), 41–43.

RASOOL, S. I., D. M. HUNTEN, AND W. M. KAULA (1977). "What the Exploration of Mars Tells Us about Earth," *Physics Today,* **23**(7), 23–32.

RAUP, D. M., AND J. J. SEPKOSKI (1984). "Periodicity of Extinctions in the Geologic Past," *Proceedings of the National Academy of Sciences of the USA,* **81,** 801–805.

REINHARD, R. (1986). "The Giotto Encounter With Comet Halley," *Nature,* **321,** 313–318.

REITSEMA, H. J. (1978). "Photometric Confirmation of the Encke's Division in Saturn's Ring," *Nature,* **272,** 601–602.

REITSEMA, H. J., B. A. SMITH, AND S. M. LARSON (1980). "A New Saturnian Satellite Near Dione's L4 Point," *Icarus,* **43,** 116–119.

REITSEMA, H. J., F. VILAS, AND B. A. SMITH (1983). "A Charge-Coupled Device Observation of Charon," *Icarus,* **56,** 75–79.

REME, H., ET AL. (1986). "Comet Halley–Solar Wind Interaction from Electron Measurements Aboard Giotto," *Nature,* **321,** 349–352.

REYNOLDS, R. T., AND P. M. CASSEN (1979). "On the Internal Structure of the Major Satellites of the Outer Planets," *Geophysical Research Letters,* **6,** 121–124.

REYNOLDS, R. T., ET AL. (1983). "On the Habitability of Europa," *Icarus,* **56,** 246–254.

RIEDLER, W., ET AL. (1986). "Magnetic Field Observations in Comet Halley's Coma," *Nature,* **321,** 288–289.

RIEKE, G. H., ET AL. (1981). "Unidentified Features in the Spectrum of Triton," *Nature,* **294,** 59–60.

ROSS, M. (1981). "The Ice Layer in Uranus and Neptune — Diamonds in the Sky?" *Nature,* **292,** 435–436.

RUSSELL, C. T. (1979). "The Martian Magnetic Field," *Physics of the Earth and Planetary Interiors,* **20,** 237–246.

RUSSELL, C. T. (1980). "Planetary Magnetism, *Reviews of Geophysics and Space Physics,* **18,** 77–106.

SAGAN, C. (1975). "The Solar System," *Scientific American,* **233**(3), 23–31.

SAGAN, C. (1979). "Sulphur Flows on Io," *Nature,* **280,** 750–753.

SAGAN, C., AND A. DRUYAN (1985). *Comet.* Random House, New York.

SAGDEEV, R. Z., ET AL. (1986). "Overview of VEGA Venus Balloon in-situ Meteorological Measurements," *Science,* **231,** 1411–1414.

SAGDEEV, R. Z., ET AL. (1986). "Vega Spacecraft Encounters with Comet Halley," *Nature,* **321,** 259–262.

SAGDEEV, R. Z., ET AL. (1986). "Television Observations of Comet Halley from Vega Spacecraft," *Nature,* **321,** 262–266.

SAITO, T., ET AL. (1986). "Interaction Between Comet Halley and the Interplanetary Magnetic Field Observed by Sakigake," *Nature,* **321,** 303–307.

SAMUELSON, R. E., ET AL. (1983). "CO_2 on Titan," *Journal of Geophysical Research,* **88**(A11), 8709–8715.

SANDEL, B. R., ET AL. (1982). "Extreme Ultraviolet Observations from the Voyager 2 Encounter with Saturn," *Science,* **215,** 548–553.

SCHARDT, A. W. (1983). "The Magnetosphere of Saturn," *Reviews of Geophysics and Space Physics,* **21,** 390–402.

SCHENK, P. M., AND W. B. McKINNON (1985). "Dark Halo Craters and the Thickness of Grooved Terrain on Ganymede," *Journal of Geophysical Research,* **90,** C775–C783.

SCHOPF, J. W. (1978). "The Evolution of the Earliest Cells," *Scientific American,* **239**(3), 110–138.

SCHRAMM, D. N., AND R. N. CLAYTON (1978). "Did a Supernova Trigger the Formation of the Solar System?" *Scientific American,* **239**(4), 124–139.

SCHROLL, A., H. F. HAUPT, AND H. M. MAITZEN (1976). "Rotation and Photometric Characteristics of Pallas," *Icarus,* **27,** 147–156.

SCHUBART, J. (1974). "The Masses of the First Two Asteroids," *Astronomy and Astrophysics,* **30,** 289–292.

SCHUBERT, G., AND C. COVEY (1981). "The Atmosphere of Venus," *Scientific American,* **245**(1), 66–74.

SCHUBERT, G., D. J. STEVENSON, AND K. ELLSWORTH (1981). "Internal Structures of the Galilean Satellites," *Icarus,* **47,** 46–59.

SCHWARTZ, R. D., AND P. B. JAMES (1984). "Periodic Mass Extinctions and the Sun's Oscillation About the Galactic Plane," *Nature,* **308,** 712–713.

SEIDELMANN, P. K., AND R. S. HARRINGTON (1984). "Dynamics of Saturn's E Ring," *Icarus,* **58,** 169–177.

SEIFF, A., ET AL. (1979). "Thermal Contrast in the Atmosphere of Venus: Initial Appraisal from Pioneer Venus Probe Data," *Science,* **205,** 46–49.

SEKANINA, Z. (1984). "Disappearance and Disintegration of Comets," *Icarus,* **58,** 81–100.

SHAPIRO, I. I., D. B. CAMPBELL, AND W. M. DeCAMPLI (1979). "Nonresonance Rotation of Venus?" *Astrophysical Journal,* **230,** L123–L126.

SHARPLESS, B. P. (1945). "Secular Accelerations in the Longitudes of the Satellites of Mars," *Astronomical Journal,* **51,** 185–186.

SHOWALTER, M. R., ET AL. (1985). "Discovery of Jupiter's 'Gossamer' Ring," *Nature,* **316,** 526–528.

SIEVER, R. (1975). "The Earth," *Scientific American,* **233**(3), 83–90.

SIMPSON, J. A., ET AL. (1986). "Dust Counter and Mass Analyzer (DUCMA) Measurements of Comet Halley's Coma from Vega Spacecraft," *Nature,* **321,** 278–280.

SINCLAIR, A. T. (1978). "The Orbits of the Satellites of Mars," *Vistas in Astronomy,* **22,** 133–140.

SISCOE, G. L., AND J. A. SLAVIN (1979). "Planetary Magnetospheres," *Reviews of Geophysics and Space Physics,* **17,** 1677–1693.

SMITH, B. A., ET AL. (1979). "The Jupiter System Through the Eyes of Voyager 1," *Science,* **204,** 951–972.

SMITH, B. A., ET AL. (1979). "The Galilean Satellites and Jupiter: Voyager 2 Imaging Science Results," *Science,* **206,** 927–950.

SMITH, B. A., ET AL. (1979). "The Role of SO_2 in Volcanism on Io," *Nature,* **280,** 738–743.

SMITH, B. A., ET AL. (1981). "Encounter with Saturn: Voyager 1 Imaging Science Results," *Science,* **212,** 163–191.

SMITH, B. A., ET AL. (1982). "A New Look at the Saturn System: The Voyager 2 Images," *Science,* **215,** 504–537.

SMITH, B. A, ET AL. (1986). "Voyager 2 in the Uranian System: Imaging Science Results," *Science,* **233,** 43–64.

SMITH, E. J., ET AL. (1963). "Mariner II: Preliminary Reports on

Measurements of Venus—Magnetic Field," *Science,* **139,** 910–911.

SMITH, G. R., ET AL. (1982). "Titan's Upper Atmosphere: Composition and Temperature from the EUV Solar Occultation Results," *Journal of Geophysical Research,* **87**(A3), 1351–1359.

SMOLUCHOWSKI, R. (1979). "Planetary Rings," *Comments on Astrophysics,* **8**(3), 69–78.

SMOLUCHOWSKI, R. (1979). "The Ring Systems of Jupiter, Saturn, and Uranus," *Nature,* **280,** 377–378.

SODERBLOM, L. A. (1979). "The Galilean Moons of Jupiter," *Scientific American,* **242**(1), 88–99.

SODERBLOM, L. A., AND T. V. JOHNSON (1982). "The Moons of Saturn," *Scientific American,* **246**(1), 100–116.

SOFFEN, G. A. (1976). "Scientific Results of the Viking Missions," *Science,* **194,** 1274–1276.

SOFFEN, G. A. (1977). "The Viking Project," *Journal of Geophysical Research,* **82,** 3959–3983.

SOFFEN, G. A., AND C. W. SNYDER (1976). "The First Viking Mission to Mars," *Science,* **193,** 759–766.

SOLOMON, S. C., AND J.W. HEAD (1980). "Lunar Mascon Basins: Lava Filling, Tectonics, and Evolution of the Lithosphere," *Reviews of Geophysics and Space Physics,* **18,** 107–141.

SOLOMON, S. C., AND J. W. HEAD (1984). "Venus Banded Terrain: Tectonic Models for Band Formation and Their Relationship to Lithospheric Thermal Structure," *Journal of Geophysical Research,* **89**(B8), 6885–6897.

SOLOMON, S. C., AND S. K. STEPHENS (1982). "On Venus Impact Basins: Viscous Relaxation of Topographic Relief," *Journal of Geophysical Research,* **87**(B9), 7763–7771.

SOMOGYI, A. J., ET AL. (1986). "First Observation of Energetic Particles Near Comet Halley," *Nature,* **321,** 285–288.

SQUYRES, S. W. (1982). "The Evolution of Tectonic Features on Ganymede," *Icarus,* **52,** 545–559.

SQUYRES, S. W., AND M. H. CARR (1986). "Geomorphic Evidence for the Distribution of Ground Ice on Mars," *Science,* **231,** 249–252.

SQUYRES, S. W., R. T. REYNOLDS, AND P. M. CASSEN (1983). "The Evolution of Enceladus," *Icarus,* **53,** 319–331.

SQUYRES, S. W., ET AL. (1983). "Liquid Water and Active Resurfacing on Europa," *Nature,* **301,** 225–226.

STEVENSON, D. J. (1981). "Models of the Earth's Core," *Science,* **214,** 611–619.

STONE, E. C., AND A. L. LANE (1979). "Voyager 1 Encounter with the Jovian System," *Science,* **204,** 945–948.

STONE, E. C., AND A. L. LANE (1979). "Voyager 2 Encounter with the Jovian System," *Science,* **206,** 925–927.

STONE, E. C. AND E. D. MINER (1981). "Voyager 1 Encounter with the Saturnian System," *Science,* **212,** 159–163.

STONE, E. C., AND E. D. MINER (1982). "Voyager 2 Encounter with the Saturnian System," *Science,* **215,** 499–504.

STONE, E. C., AND E. D. MINER (1986). "The Voyager 2 Encounter with the Uranian System," *Science,* **233,** 39–43.

STONE, E. C., ET AL. (1986). "Energetic Charged Particles in the Uranian Magnetosphere," *Science,* **233,** 93–97.

STROBEL, D. F. (1979). "The Ionospheres of the Major Planets," *Reviews of Geophysics and Space Physics,* **17,** 1913–1922.

STROM, R. G. (1979). "Mercury: A Post-Mariner 10 Assessment," *Space Science Reviews,* **24,** 3–70.

STROM, R. G., A. WORONOW, AND M. GURNIS (1981). "Crater Populations on Ganymede and Callisto," *Journal of Geophysical Research,* **86**(A10), 8659–8674.

STROM, R. G., ET AL. (1979). "Volcanic Eruption Plumes on Io," *Nature,* **280,** 733–736.

STROM, R. G., ET AL. (1981). "Volcanic Eruptions on Io," *Journal of Geophysical Research,* **86**(A10), 8593–8620.

SYNNOTT, S. P. (1980). "1979J2: The Discovery of a Previously Unknown Jovian Satellite," *Science,* **210,** 786–788.

SYNNOTT, S. P. (1981). "1979J3: Discovery of a Previously Unknown Satellite of Jupiter," *Science,* **212,** 1392.

SYNNOTT, S. P. (1984). "Orbits of the Small Inner Satellites of Jupiter," *Icarus,* **58,** 178–181.

SYNNOTT, S. P., ET AL. (1981). "Orbits of the Small Satellites of Saturn," *Science,* **212,** 191–192.

TAYLOR, F. W. (1980). "Structure and Meteorology of the Middle Atmosphere of Venus: Infrared Remote Sensing from the Pioneer Orbiter," *Journal of Geophysical Research,* **85**(A13), 7963–8006.

TAYLOR, S. R. (1979). "Structure and Evolution of the Moon," *Nature,* **281,** 105–110.

TAYLOR, S. R. (1982). *Planetary Science: A Lunar Perspective.* Lunar and Planetary Institute, Houston.

TEDESCO, E. F. (1980). "Rotational Properties of Asteroids: Correlations and Selection Effects," *Icarus,* **43,** 33–50.

THOMAS, P. (1979). "Surface Features of Phobos and Deimos," *Icarus,* **40,** 223–243.

THOMAS, P., AND J. VEVERKA (1985). "Hyperion: Analysis of Voyager Observations," *Icarus,* **64,** 414–424.

THOMAS, P., ET AL. (1983). "Phoebe: Voyager 2 Observations," *Journal of Geophysical Research,* **88**(A11), 8736–8742.

THOMAS, P., ET AL. (1983). "Saturn's Small Satellites: Voyager Imaging Results," *Journal of Geophysical Research,* **88**(A11), 8743–8754.

THOMAS, P., ET AL. (1984). "Hyperion: 13-day Rotation From Voyager Data," *Nature,* **307,** 716–717.

TOKSOZ, M. N. (1979). "Planetary Seismology and Interiors," *Reviews of Geophysics and Space Physics,* **17,** 1641–1655.

TORBETT, M., AND R. SMOLUCHOWSKI (1979). "The Structure and the Magnetic Field of Uranus," *Geophysical Research Letters,* **6,** 675–676.

TORBETT, M. V., AND R. SMOLUCHOWSKI (1984). "Orbital Stability of the Unseen Solar Companion Linked to Periodic Extinction Events," *Nature,* **311,** 641–642.

TOWE, K. M. (1978). "Early Precambrian Oxygen: A Case Against Photosynthesis," *Nature,* **274,** 657–661.

TRAFTON, L. (1980). "Does Pluto Have a Substantial Atmosphere?" *Icarus,* **44,** 53–61.

TRAFTON, L. (1981). "The Atmospheres of the Outer Planets and Satellites," *Reviews of Geophysics and Space Physics,* **19,** 43–89.

TRAFTON, L. (1984). "Large Seasonal Variations in Triton's Atmosphere," *Icarus,* **58,** 312–324.

TRAFTON, L., AND S. A. STERN (1983). "On the Global Distribution of Pluto's Atmosphere," *Astrophysical Journal,* **267,** 872–881.

TYLER, G. L., ET AL. (1981). "Radio Science Investigations of the Saturn System with Voyager 1: Preliminary Results,'' *Science*, **212,** 201–206.

TYLER, G. L., ET AL. (1982). ''Radio Science With Voyager 2 at Saturn: Atmosphere and Ionosphere and the Masses of Mimas, Tethys, and Iapetus,'' *Science*, **215,** 553–558.

TYLER, G. L., ET AL. (1986). ''Voyager 2 Radio Science Observations of the Uranian System: Atmosphere, Rings, and Satellites,'' *Science*, **233,** 79–84.

VAISBERG, O. L., ET AL. (1986). ''Dust Coma Structure of Comet Halley From SP-1 Detector Measurements,'' *Nature*, **321,** 274–276.

VALENTINE, J. W. (1978). ''The Evolution of Multicellular Planets and Animals,'' *Scientific American*, **239**(3), 141–158.

VAN ALLEN, J. A., ET AL. (1980). ''Saturn's Magnetosphere, Rings, and Inner Satellites,'' *Science*, **207,** 415–421.

VEVERKA, J. (1977). ''Phobos and Deimos,'' *Scientific American*, **236**(2), 30–37.

VEVERKA, J., AND J. A. BURNS (1980). ''The Moons of Mars,'' *Annual Reviews of Earth and Planetary Science*, **8,** 527–558.

VEVERKA, J., AND T. C. DUXBURY (1977). ''Viking Observations of Phobos and Deimos: Preliminary Results,'' *Journal of Geophysical Research*, **82,** 4213–4223.

VEVERKA, J., P. THOMAS, AND S. SYNOTT (1982). ''The Inner Satellites of Jupiter,'' *Vistas in Astronomy*, **25,** 245–262.

VEVERKA, J., ET AL. (1981). ''Amalthea: Voyager Imaging Results,'' *Journal of Geophysical Research*, **86**(A10), 8675–8692.

VOGT, R. E., ET AL. (1979). ''Voyager 1: Energetic Ions and Electrons in the Jovian Magnetosphere,'' *Science*, **204,** 1003–1007.

VOGT, R. E., ET AL. (1979). ''Voyager 2: Energetic Ions and Electrons in the Jovian Magnetosphere,'' *Science*, **206,** 984–987.

WALDROP, M. (1978). ''U.S., U.S.S.R. Probes Begin Study of Venus,'' *Chemical and Engineering News*, December 14, 19–24.

WALDROP, M. (1986). ''Voyage to a Blue Planet,'' *Science*, **231,** 916–918.

WALLACE, L. (1980). ''The Structure of the Uranus Atmosphere,'' *Icarus*, **43,** 231–259.

WALLIS, M. K., AND W.-H. IP (1982). ''Atmospheric Interactions of Planetary Bodies with the Solar Wind,'' *Nature*, **298,** 229–234.

WARWICK, J. W., ET AL. (1979). ''Voyager 1 Planetary Radio Astronomy Observations Near Jupiter,'' *Science*, **204,** 995–998.

WARWICK, J. W., ET AL. (1979). ''Planetary Radio Astronomy Observations from Voyager 2 Near Jupiter,'' *Science*, **206,** 991–995.

WARWICK, J.W., ET AL. (1986). ''Voyager 2 Radio Observations of Uranus,'' *Science*, **233,** 102–106.

WASHBURN, M. (1981). *In the Light of the Sun*. Harcourt Brace Jovanovich, New York.

WASHBURN, M. (1983). *Distant Encounters — The Exploration of Jupiter and Saturn*. Harcourt Brace Jovanovich, New York.

WASHBURN, S. L. (1978). ''Evolution of Man,'' *Scientific American*, **239**(3), 194–207.

WEIDENSCHILLING, S. J. (1983). ''Progress Toward the Origin of the Solar System,'' *Reviews of Geophysics and Space Physics*, **21,** 206–213.

WETHERILL, G. W. (1981). ''Apollo Objects,'' *Scientific American*, **240**(3), 54–65.

WETHERILL, G. W. (1981). ''The Formation of the Earth from Planetismals,'' *Scientific American*, **244**(6), 163–174.

WETHERILL, G. W. (1985). ''Occurrence of Giant Impacts During the Growth of the Terrestrial Planets,'' *Science*, **228,** 877–879.

WHIPPLE, F. L. (1980). ''The Spin of Comets,'' *Scientific American*, **242**(3), 124–134.

WHITMIRE, D. P., AND A. A. JACKSON (1984). ''Are Periodic Mass Extinctions Driven by a Distant Solar Companion?'' *Nature*, **308,** 713–715.

WILSON, K. T. (1985). ''The CRAF Mission,'' *Spaceflight*, **27,** 452–453.

WISDOM, J., S. J. PEALE, AND F. MIGNARD (1984). ''The Chaotic Rotation of Hyperion,'' *Icarus*, **58,** 137–152.

WOOD, J. A. (1975). ''The Moon,'' *Scientific American*, **233**(3), 93–102.

WOODS, D. R. (1976). ''A Review of the Soviet Lunar Exploration Programme,'' *Spaceflight*, **18,** 273–290.

WOODWELL, G. M. (1978). ''The Carbon Dioxide Question,'' *Scientific American*, **238**(1), 34–43.

WYATT, S. P. (1977). *Principles of Astronomy*. Allyn & Bacon, Rockleigh, N.J.

WYLLIE, P. J. (1975). ''The Earth's Mantle,'' *Scientific American*, **232**(3), 50–63.

YOUNG, A., AND L. YOUNG (1975). ''Venus,'' *Scientific American*, **233**(3), 71–78.

ZEBKER, H. A., AND G. L. TYLER (1984). ''Thickness of Saturn's Rings Inferred from Voyager 1 Observations of Microwave Scatter,'' *Science*, **223,** 396–398.

ZEILIK, M. (1976). *Astronomy: The Evolving Universe*. Harper & Row, New York.

Index